# Ecological Studies

Analysis and Synthesis

---

Edited by

W. D. Billings, Durham (USA) F. Golley, Athens (USA)
O. L. Lange, Würzburg (FRG) J. S. Olson, Oak Ridge (USA)



Volume 17

# Fennoscandian Tundra Ecosystems

## Part 2 Animals and Systems Analysis

Edited by F. E. Wielgolaski

Editorial Board

P. Kallio, H. Kauri, E. Østbye, and T. Rosswall

With 81 Figures and 97 Tables

Springer-Verlag Berlin Heidelberg New York 1975

The picture on the cover is a simplified version of Fig. 1 on p. 4 of Ecological Studies, Vol. 16: Fennoscandian Tundra Ecosystem, Part 1.

ISBN-13: 978-3-642-66278-2 e-ISBN-13: 978-3-642-66276-8
DOI: 10.1007/978-3-642-66276-8

Distributed in the British Commonwealth Market by Chapman & Hall Limited, London.

Library of Congress Cataloging in Publication Data. (Revised). Main entry under title: Fennoscandian tundra ecosystems. (Ecological studies; v. 16–17). Bibliography: p. Includes index. CONTENTS: pt. 1. Plants and microorganisms. – pt. 2. Animals and systems analysis. 1. Tundra ecology – Scandinavia. 2. Tundra ecology – Finland. I. Wielgolaski, F. E., 1931 – II. Kallio, Paavo. III. Rosswall, Thomas. IV. Series. QH164.F45 574.5'264 75–4809



Softcover reprint of the hardcover 1st edition 1975

# Preface

Animals are important components of any ecosystem and it is impossible to describe structure and functioning of the Fennoscandian tundra ecosystems without including this part of the system. However, the strong diversity between functionally highly different fauna groups makes it impossible to study all groups in great detail with the funds and expertise available in Fennoscandia.

Relatively few productivity studies were carried out on animals in Fennoscandian tundra regions before the IBP projects started in 1969–1970. Within IBP, more fauna groups have been investigated in the Norwegian tundra project than in the other countries, due to better financial support. Even in Norway, however, only a limited number of invertebrate and vertebrate taxons is studied.

General lack of information on the same animals in various parts of the Fennoscandian tundra makes it more difficult to present structural and functional comparisons of fauna for the whole region than in plants and microorganisms (see Part 1 of Fennoscandian Tundra Ecosystems). Brief surveys on faunal problems within the area are given in the introduction to the animal section, and general aspects are also discussed in the four first papers of the volume.

Fennoscandian tundra ecosystems are influenced by man through tourism, hydroelectric damming and domestic animals. Native animals are also important through herbivory and trampling by mammals as reindeer; and invertebrates sometimes cause ecocatastrophic effects. All these factors are important in multiple use discussions e. g. on stability of the systems as discussed in the present volume. Within the conservation section (CT) of IBP in Norway, vegetation mapping by remote sensing has been carried out in alpine tundra areas, a method which is also useful in studies of influences by man and other animals.

One way of synthesizing biological data is to use modeling. Mathematical models may be used for integration and predictive purposes. The models may either be general and then usually with few details and low precision, or on specific and more detailed problems. Examples of both types are given in the volume, but data are still missing for many parts of the models. Another type of model is a "word" model which describes the problems in text form. The last paper in this volume is an example of a "word" model trying to summarize and synthesize the data presented in both volumes of Fennoscandian Tundra Ecosystems.

The reader is referred to the preface of Ecological Studies Volume 16 for more details on the background of Fennoscandian tundra studies carried out through the latest years, and forming the bulk of data presented in the two Fennoscandian Tundra Ecosystem volumes. The Editor and the Board express their thanks to all authors for their valuable contributions and to Springer-Verlag for publication of the Ecological Studies series. The Editor also wishes to thank Mrs. M. ESPELAND for correcting the English style in most of the manuscripts and to Mrs. I. LAUVSTAD who, as secretary, has had responsibility for the daily editorial work in preparation of the volume.

Oslo, September 1975 F. E. WIELGOLASKI

# Contents of Part 2

## 1. Animals

KAURI, H., ØSTBYE, E.: Introduction . . . . . . . . . . . . . . . . . 3
KAURI, H.: Area Size and Niche Breadth at Higher Latitudes . . . . . . 8

### 1.1. Faunal Structure of Research Areas

HAUKIOJA, E., KOPONEN, S.: Faunal Structure of Investigated Areas at Kevo, Finland . . . . . . . . . . . . . . . . . . . . . . . . . . 19
SOLHØY, T., ØSTBYE. E., KAURI, H., HAGEN, A., LIEN, L., SKAR, H.-J.: Faunal Structure of Hardangervidda, Norway . . . . . . . . . . . . . . . 29
JONSSON, S.: Faunal Structure of the IBP Tundra Site and Its Surroundings, Abisko, Sweden . . . . . . . . . . . . . . . . . . . . . . 46

### 1.2. Organization and Dynamics of Populations

SOLHØY, T.: Dynamics of Enchytraeidae Populations on Hardangervidda . 55
SOLHØY, T.: Dynamics of Oribatei Populations on Hardangervidda . . . . 60
KOPONEN, S.: Spider Populations in a Subalpine Birch Forest . . . . . . 66
FJELLBERG, A.: Organization and Dynamics of Collembola Populations on Hardangervidda . . . . . . . . . . . . . . . . . . . . . . . 73
KJELLSEN, E. K.: Dynamics of Thysanoptera Populations on Hardangervidda 80
TAMBS-LYCHE, H.: Dynamics of Aphididae Populations on Hardangervidda 84
HÅGVAR, S., ØSTBYE, E.: Occurrence and Role of Different Invertebrate Groups in Alpine *Salix herbacea* Snowbeds at Hardangervidda . . . . . 88
HOGSTAD, O.: Structure of Small Passerine Communities in Subalpine Birch Forests in Fennoscandia . . . . . . . . . . . . . . . . . . . 94
LIEN, L., ØSTBYE, E., HAGEN, A., SKAR, H.-J., SVALASTOG, D.: Density Variations of Bird Populations in High Mountain Habitats, Hardangervidda . . 105

### 1.3. Bioenergetics

SOLHØY, T., SKARTVEIT, A.: Influence of Various Temperature Approximations on Estimates of Field Respiration in a Mountain Community 111
KAURI, H., MOLDUNG, T. J., FJELLBERG, A.: Respiration Rates, Winter and Summer Activity in Collembola on Hardangervidda . . . . . . . . . . 117
STEIGEN, A. L.: Respiratory Rates and Respiratory Energy Loss in Terrestrial Invertebrates from Hardangervidda . . . . . . . . . . . . . . . 122

STEIGEN, A. L.: Energetics in a Population of *Pardosa palustris* (L.) (Araneae, Lycosidae) on Hardangervidda . . . 129
HOFSVANG, T.: Life Cycle and Energy Budget of *Tipula excisa* (Schum.) (Diptera, Tipulidae) . . . 145
NÆSS, S. J., STEIGEN, A. L., SOLHØY, T.: Standing Crop and Calorific Content in Invertebrates from Hardangervidda . . . 151
SKAR, H.-J., HÅGVAR, S., HAGEN, A., ØSTBYE, E.: Food Habits and Body Composition of Adult and Juvenile Meadow Pipit [*Anthus pratensis* (L.)] . 160
HAGEN, A., ØSTBYE, E., SKAR, H.-J.: Energy Budget of a Population of the Root Vole [*Microtus oeconomus* (Pall.)] in a High Mountain Habitat, Hardangervidda . . . 170

### 1.4. Herbivory Aspects

KALLIO, P., LEHTONEN, J.: On the Ecocatastrophe of Birch Forests Caused by *Oporinia autumnata* (Bkh.) and the Problem of Reforestation . . . 174
HAUKIOJA, E., KOPONEN, S.: Birch Herbivores and Herbivory at Kevo . . . 182
HAUKIOJA, E.: Importance Criteria in Birch Defoliators . . . 189
GAARE, E., SKOGLAND, T.: Wild Reindeer Food Habits and Range Use at Hardangervidda . . . 195
GAARE, E., THOMSON, B. R., KJOS-HANSSEN, O.: Reindeer Activity on Hardangervidda . . . 206
WIELGOLASKI, F. E.: Grazing by Sheep . . . 216

## 2. Conservation and Recreation in Tundra Ecosystems

DAHL, E.: Stability of Tundra Ecosystems in Fennoscandia . . . 231
HESJEDAL, O., LARSSON, J.: Remote Sensing of Vegetation in Conservation of Tundra Landscapes . . . 237

## 3. Models for Integration and Prediction

WIELGOLASKI, F. E.: Principles in the Use of Wide-Scale Models on Tundra Data . . . 245
SANDHAUG, A., KJELVIK, S., WIELGOLASKI, F. E.: A Mathematical Simulation Model for Terrestrial Tundra Ecosystems . . . 251
GYLLENBERG, G.: Systems Analysis on Invertebrates . . . 267
LIESTØL, K., ØSTBYE, E., SKAR, H.-J., SWARTZMAN, G.: A Simulation Model of a Small Rodent Population . . . 273
STENSETH, N. C.: Energy Model for Individual Small Rodents and Its Significance in General Population Theory . . . 283
WIELGOLASKI, F. E.: Functioning of Fennoscandian Tundra Ecosystems . . 300

**Subject Index** . . . 327

# Contents of Part 1


**1. Site Description**

SONESSON, M., WIELGOLASKI, F. E., KALLIO, P.: Description of Fennoscandian Tundra Ecosystems

**2. Abiotic Variables**

HINNERI, S., SONESSON, M., VEUM, A. K.: Soils of Fennoscandian IBP Tundra Ecosystems

SKARTVEIT, A., RYDÉN, B. E., KÄRENLAMPI, L.: Climate and Hydrology of Some Fennoscandian Tundra Ecosystems

**3. Primary Producers**

WIELGOLASKI, F. E., KALLIO, P.: Introduction

**3.1. Plant Structure**

DAHL, E.: Flora and Plant Sociology in Fennoscandian Tundra Areas

LYE, K. A.: Survey of the Main Plant Communities on Hardangervidda

HESJEDAL, O.: Vegetation Mapping at Hardangervidda

ELVEN, R., RYVARDEN, L.: Dispersal and Primary Establishment of Vegetation

WIELGOLASKI, F. E.: Comparison of Plant Structure on Grazed and Ungrazed Tundra Meadows

WIELGOLASKI, F. E., KÄRENLAMPI, L.: Plant Phenology of Fennoscandian Tundra Areas

BERG, A., KJELVIK, S., WIELGOLASKI, F. E.: Measurement of Leaf Areas and Leaf Angles of Plants at Hardangervidda, Norway

**3.2. Biomass and Production**

KJELVIK, S., KÄRENLAMPI, L.: Plant Biomass and Primary Production of Fennoscandian Subarctic and Subalpine Forests and of Alpine Willow and Heath Ecosystems

WIELGOLASKI, F. E.: Primary Productivity of Alpine Meadow Communities

FLOWER-ELLIS, J. G. K.: Growth in Populations of *Andromeda polifolia* on a Subarctic Mire

**3.3. Plant Processes**

KÄRENLAMPI, L., TAMMISOLA, J., HURME, H.: Weight Increase of Some Lichens as Related to Carbon Dioxide Exchange and Thallus Moisture

KALLIO, P., HEINONEN, S.: $CO_2$ Exchange and Growth of *Rhacomitrium lanuginosum* and *Dicranum elongatum*

KALLIO, P., VALANNE, N.: On the Effect of Continuous Light on Photosynthesis in Mosses

NYGAARD, R. T.: Acclimatization Effect in Photosynthesis and Respiration

SKRE, O.: $CO_2$ Exchange in Norwegian Tundra Plants Studied by Infrared Gas Analyzer Technique

KJELVIK, S., WIELGOLASKI, F. E., JAHREN, A.: Photosynthesis and Respiration of Plants Studied by Field Technique at Hardangervidda, Norway

JOHANSSON, L.-G., LINDER, S.: The Seasonal Pattern of Photosynthesis of Some Vascular Plants on a Subarctic Mire

WIELGOLASKI, F. E., KJELVIK, S.: Energy Content and Use of Solar Radiation of Fennoscandian Tundra Plants

BERG, A., KJELVIK, S., WIELGOLASKI, F. E.: Distribution of $^{14}C$ Photosynthates in Norwegian Alpine Plants

BERG, A.: Pigment Structure of Vascular Plants, Mosses, and Lichens at Hardangervidda, Norway

**4. Decomposers**

ROSSWALL, T.: Introduction

GOKSØYR, J.: Decomposition, Microbiology, and Ecosystem Analysis

HANSSEN, J. F., GOKSØYR, J.: Biomass and Production of Soil and Litter Fungi at Scandinavian Tundra Sites

HAYES, A. J., RHEINBERG, P.: Microfungal Populations of the Abisko Area, Northern Sweden

CLARHOLM, M., LID-TORSVIK, V., BAKER, J. H.: Bacterial Populations of Some Fennoscandian Tundra Soils

BERG, B., KÄRENLAMPI, L., VEUM, A. K.: Comparisons of Decomposition Rates Measured by Means of Cellulose

ROSSWALL, T., VEUM, A. K., KÄRENLAMPI, L.: Plant Litter Decomposition at Fennoscandian Tundra Sites

SVENSSON, B. H., VEUM, A. K., KJELVIK, S.: Carbon Losses from Tundra Soils

**5. Nutrient Cycling**

KALLIO, P., WIELGOLASKI, F. E.: Introduction

KALLIO, S., KALLIO, P.: Nitrogen Fixation in Lichens at Kevo, North-Finland

GRANHALL, U., LID-TORSVIK, V.: Nitrogen Fixation by Bacteria and Free-Living Blue-Green Algae in Tundra Areas

Wielgolaski, F. E., Kjelvik, S., Kallio, P.: Mineral Content of Tundra and Forest Tundra Plants in Fennoscandia

Kallio, P., Veum, A. K.: Analysis of Precipitation at Fennoscandian Tundra Sites

Skre, O., Berg, A., Wielgolaski, F. E.: Organic Compounds in Alpine Plants

Larsson, P., Tangen, K.: The Input and Significance of Particulate Terrestrial Organic Carbon in a Subalpine Freshwater Ecosystem

**Subject Index**

# Contributors Part 2


| | |
|---|---|
| DAHL, E. | Botanical Institute, Agricultural College, 1432 Ås-NLH, Norway |
| FJELLBERG, A. | Zoological Museum, Department of Animal Ecology and Systematics, University of Bergen, 5014 Bergen, Norway |
| GYLLENBERG, G. | Zoological Museum, University of Bergen, Norway. Present address: Department of Zoology, University of Helsinki, Rautatiekatu 13, 00100 Helsinki 10, Finland |
| GAARE, E. | Norwegian State Game Research Institute, Elgesetergate 10, 7000 Trondheim, Norway |
| HAGEN, A. | Zoological Institute, University of Oslo, Blindern, Oslo 3, Norway |
| HÅGVAR, S. | Zoological Institute, University of Oslo. Present address: NISK, 1430 Ås, Norway |
| HAUKIOJA, E. | Department of Zoology, University of Turku, 20500 Turku 50, Finland |
| HESJEDAL, O. | Department of Agriculture, Division of Land Registration (DLR), Ås. Present address: Telemark Distriktshøgskole, 3800 Bø i Telemark, Norway |
| HOFSVANG, T. | Zoological Institute, University of Oslo. Present address: Department of Zoology, Agricultural College, 1432 Ås-NLH, Norway |
| HOGSTAD, O. | Zoological Institute, University of Trondheim, 7000 Trondheim, Norway |
| JONSSON, S. | Department of Entomology, University of Uppsala, 75122 Uppsala, Sweden |
| KALLIO, P. | Department of Botany, University of Turku, 20500 Turku 50, Finland |
| KAURI, H. | Zoological Museum, Department of Animal Zoology and Systematics, University of Bergen, 5014 Bergen, Norway |
| KJELLSEN, E.K. | Zoological Museum, Department of Animal Ecology and Systematics, University of Bergen. Present address: 9300 Finnsnes, Norway |
| KJELVIK, S. | Botanical Laboratory, University of Oslo. Present address: Botanical Institute, Norwegian Agricultural College, 1432 Ås-NLH, Norway |
| KJOS-HANSSEN, O. | Norwegian State Game Research Institute, Elgesetergate 10, 7000 Trondheim, Norway |
| KOPONEN, S. | Department of Zoology, University of Turku, 20500 Turku 50, Finland |
| LARSSON, J. | Department of Agriculture, Division of Land Registration (DLR), 1430 Ås, Norway |
| LEHTONEN, J. | Department of Botany, University of Turku, 20500 Turku 50, Finland |
| LIEN, L. | Zoological Museum, University of Oslo, Sarsgt. 1, Oslo 5, Norway |

LIESTØL, K. — Department of Physics, University of Oslo. Present address: Bjerkåsen 74, 1310 Blommenholm, Norway

MOLDUNG, T. J. — Zoological Museum, Department of Animal Ecology and Systematics, University of Bergen, 5014 Bergen, Norway

NÆSS, S. J. — Zoological Museum, Department of Animal Ecology and Systematics, University of Bergen. Present address: 5300 Kleppestø, Norway

ØSTBYE, E. — Zoological Institute, University of Oslo, Blindern, Oslo 3, Norway

SANDHAUG, A. — Botanical Laboratory, University of Oslo. Present adress: Norsk Medisinaldepot, POB 100 Veitvedt, Oslo 9, Norway

SKAR, H.-J. — Zoological Institute, University of Oslo, Blindern, Oslo 3, Norway

SKARTVEIT, A. — Department of Geophysics, Section B, University of Bergen, 5014 Bergen, Norway

SKOGLAND, T. — Norwegian State Game Research Institute, Elgesetergate 10, 7000 Trondheim, Norway

SOLHØY, T. — Zoological Museum, Department of Animal Ecology and Systematics, University of Bergen, 5014 Bergen, Norway

STEIGEN, A. L. — Zoological Museum, Department of Animal Ecology and Systematics, University of Bergen, 5014 Bergen, Norway

STENSETH, N. C. — Zoological Institute, University of Oslo, Blindern, Oslo 3, Norway

SVALASTOG, D. — Zoological Institute, University of Oslo. Present address: NIVA, Gaustadalléen 25, Oslo 3, Norway

SWARTZMAN, G. — University of Washington, Seattle, WA. 98195, USA

TAMBS-LYCHE, H. — Zoological Museum, Department of Animal Ecology and Systematics, University of Bergen. Present address: Malmosvej 83 A, Virum, Denmark

THOMSON, B. R. — Norwegian State Game Research Institute, Trondheim. Present address: The Scottish Landowner's Federation, 26 Ruthland Square, Edinburgh EH1 2BT, U.K.

WIELGOLASKI, F. E. — Botanical Laboratory, University of Oslo, Blindern, Oslo 3, Norway

# 1. Animals

# Introduction

H. KAURI and E. ØSTBYE

The alpine zone of the Scandinavian mountains has many characteristics in common with the arctic tundra: the pronounced seasonal changes of the climate, and the short turnover intervals between winter and summer. In the treeless plateau e.g. of Hardangervidda at 60° N in southwestern Norway (SONESSON et al., 1975), particularly the spring turnover is of short duration. A long winter, a short summer (2.5–3.5 months) accompanied by lower temperatures characterize to a greater or lesser extent both arctic and alpine regions. Considering these and other characteristics, one can recognise a similarity in environmental conditions and habitats between the alpine zone in the Scandinavian mountains and the arctic tundra. A lower diversity and a smaller range in the true species of tundra and of alpine zone are characteristics of both regions (KAURI, 1975). A great number of animal species are common to both the arctic tundra and the alpine zone, for example in the Norwegian IBP tundra research area at Hardangervidda, including numerous invertebrates and some 50% of the birds and mammals. It therefore seems reasonable to link the alpine belt of the Scandinavian mountains and the arctic tundra together into an arcto-alpine tundra biome (in the sense of LATTIN, 1967) or simply a tundra biome (KENDEIGH, 1961; HOFFMANN, 1974).

The differences between the areas are explained by their geographical location. In southwestern Fennoscandia there are great variations in the day length compared to the everlasting day of the arctic summer. Correspondingly, the daily fluctuations of summer temperature are greater on e.g. the mountain plateau of Hardangervidda and the temperatures higher than in the arctic tundra. Somewhat thinner atmosphere in the mountains further accentuates the fluctuations of temperature. The precipitation is heavier on Hardangervidda than is usual in arctic tundra and provides a thick sheltering cover of snow throughout the winter (SKARTVEIT et al., 1975).

The subalpine birch forest region is a characteristic feature of Fennoscandia and with few exceptions it makes a more or less distinct zone between the coniferous forest and the alpine region (SJÖRS, 1967). It has some characteristics in common with forest tundra (BLÜTHGEN, 1970). Birch *(Betula tortuosa)* is the dominant primary producer of the subalpine zone in northern Finland (KÄRENLAMPI and KAUHANEN, 1972) and this zone has a large coverage in northernmost Fennoscandia (HÄMET-AHTI, 1963).

The Finnish IBP tundra research area in Kevo at 70° N is associated with the subalpine birch forest, intermingled with patches of pine. In the IBP birch site at

Kevo the birch represents 57% of the total above-ground biomass and a large proportion of the below-ground biomass. The Stordalen mire, the Swedish IBP tundra site at Abisko 68° N, has a similar location in connection with the belt of birch forest (SONESSON et al., 1975).

The three IBP research areas are climatically different, and indicate a gradient, reaching from the arctic-continental Kevo to the atlantic-maritime Hardangervidda. Along the gradient, there occurs a decrease of the diversity as illustrated on the spiders (KAURI, unpubl.), springtails (FJELLBERG, 1975), stoneflies (LILLEHAMMER, 1974) and ichneumonid wasps (JUSSILA, 1965, 1973). A great part of the invertebrates of the Hardangervidda is also common in the adjacent lowland habitats. A passive aerial transport occurs (GREVE, 1969). Dominating taxons among the invertebrates at Hardangervidda are the microarthropods Collembola and Acari, Enchytraeidae and some taxons of Diptera, the prevailing part of which is living in the litter and the edaphic layer—in an environment with moderate microclimatical fluctuations. An important characteristic feature of the invertebrate composition is the absence of several taxons important in the decomposition cycle in the lowland habitats: isopods, millipedes, centipedes and ants. The scantiness of earthworms and snails is also pronounced compared with the adjacent lowlands, e.g. west of the mountain plateau (STØP-BOWITZ, 1969, 1974; SOLHØY, unpubl.).

Certain adaptations to the alpine environment have been recognized. The relatively higher respiratory rates for e.g. Collembola and Araneae *(Pardosa palustris)* have been considered as compensatory to the low temperatures (STEIGEN, 1975a). On the higher altitudes, a great number of species have a prolonged life cycle (FJELLBERG, 1975; STEIGEN, 1975a; SOLHØY, 1975; KAURI, unpubl.). The ecological efficiency (the index of production/consumption) of e.g. *P. palustris* however, seems to have about the same value as in species from lowland habitats (STEIGEN, 1975b). The prolonged life cycle thus seems to be dependent on the qualities of the environment, especially on the short summer season.

Many vertebrate species have their main distribution range on the Eurasian continent, with a Fennoscandian outlet having its southern distribution limit at the southern borders of Hardangervidda in Norway, e.g. the arctic fox, the grey-sided vole (SIIVONEN, 1968) and the snowy owl (HAFTORN, 1971).

In the latter decades, some species have extended their distribution range, e.g. the moose has expanded its territory from the coniferous forests at the eastern border of Hardangervidda, over the mountain plateau, to the forests on the west-coast of Norway. Other species have also extended their distribution in Fennoscandian tundra areas, e.g. the common gull, redwing, skylark and American mink. A survey of the number of species breeding within the Fennoscandian IBP tundra areas is given in Table 1. More details on species composition are given in the faunal structure papers in the present volume (HAUKIOJA and KOPONEN, 1975; SOLHØY et al., 1975; JONSSON, 1975).

Fennoscandian alpine regions and adjacent northern coniferous forests are characterized by pronounced 3–4 year cycles in productivity of certain vertebrate animal groups, for instance small rodents (KALELA, 1962). The degree of synchronism of rodent production is variable between different geographical areas in Fennoscandia. During certain cycles, the rodent biomass is in the same phase

Table 1. Vertebrate species breeding within Fennoscandian tundra areas

| | Hardanger-vidda | Kevo | Stordalen |
|---|---|---|---|
| Pisces | 4 | — | 10 |
| Amphibia | 1 | 1 | 1 |
| Reptilia | 1 | 1 | 1 |
| Aves | 61 | 48 | 112 |
| Mammalia | 17 | 15 | 23 |
| Total | 84 | 65 | 147 |

over larger areas. During other 3–4 year periods, rodent production over smaller or larger areas displays a 1–2 years lag phase resulting in an asynchronous and complex productivity pattern, probably partly reflecting the high ecological diversity of Fennoscandia (HAGEN, 1956; MYRBERGET, 1965; 1973; MYSTERUD, 1970).

Some information on density variations and population energetics in small rodents in Fennoscandia exists, e.g. from Hardangervidda (LIEN et al., 1975; HAGEN et al., 1975; ØSTBYE et al., 1975). A simulation of the cyclic pattern of a small rodent population at Hardangervidda was the intention for a model presented in this volume (LIESTØL et al., 1975).

One of the dominant passerine bird species in Fennoscandian high mountain areas is the meadow pipit *(Anthus pratensis)*. Both this species and other passerines at Hardangervidda show great yearly variations in density coinciding with small rodent fluctuations (LIEN et al., 1975). The food habits of the meadow pipit reveal that this species finds its food over a large part of the nutritional specter available (SKAR et al., 1975).

Despite great annual fluctuations of the density of most of the birds inhabiting the subalpine birch forests, the passerine communities found in the different Fennoscandian study areas show similarities. They seem to have qualitative as well as quantitative patterns, characteristic of the sub-alpine birch forest as a biome (HOGSTAD, 1975). Birch leaves and sap are the resources most widely used by herbivores. Vertebrates graze on leaves, buds and small twigs and shoots, but are not dependent on the birch as their sole food. Reindeer use it as a summer diet (cf. HAUKIOJA and HEINO, 1974; GAARE and SKOGLAND, 1975) and voles also take a lot during peak years. However, birch represents only a small part of their total annual energy intake when compared with that of invertebrate herbivores. Many or most of these are monophagous, the birch being their host plant and their specialization to a certain resource of the host is strict. Sometimes birch trees are fully defoliated by invertebrates (KALLIO and LEHTONEN, 1975). Three main functional groups can be distinguished according to feeding habits in herbivore invertebrates (HAUKIOJA and KOPONEN, 1975): free-living leaf grazers (e.g. lepidopteran and hymenopteran larvae), free-living sap suckers (e.g. psyllids and aphides) and intraleaf (in mines or galls) living forms (e.g. lepidopteran and dipteran larvae).

## References

BLÜTHGEN, J.: Problems of definition and geographical differentiation of the Subarctic with special regard of northern Europe. In: ECKHARDT, F. E. (Ed.): Ecology of the subarctic regions, pp. 11–33. Paris: UNESCO 1970.

FJELLBERG, A.: Organisation and dynamics of Collembola populations at Hardangervidda. In: WIELGOLASKI, F. E. (Ed.): Fennoscandian tundra ecosystems, Part 2: Animals and systems analysis, pp. 73–79. Berlin-Heidelberg-New-York: Springer 1975.

GAARE, E., SKOGLAND, T.: Wild reindeer food habits and range use at Hardangervidda. In: WIELGOLASKI, F. E. (Ed.): Fennoscandian tundra ecosystems, Part 2: Animals and systems analysis, pp. 195–205. Berlin-Heidelberg-New York: Springer 1975.

GREVE, L.: An aerial-drift of Neuroptera from Hardangervidda, western Norway. Årb. Univ. Bergen. Mat.-Nat. Ser., **2**, 1–15 (1969).

HAFTORN, S.: Norges fugler. Oslo-Bergen-Tromsø: Universitetsforlaget 1971.

HAGEN, A., ØSTBYE, E., SKAR, H. J.: Energy budget of a population of the root vole, [*Microtus oeconomus* (Pall.)], in a high mountain habitat, Hardangervidda. In: WIELGOLASKI, F. E. (Ed.): Fennoscandian tundra ecosystems. Part 2: Animals and systems analysis, pp. 170–173. Berlin-Heidelberg-New York: Springer 1975.

HAGEN, Y.: The irruption of hawk-owls [*Surnia ulula* (L.)] in Fennoscandia 1950–51. Sterna, Opusc. Ser. Zool. **24**, 1–22 (1956).

HÄMET-AHTI, L.: Zonation of the mountain birch forests in northernmost Fennoscandia. Ann. Bot. Soc. "Vanamo" **34**, 1–127 (1963).

HAUKIOJA, E., HEINO, J.: Birch consumption by reindeer *(Rangifer tarandus)* in Finnish Lapland. Rep. Kevo Subarctic Res. Stat. **11**, 22–25 (1974).

HAUKIOJA, E., KOPONEN, S.: Faunal structure of investigated areas at Kevo, Finland. In: WIELGOLASKI, F. E. (Ed.): Fennoscandian tundra ecosystems. Part 2: Animals and systems analysis, pp. 19–28. Berlin-Heidelberg-New York: Springer 1975.

HOFFMANN, R. S.: Terrestrial vertebrates. In: IVES, J. D., BARRY, R. G. (Eds.): Arctic and alpine environments, pp. 475–568. London: Methuen 1974.

HOGSTAD, O.: Structure of small passerine communities in subalpine birch forests in Fennoscandia. In: WIELGOLASKI, F. E. (Ed.): Fennoscandian tundra ecosystems. Part 2: Animals and systems analysis, pp. 94–104. Berlin-Heidelberg-New York: Springer 1975.

JONSSON, S.: Faunal structure of the IBP tundra site and its surroundings, Abisko, Sweden. In: WIELGOLASKI, F. E. (Ed.): Fennoscandian tundra ecosystems. Part 2: Animals and systems analysis, pp. 46–54. Berlin-Heidelberg-New York: Springer 1975.

JUSSILA, R.: The Ichneumonidae of the Kevojoki area in Inari Lapland (Finland). Ann. Univ. Turku A, **11**, 1–186 (1965).

JUSSILA R.: Ichneumonidae from Hardangervidda. Fauna of Hardangervidda, Vol. 2, 1–50. Oslo-Bergen: Universitetsforlaget 1973.

KALELA, O.: On the fluctuations in the numbers of arctic and boreal small rodents as a problem of production bioloby. Ann. Acad. Sci. Fenn. Ser. A IV Biol. 66 (1962).

KALLIO, P., LEHTONEN, J.: On the ecocatastrophe of birch forests caused by *Oporinia autumnata* (Bkh.) and the problem of reforestation. In: WIELGOLASKI, F. E. (Ed.): Fennoscandian tundra ecosystems. Part 2: Animals and systems analysis, pp. 174–180. Berlin-Heidelberg-New York: Springer 1975.

KÄRENLAMPI, L., KAUHANEN, H.: A direct gradient analysis of the vegetation of the surroundings of the Kevo Subarctic Station. Rep. Kevo Subarct. Res. Stat. **9**, 82–98 (1972).

KAURI, H.: Area size and niche breadth at higher latitudes. In: WIELGOLASKI, F. E. (Ed.): Fennoscandian tundra ecosystems. Part 2: Animals and systems analysis, pp. 8–18. Berlin-Heidelberg-New York: Springer 1975.

KENDEIGH, S. C.: Animal ecology. Englewood Cliffs: Prentice-Hall. 1961.

LATTIN, G. DE: Grundriß der Zoogeographie. Stuttgart: Gustav Fischer 1967.

LIEN, L., ØSTBYE, E., HAGEN, A., SKAR, H. J., SVALASTOG, D.: Density variations of bird populations in high mountain habitats, Hardangervidda. In: WIELGOLASKI, F. E. (Ed.): Fennoscandian tundra ecosystems. Part 2: Animals and systems analysis, pp. 105–110. Berlin-Heidelberg-New York: Springer 1975.

LIESTØL, K., ØSTBYE, E., SKAR, H. J., SWARTZMAN, G.: A simulation model of a small rodent population. In: WIELGOLASKI, F. E. (Ed.): Fennoscandian tundra ecosystems. Part 2: Animals and systems analysis, pp. 273–282. Berlin-Heidelberg-New York: Springer 1975.

LILLEHAMMER, A.: Norwegian stoneflies, II. Distribution and relationship to the environment. Norwegian J. Ent. **21**, 195–250 (1974).

MYRBERGET, S.: Vekslinger i bestandstørrelsen hos norske smågnagere i årene 1946–60. Meddr. St. Viltunders. Ser. 2 (19), 1–54 (1965).

MYRBERGET, S.: Geographical synchronism of cycles of small rodents in Norway. Oikos **24**, 220–224 (1973).

MYSTERUD, I.: Hypothesis concerning characteristics and causes of population movements in Tengmalm's Owl [*Aegolius funereus* (L.)]. Norw. J. Zool. **18**, 49–74 (1970).

ØSTBYE, E. (ED.), BERG, A., BLEHR, O., ESPELAND, M., GAARE, E., HAGEN, A., HESJEDAL, O., HÅGVAR, S., KJELVIK, S., LIEN, L., MYSTERUD, I., SANDHAUG, A., SKAR, H. J., SKARTVEIT, A., SKRE, O., SKOGLAND, T., SOLHØY, T., STENSETH, N. C., WIELGOLASKI, F. E.: Hardangervidda. In: ROSSWALL, T., HEAL, O. W. (Eds.): Structure and function of tundra ecosystems. Ecol. Bull. **20**: 225–264. Stockholm: Swed. Nat. Sci. Res. Council 1975.

SIIVONEN, L.: Nordeuropas däggdjur. Stockholm: Norstedt & Søner 1968.

SJÖRS, H.: Nordisk växtgeografi. Stockholm: Svenska Bokförlaget 1967.

SKAR, H. J., HÅGVAR, S., HAGEN, A., ØSTBYE, E.: Food habits and body composition of adult and juvenile meadow pipit [*Anthus pratensis (L.)]. In:* WIELGOLASKI, F. E. (Ed.): Fennoscandian tundra ecosystems. Part 2: Animals and systems analysis, pp. 160–169. Berlin-Heidelberg-New York: Springer 1975.

SKARTVEIT, A., RYDÉN, B. E., KÄRENLAMPI, L.: Climate and hydrology of some Fennoscandian tundra ecosystems. In: WIELGOLASKI, F. E. (Ed.): Fennoscandian tundra ecosystems. Part 1: Plants and microorganisms, pp. 41–53. Berlin-Heidelberg-New York: Springer 1975.

SOLHØY, T.: Dynamics of Oribatei populations on Hardangervidda. In: WIELGOLASKI, F. E. (Ed.): Fennoscandian tundra ecosystems. Part 2: Animals and systems analysis, pp. 60–65. Berlin-Heidelberg-New York: Springer 1975.

SOLHØY, T., KAURI, H., ØSTBYE, E., HAGEN, A., LIEN, L., SKAR, H. J.: Faunal structure of Hardangervidda, Norway. In: WIELGOLASKI, F. E. (Ed.): Fennoscandian tundra ecosystems. Part 2: Animals and systems analysis, pp. 29–45. Berlin-Heidelberg-New York: Springer 1975.

SONESSON, M., WIELGOLASKI, F. E., KALLIO, P.: Description of Fennoscandian tundra ecosystems. In: WIELGOLASKI, F. E. (Ed.): Fennoscandian tundra ecosystems. Part 1: Plants and microorganisms, pp. 3–28. Berlin-Heidelberg-New York: Springer 1975.

STEIGEN, A. L.: Respiratory rates and respiratory energy loss in terrestrial invertebrates from Hardangervidda. In: WIELGOLASKI, F. E. (Ed.): Fennoscandian tundra ecosystems. Part 2: Animals and systems analysis, pp. 122–128. Berlin-Heidelberg-New York: Springer 1975a.

STEIGEN, A. L.: Energetics in a population of *Pardosa palustris* (L.) (Araneae, Lycosidae) at Hardangervidda. In: WIELGOLASKI, F. E. (Ed.): Fennoscandian tundra ecosystems. Part 2: Animals and systems analysis, pp. 129–144. Berlin-Heidelberg-New York: Springer 1975b.

STØP-BOWITZ, C.: A contribution to our knowledge of the systematics and zoogeography of Norwegian earthworms (Annelida, Oligochaeta, Lumbricidae). Nytt Mag. Zool. **17**, 169–280 (1969).

STØP-BOWITZ, C.: Earthworms (Annelida, Oligochaeta, Lumbricidae). Fauna of Hardangervidda, **6**, 37–46. Zool. Mus. Univ. Bergen 1974.

# Area Size and Niche Breadth at Higher Latitudes

H. KAURI

Fewer plant and animal species, lower production, and a slower rate of decomposition have been considered aspects of the higher latitudes. An increase in niche breadth (KLOPFER and MACARTHUR, 1960; MACARTHUR, 1969; WILLSON, 1969) and a decrease in the degree of specialization (SCRIBER, 1973) have also been observed. A lower level of competition over a resource gradient has been presumed to apply to the unpredictable environment of temperature regions (SLOBODKIN and SANDERS, 1969). The arctic and subarctic regions however have hitherto not been included in this discussion.

The extent to which species have established themselves geographically may yield some new information about the functioning of ecological systems. The size of the range of a species would provide a convenient measurable unit.

Such a study has been attempted in connection with the IBP Tundra Biome investigations in Norway.

There are only a few terrestrial animal groups which can be treated in this way. Sufficient knowledge of their distribution area exists only for birds, mammals, some amphibians, reptiles, and a few invertebrate groups, perhaps only for butterflies, some beetles and spiders. For the investigation of a sequence of range sizes, the species studied should share a common resource gradient. The family as a taxonomic category seems to be a suitable group of species for such studies. The related niches of species in a family taxon can be considered the result of a common trend in ecological radiation.

The families Sylviidae, Fringillidae and Emberizidae (Aves) were used. Birds of the first family are mainly secondary consumers (predators), the others primary consumers (herbivores).

## Method

The area between 0–20° E longitude, 0–80° N and 0–40° S latitude was chosen for this examination (from 70–80° N the quadrilateral 0–20° E was replaced by 10–30° E). Within this area the sizes of the ranges for all species within every quadrilateral (20° longitude by 10° latitude) were measured. For every square the mean range of the inhabiting species was calculated. The same procedure was used for Sylviidae south of the Equator. The maps given by VOOUS (1962) were used, which means that winter and transitory territories were not included. The areas in Africa were treated according to MACKWORTH-PRAED and GRANT (1960, 1963, 1973) and in some cases for desert species according to ETCHÉCOPAR and HÜE (1967), ALI and RIPLEY (1973) and DEMENT'EV and GLADKOV (1968). All the areas in Africa were transformed to the scale of VOOUS maps. Some corrections were made to the limits of distribution in Scandinavian countries (see HAFTORN, 1971) and in Asia (see DEMENT'EV and GLADKOV, 1968, 1970).

*Erythropygia galactotes* (Temminck), in VOOUS (1962) included in the Sylviidae family, was excluded from this study as it is classified in the Turdidae family.

Compared with VOOUS maps, those of MACKWORTH-PRAED and GRANT are schematic. Nevertheless they express the known range of African birds. Any faults that may occur in the limits have been somewhat magnified by the transformation to VOOUS scale, but the relations of the range between species have generally been unaffected.

Equiareal maps would have made direct comparison possible, but the maps of VOOUS and MACKWORTH-PRAED and GRANT are not equiareal. Corrections were therefore necessary for both latitude and longitude. The Smithsonian Tables given by ROBINSON and SALE (1969) were used in calculations. Measurements of the ranges on the maps were made by means of a dotted slide. For species with disjunct distributions, a summation was made of the various parts. Temporary expansions and contractions of the range are generally small and of little importance in this treatment.

## Number of Species

The total number of species in the northern gradient is 104 of Sylviidae. Between 30 and 80° N there are 17 species of Fringillidae and 12 of Emberizidae. In the southern gradient there are 41 species of the Sylviidae. The number of species decreases from the tropics towards higher latitudes. The highest number of Sylviidae species occurs in tropical rain forest, where there are 65 species (square 0–10° N), and the lowest in the arctic-alpine tundra in northern Norway, with 3 species (Table 1). The decrease in the number of species between the tropical maximum and the minimum in the tundra is not regular, and large divergences due to changes in the ecological environment exist; between the low number in the Sahara (4 species) and in the tundra there is a peak in numbers in the square 40–50° N. This square is characterized by considerable habitat diversity, along with the Alps, Pyrenees, Mediterranean summer-dry habitats and Central European landscapes.

South of the Equator, in Africa, the sequence of the biomes is more homogeneous and the decrease in the number of species follows a regular pattern (Table 1). The highest number of species (36) occurs in the quadrilateral 0–10° S, the lowest (6) in the quadrilaterals 20–30–40° S. It must be mentioned that the diversity on the two sides of the Equator differs; the southern quadrilateral (0–10° S) consists to a large extent of savannah and dry forest with little rain forest, and it contains fewer species than the neighboring area of rain forest to the north.

Table 1. Mean range size values in various quadrilaterals for species of Sylviidae, Fringillidae and Emberizidae

| Quadri-laterals | No. of species | Mean range size $km^2 \cdot 10^6$ | Quadri-laterals | No. of species | Mean range size $km^2 \cdot 10^6$ |
|---|---|---|---|---|---|
| Sylviidae | | | Fringillidae | | |
| 70–80° N | 3 | 13.79 | 70–80° N | 3 | 13.29 |
| 60–70° | 12 | 13.06 | 60–70° | 12 | 13.54 |
| 50–60° | 21 | 10.52 | 50–60° | 14 | 12.87 |
| 40–50° | 33 | 8.21 | 40–50° | 12 | 11.95 |
| 30–40° | 26 | 7.00 | 30–40° | 8 | 11.77 |
| 20–30° | 4 | 8.02 | | | |
| 10–20° | 30 | 7.12 | Emberizidae | | |
| 0–10° | 65 | 4.46 | 70–80° N | 3 | 15.58 |
| 0–10° S | 36 | 6.32 | 60–70° | 7 | 13.43 |
| 10–20⁰ | 22 | 8.06 | 50–60° | 6 | 14.25 |
| 20–30° | 6 | 10.99 | 40–50° | 7 | 12.45 |
| 30–40° | 6 | 11.59 | 30–40° | 6 | 10.39 |

MACARTHUR and WILSON (1967) suggested that the low diversity on peninsulas is a result of reduced immigration resulting from the absence of source regions around most of their boundaries. It is probably true that impoverishment of the islands and peninsulas in the Mediterranean (square 30–40° N) is a result of this, but this is not the case with the low number of species further south in the Sahara, and in the arid environments in southern Africa. The severe physical conditions, particularly the lack of water, are the real reason for the low diversity. There is some similarity between desert and tundra: rigorous environment, low diversity, extended range, difficulties for organisms in utilizing available resources etc. It is understandable therefore that the same model has proved applicable to both ecosystems (WIELGOLASKI, 1975). Recently KARR (1975) has shown that diversity is proportional to the number of resource "packages" in the community, and not so much to the increase in general productivity. Tundra and desert ecosystems, with their low diversity and the relative homogeneity of the environment, fit well into KARR's model.

The latitudinal variations in the number of species of Fringillidae and Emberizidae are similar to those for the Sylviidae (Table 1). The same pattern was found by SCRIBER (1973) in the family Papilionidae (Lepidoptera).

## Range Size

The range of a species is principally dependent on the presence of the necessary resources, the time-span during which the species has been able to colonize the resource area, and its success in utilizing the resource. The dispersal capacity of the species becomes the decisive factor (EKMAN, 1922; GISLÉN, 1940). The regions of Northern Europe once covered by ice cause some uncertainty, but it may be supposed that for animals which spread as easily as birds, colonization has reached its natural limits. Further changes are due to environmental and/or evolutionary factors and not to historical developments.

The range increases in a linear regression towards higher latitudes. The correlation coefficients are high: 0.94 for Sylviidae in the north and 0.97 in the south; 0.92 for Fringillidae and 0.91 for Emberizidae in the north.

It seems possible to describe the range size *versus* degree of latitude by means of the logarithmic function

$$\Phi = \frac{K}{1+ae^{-rt}} \quad \text{(Kauri and Lie, in press)},$$

where the symbols are: $t = \frac{\varphi}{5}$; $\varphi$ = latitude; $\Phi$ = range size; $K$ = upper asymptote; $a$ and $r$ are constants. Thus the equation will be transcribed as

$$\Phi = \frac{K}{1+ae^{-r\frac{\varphi}{5}}}$$

As examples the following equations have been calculated for Sylviidae:

(1) The northern regression, with the upper asymptote in the tundra

$$\Phi_A = \frac{17.53}{1+1.55 \cdot e^{-0.28 \frac{\varphi_{\text{NORTH-40}}}{5}}}.$$

(2) The equation of the Saharan regression

$$\Phi_B = \frac{8.26}{1+1.97 \cdot e^{-0.84 \frac{\varphi_{\text{NORTH}}}{5}}}.$$

(3) The southern regression

$$\Phi_C = \frac{11.70}{1+8.37 \cdot e^{-0.84 \frac{\varphi_{\text{SOUTH}}}{5}}}.$$

As stated above, the range of the species increases with the distance from the Equator. Upper asymptotes are established in the tundra and the deserts (Figs. 1, 2). Of particular interest is the Saharan asymptote, since the Sahara is an inland desert not influenced by coastal effects.

For further discussion of the regressions of the range, some considerations are necessary:

(1) There may be a threshold value of range size as a site for the maintenance of optimal population size. The lower asymptotes observed near the Equator can be explained in this way.

(2) If natural selection at the higher latitudes is basically controlled by the physical environment and not by biological interactions (Dobzhansky, 1950; Williams, 1964; Slobodkin and Sanders, 1969) and interspecific competition is

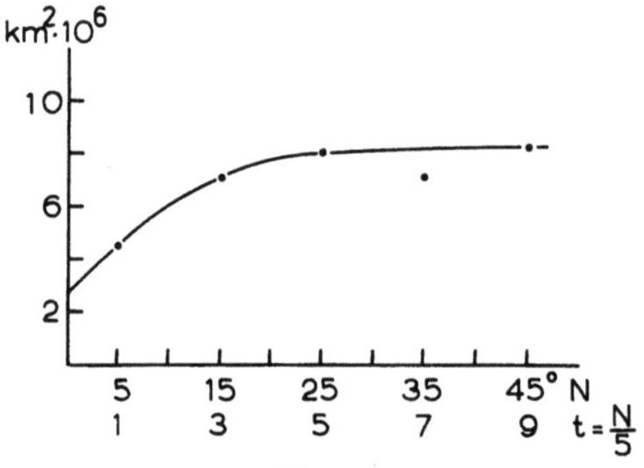


Fig. 1

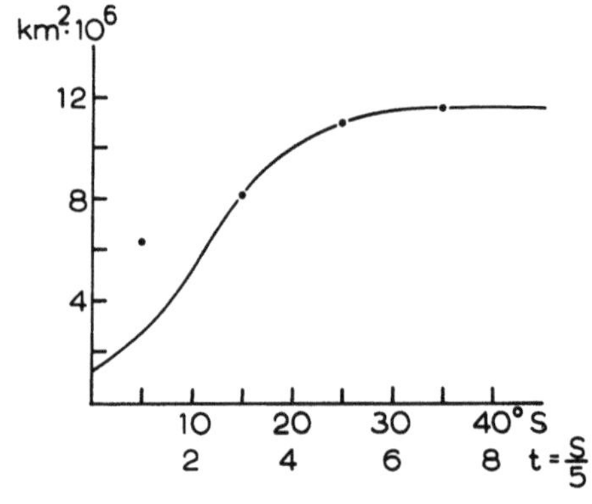


Fig. 2

Fig. 1. The species range in Sylviidae from the tropics to the Sahara plotted against the degree of latitude

Fig. 2. The species range in Sylviidae from the tropics to the Cape of Good Hope plotted against the degrees of latitude

reduced, the boundaries of the range will be determined by the abiotic environmental factors. Changes in the boundaries associated with climatic changes in northern Europe during this century are good examples of the strong influence of abiotic environmental factors on the distribution of animals (SIIVONEN and KALELA, 1937; KALELA, 1952; SIIVONEN, 1952).

Experimental research work directly related to field observations has greatly helped in verifying these ideas. The literature on these problems is vast; here only some outstanding Finnish and Swedish authors can be referred to, namely KROGERUS (1932), LINDROTH (1949), KOSKIMIES and LAHTI (1964).

(3) Interactions between individuals of the same species become more intense with increasing abundance, and consequently intraspecific competition increases. LEVINS (1968) and ROTHSTEIN (1973) suggest the idea of increase of niche breadth for abundant species. The increasing intraspecific competition and greater abundance as complementary factors lead to wider distribution.

(4) Species in temperate regions tend to have broader niches than in tropical regions. KLOPFER and MACARTHUR (1960) and LEVINS (1968) regarded this as an adaptation to the unstable environment. The amplitude of temperature tolerance of the European anurans, plotted against the latitudes of distribution, is an example of this relationship (Table 2). The temperature tolerance in relation to niche space is given by HUTCHINSON (1968). The amplitude of tolerance increases towards the higher latitudes, but the rate decreases (Fig. 3).

The six anuran species in Table 2 belong to four families. A comparison of temperature tolerance and range size in this sample within families is only possible in the Ranidae and Bufonidae, each of which is represented by two species. In both families the species with the wider temperature tolerance has the greater range; the ranges of *Rana temporaria* L. and *Bufo bufo* (L.) are greater than those of *R. arvalis* Nilsson and *Bufo viridis* Laurenti. However, these relations must be regarded critically. The difference in temperature tolerance between *R. tempo-*

Table 2. The temperature tolerance and range of some anuran species

| Species | Northern boundary of distribution | Temperature tolerance[a] | Range size $km^2 \cdot 10^6$ |
|---|---|---|---|
| *R. temporaria* | 71° | 30.2 | 12.9 |
| *R. arvalis* | 68° 45′ | 30.1 | 10.9 |
| *B. bufo* | 66° | 29.5 | 18.2 |
| *P. fuscus* | 59° | 29.1 | |
| *B. viridis* | 58° 20′ | 28.5 | 17.2 |
| *H. arborea* | 57° | 27.7 | |

[a] The temperature tolerance is regarded as the interval between the temperature of heat collapse of the larvae and the temperature at which embryonic development stops.

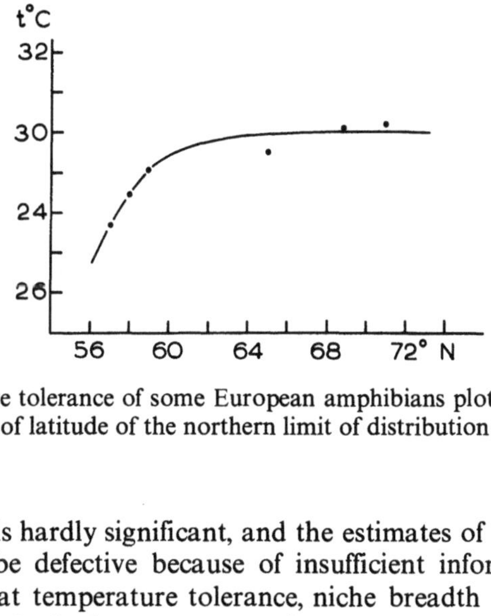


Fig. 3. The temperature tolerance of some European amphibians plotted against the degree of latitude of the northern limit of distribution

*raria* and *R. arvalis* is hardly significant, and the estimates of the range of *B. bufo* and *B. viridis* may be defective because of insufficient information. Generally, though, it seems that temperature tolerance, niche breadth and range increase towards higher latitudes in the temperate regions.

SCRIBER (1973) has studied the specialization of larval feeding in the family Papilionidae and found that the percentage of generalized feeders was significantly larger in the higher latitudes (Fig. 4). The trophic niche of the Papilionidae is thus broader in the higher latitudes, which provides possibilities for an increase in range.

(5.) Consequently, it seems that distributional limits are dependent on niche breadth, which is related to the available food resource and the physical/chemical qualities of the habitat. It is concluded that range is one of the dimensions of the fundamental niche and is an integral part of the concept of niche-phenotype model as proposed by MACARTHUR (1968) (see also VAN VALEN, 1965).

The increase of range towards higher latitudes is general but not universal. Actually the increase occurs towards the "impoverished environments", as stated above (KARR, 1975) or, expressed in the terms of MACARTHUR (1969, 1972), towards environments with a smaller resource spectrum.

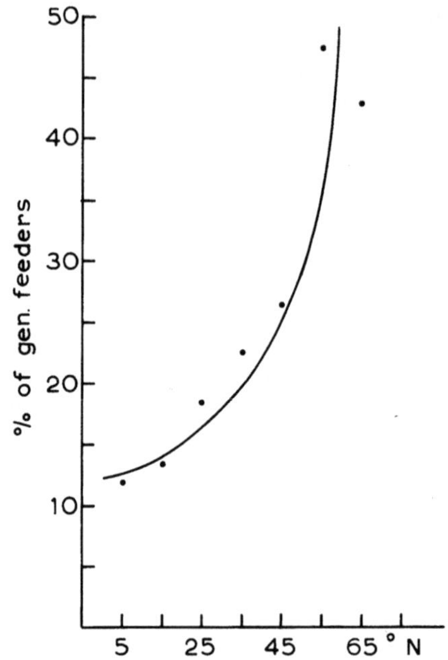


Fig. 4. The percentage of general feeders in the family Papilionidae plotted against the degree of latitude (According to data given by SCRIBER, 1973)

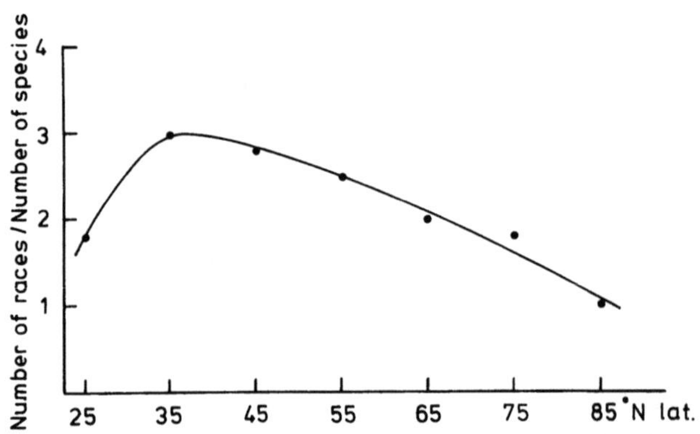


Fig. 5. Number of races per species (Ra/$S$) plotted against the degree of latitude

This is, as mentioned, the case concerning the desert environment, e.g. the Sahara. From the Equator to 20–30° N there is a steep gradient of decreasing number of species and increasing range size. This trend certainly has many causes, but the main factor seems to be the reduced supply of water (moisture), a physical component of the environment. From the Sahara northwards the number of species increases again and the range size decreases up to 30–40° N. Further north towards lower resource diversity, the range size increases while the number of species decreases.

The "unpredictability" of the climatic conditions in the temperate regions is also important, particularly for the occurrence of clines and of phenotypic variability. Variability is already at its maximum between 30° and 40° N and decreases further northwards towards its lowest value in the tundra. This is shown by the variability factor in Fringillidae expressed as the number of races per number of species (Ra/S), calculated for latidudinal squares (Fig. 5) (see also SCHWARZ, 1971).

The range as used in this study is expressed as an average value. This means that an increase in average range is the result of fewer species with smaller ranges and/or more species with larger ranges. The composition of species changes continuously along the gradient, and those best adapted to the environment tend to dominate.

Accompanying the decline of the resource spectrum there is, as we have seen, a widening of niche breadth which is followed by an increase in range. This may be interpreted as compensation enabling maintenance of an optimal population size.

## Range in the Tundra and Arctic-Alpine Tundra

Nine species of the three families, Sylviidae, Emberizidae and Fringillidae, belong to the tundra fauna between 70 and 80° N and 10 and 30° E. They are the true tundra species, *Calcarius lapponicus* (L.), *Plectrophenax nivalis* (L.) and to a lesser degree, *Acanthis flavirostris* (L.); the boreal forest (taiga) species, *Phylloscopus trochilus* (L.), *Phylloscopus borealis* (Blas.), *Acanthis flammea* (L.) and *Fringilla montifringilla* (L.); and finally *Acrocephalus schoenobaenus* (L.) and *Emberiza schoeniclus* (L.), two species associated with water and wet meadows. *Acrocephalus schoenobaenus* only penetrates as far as 70° N in Scandinavia. The range of these ecological groups is seen in Table 3.

The species of the tundra group are characterized by the smallest range. They are adapted to rigorous environments and to special, well-defined habitats (HAFTORN, 1971; DEMENT'EV and GLADKOV, 1968).

The boreal forest (taiga) species penetrate the forest-shrub and dwarf-shrub tundra and show a greater diversity of habitats and environments (DEMENT'EV and GLADKOV, 1968; HAFTORN, 1971). Their range is almost twice that of the former group.

Table 3. The range size of tundra species

| True tundra species | | Forest, shrub tundra | | Water-near habitats | |
|---|---|---|---|---|---|
| Species | Range $km^2 \cdot 10^6$ | Species | Range $km^2 \cdot 10^6$ | Species | Range $km^2 \cdot 10^6$ |
| *C. lapponicus* | 9.5 | *P. trochilus* | *14.3* | *A. schoenobaenus* | 13.4 |
| *P. nivalis* | 8.8 | *P. borealis* | *13.7* | *E. schoeniclus* | 28.5 |
| *A. flavirostris* | 7.3 | *F. montifringilla* | *12.0* | | |
| | | *A. flammea* | *20.6* | | |
| Mean | $\bar{x} = 8.5$ | | $\bar{x} = 15.1$ | | $\bar{x} = 20.9$ |

Only two species belong to the third group. Along the water courses they penetrate a vast area and inhabit different habitats from the steppe and semi-desert to the taiga and shrub tundra. Their range is great.

It is obvious that the niche breadth of the true tundra species is smaller than that of the boreal forest species. It must be noted, however, that this applies to the spatial niche, not the nutritional. There is a multitude of food resources even in the tundra, but most of the "packages" lie beneath or near the threshold level and/or fluctuate and provide an insufficient basis for specialization. The snow bunting, lapland bunting and twite claim particular habitats and nesting sites but are hardly distinguishable in their choice of food (LÖVENSKJOLD, 1964; DEMENT'EV and GLADKOV, 1970; HAFTORN, 1971). This applies only to the nominate race of the twite, *Acanthis f. flavirostris*. This species has a great number of races, distributed in the various parts of its disjunct range. In the Caucasus in Central Asia it is almost a species of mountain environments.

As mentioned above, the range of the specialized tundra species is smaller than that of the boreal forest species occurring in tundra. The trend of increasing range and widening of niche breadth observable in temperate regions is reversed in the Arctic with increasing latitude. The cause seems to be a strong adaptation to the special environmental conditions and a loss of ecological plasticity.

This explanation is also useful for the alpine tundra, e.g., Hardangervidda. The difference in range size compared with arctic lowland tundra is slight. The average range for nine species of the same families is $14.2 \cdot 10^6$ km$^2$, and that for the six species found on Hardangervidda is $14.6 \cdot 10^6$ km$^2$. The composition of species in both biomes is almost the same; only three of the species occurring in the tundra are absent from the Hardangervidda, namely *P. borealis*, *A. schoenobaenus* and *F. montifringilla*.

## Conclusions

1. An increase in range towards environments with lower resource/diversity has been demonstrated in passerine birds.
2. The increase in range from tropical rain forest over a resource gradient towards desert and tundra can be described as a logarithmic function.
3. The range size is regarded as one of the dimensions of the fundamental niche.
4. The increase in range is associated with a decrease in the number of species, and increase of niche breadth.
5. Rigorous environmental conditions as in the tundra and desert demand special adaptations which cause reductions in range and in niche breadth.

*Acknowledgements.* I am grateful to Dr. T. OUREN, of the Norwegian School of Economic and Business Administration, Bergen, for instructions and for assistance in the use of cartographic methods. I am indebted to Dr. R. LIE of the Zoological Museum, University of Bergen, for the calculation of the logarithmic equations.

Many years ago I had the pleasure of discussing the ideas developed in this paper with friends and colleagues at the Department of Zoology and Animal Ecology, University of Lund, Sweden, especially Dr. P. BRINCK, Dr. E. DAHL and Dr. C.H. LINDROTH. My friends Dr. S.-A. BENGTSON, Zoological Museum, Bergen and Dr. G. RUDEBECK, Department of

Zoology, University of Lund, have generously helped me with valuable information and guided me in the literature on African birds. I would add my thanks to J. B. L. MATTHEWS, Ph. D., for correcting the language.

## References


ALI, S., RIPLEY, S. D.: Handbook of the Birds of India and Pakistan. Vol. 8. Bombay-London-New York: Oxford University Press 1973.

DEMENT'EV, G. P., GLADKOV, N. A.: Birds of the Soviet Union. Vols. 5 and 6. Jerusalem: Israel Program for Scientific Translations. 1968, 1970.

DOBZHANSKY, T.: Evolution in the tropics. Am. Sci. **38**, 209–221 (1950).

EKMAN, S.: Djurvärldens utbredningshistoria på Skandinaviska halvön. Stockholm: Albert Bonniers Förlag 1922.

ETCHÉCOPAR, R. D., HÜE, E.: The Birds of North Africa. Edinburgh-London: Oliver & Boyd 1967.

GISLÉN, T.: The number of animal species in Sweden with remarks on some rules of distribution especially of the microfauna. Lunds Universitets Årsskrift. N.F. Ser. 2. **36**, 1–23 (1940).

HAFTORN, S.: Norges fugler. Oslo: Universitetsforlaget 1971.

HUTCHINSON, G. E.: When are species necessary? In: LEWONTIN, R. C. (Ed.): Population Biology and Evolution, pp. 177–186. Syracuse: Univ. Press 1968.

KALELA, O.: Changes in the geographic distribution of Finnish birds and mammals in relation to recent changes in climate. Fennia **75**, 38–51 (1952).

KARR, J. R.: Production, energy pathways, and community diversity in forest birds. In: GOLLEY, F. B., MEDINA, E. (Eds.): Tropical ecological systems. New York-Heidelberg-Berlin: Springer 1975.

KAURI, H., LIE, R.: Latitudinal changes of the area-sizes (in press).

KLOPFER, P. H., MACARTHUR, R. H.: Niche size and faunal diversity. Am. Naturalist **5**, 293–300 (1960).

KOSKIMIES, J., LAHTI, L.: Cold-hardiness of the newly hatched young in relation to ecology and distribution in the species of European ducks. Auk **81**, 281–307 (1964).

KROGERUS, R.: Über die Ökologie und Verbreitung der Arthropoden der Triebsandgebiete an den Küsten Finnlands. Acta Zool. Fennica **12**, 1–308 (1932).

LEVINS, R.: Evolution in changing environments. Princeton: Univ. Press 1968.

LINDROTH, C. H.: Die fennoskandischen Carabidae. III. 1–910. Göteborg, K. Vet. and Vitt. Samh. Handl. B **4:3** (1949).

LÖVENSKJOLD, H. L.: Avifauna Svalbardensis. Norsk Polarinstitutt, Skrifter **129**, 1–460 (1964).

MACARTHUR, R. H.: The theory of the niche. In: LEWONTIN, R. C. (Ed.): Population Biology and Evolution, pp. 159–176. Syracuse: Univ. Press 1968.

MACARTHUR, R. H.: Species packing or what competition minimizes. Proc. Nat. Acad. Sci. U.S. **64**, 1369–1371 (1969).

MACARTHUR, R. H.: Geographical ecology. Patterns in the distribution of species. New York: Harper and Row 1972.

MACARTHUR, R. H., WILSON, E. O.: The Theory of Island Biogeography. Princeton: Princeton University Press 1967.

MACKWORTH-PRAED, C. W., GRANT, C. H. B.: Birds of Eastern and North Eastern Africa. African Handbook of Birds, Ser. I., Vol. 2. London: Longmans 1960.

MACKWORTH-PRAED, C.W., GRANT, C.H.B.: Birds of Southern Third of Africa. African Handbook of Birds. Ser. II, Vol. 2. London: Longmans 1963.

MACKWORTH-PRAED, C.W., GRANT, C.H.B.: Birds of West Central and Western Africa. African Handbook of Birds. Ser. III, Vol. 2. London: Longman 1973.

ROBINSON, A. H., SALE, R. D.: Elements of Cartography. New York: Wiley 1969.

ROTHSTEIN, S. I.: The niche-variation model—is it valid? The Am. Naturalist **107**, 598–620 (1973).

Schwarz, S. S.: Population structure of biogeocenoses of south tundra, their productivity and stability. In: Wielgolaski, F. E., Rosswall, T. (Eds.): International Biological Programme: Tundra Biome. Proc. IV. Int. Meeting on the Biological Production of Tundra, pp. 48–51. Stockholm: IBP Tundra Biome Steering Comm. 1972.

Scriber, J. M.: Latitudinal gradients in larval feeding specialization of the world. Papilionidae (Lepidoptera). Psyche **80**, 355–373 (1973).

Siivonen, L.: On the influence of the climatic variations of recent decades on game economy. Fennia **75**, 77–88 (1952).

Siivonen, L., Kalela, O.: Über die Veränderungen in der Vogelfauna Finnlands während der letzten Jahrzehnte und die darauf einwirkenden Faktoren. Acta Soc. Fauna Flora Fenn. **60**, 606–634 (1937).

Slobodkin, L. B., Sanders, A. L.: On the contribution of environmental predictability to species diversity. In: Diversity and Stability in Ecological Systems, pp. 82–95. Brookhaven Symp. Biol. **22** (1969).

Van Valen, L.: Morphological variation and width of ecological niche. Amer. Naturalist **993**, 377–390 (1965).

Voous, K. H.: Die Vogelwelt Europas und ihre Verbreitung. Hamburg: Paul Parey 1962.

Wielgolaski, F. E.: Principles in use of wide scale models on tundra data. In: Wielgolaski, F. E. (Ed.): Fennoscandian tundra ecosystems. Part 2: Animals and systems analysis, pp. 245–250. Berlin-Heidelberg-New York: Springer 1975.

Williams, C. B.: Patterns in the balance of nature and related problems in quantitative ecology. London: Academic Press 1964.

Willson, M. F.: Avian niche size and morphological variation. Am. Naturalist **103**, 531–542 (1969).

# 1.1. Faunal Structure of Research Areas

## Faunal Structure of Investigated Areas at Kevo, Finland

E. HAUKIOJA and S. KOPONEN

The Finnish IBP tundra sites consist of two forest habitats, the birch site and the pine site, and of a site on the border between the subalpine and alpine region with scattered birch formations (SONESSON et al., 1975).

Special features of the fauna are based on the vegetation of the sites (KÄRENLAMPI and KAUHANEN, 1972) and also the low altitude and rather continental climate.

The general features of the diversity of the fauna are given in Table 1, including the number of species of some well-known invertebrate groups. The number of mammalian species regularly found in the areas is about ten, and that of commonly breeding birds roughly 20, but the total number of vertebrates is much higher, as seen in Table 2. Vertebrates are presented in systematic order in this presentation, but invertebrates are discussed according to collecting methods and vegetation layers.

Table 1. Species numbers of some invertebrate groups at Kevo

| Group | Birch forest | Pine forest | Low alpine Heath | Total |
|---|---|---|---|---|
| Araneae | 71 | 54 | 37 | 93 |
| Coleoptera (ground) | 61 | 67 | 42 | 110 |
| Lepidoptera | 91 | not counted | not counted | >120 |
| Homoptera: Cicadina | 9 | 4 | 5 | 11 |
| Mollusca: Gastropoda | 5 | 2 | 0 | 5 |
| Phalangida | 1 | 1 | 1 | 1 |
| Chilopoda | 1 | 1 | 1 | 1 |

## Vertebrates

**Mammals.** The semi-domesticated reindeer *(Rangifer tarandus)* is the most conspicuous mammal of the area and may be encountered at each of the study sites. Their density is of the order of four animals per $km^2$.

Among rodents *Clethrionomys rufocanus* and *C.rutilus* occur in the birch zone, the latter preferring higher areas but both occurring at the birch site (for densities, see HAUKIOJA and KOPONEN, 1975). *C.rutilus* is also found at the alpine site.

Table 2. Vertebrate species on the three IBP sites at Kevo, usually breeding within the area. $H_g$ – granivore (seeds), $H_h$ – grazer, browser, H – herbivores, $C_i$ – carnivores on invertebrates, $C_v$ – carnivores on vertebrates, O – omnivores: A – abundant, C – common, U – uncommon, R – rare

| Ecosystem | | Relative abundance | | |
|---|---|---|---|---|
| Component | Function | Pine forest | Birch forest | Heath |
| Amphibia | | | | |
| *Rana temporaria* | $C_i$ | R | C | |
| Reptilia | | | | |
| *Lacerta vivipara* | $C_i$ | U | R | |
| Aves | | | | |
| *Aquila chrysaëtos* | $C_v$ | R | R | |
| *Buteo lagopus* | $C_v$ | U | C | C |
| *Accipiter gentilis* | $C_v$ | R | | |
| *Pandion haliaëtus* | $C_v$ | U | U | |
| *Falco rusticolus* | $C_v$ | | | R |
| *Falco columbarius* | $C_v$ | C | U | R(?) |
| *Falco tinnunculus* | $C_v$ | U | | |
| *Lagopus lagopus* | $H_g$/H | R | C | U |
| *Lagopus mutus* | $H_g$/H | | | U |
| *Tetrao urogallus* | $H_g$/H | U | | |
| *Pluvialis apricaria* | $C_i$ | | | C |
| *Eudromias morinellus* | $C_i$ | | | C |
| *Numenius phaeopus* | $C_i$ | | R(?) | U |
| *Stercorarius longicaudus* | $C_v$ | | | C |
| *Cuculus canorus* | $C_i$ | | U | |
| *Surnia ulula* | $C_v$ | R(?) | U | |
| *Dendrocopos major* | $C_i$ | R | | |
| *Dendrocopos minor* | $C_i$ | | R | |
| *Picoides tridactylus* | $C_i$ | | R | |
| *Corvus corax* | O | C | C | U |
| *Perisoreus infaustus* | O | C | U(?) | |
| *Parus major* | $C_i$/O | C | C | |
| *Parus cinctus* | $C_i$/O | C | C | |
| *Parus montanus* | $C_i$/O | C | C | |
| *Turdus pilaris* | O | C | C | |
| *Turdus philomelos* | O | R(?) | R | |
| *Turdus iliacus* | O | C(?) | A | |
| *Oenanthe oenanthe* | $C_i$ | | | A |
| *Phoenicurus phoenicurus* | $C_i$ | C | C | |
| *Luscinia svecica* | $C_i$ | | A | |
| *Erithacus rubecula* | $C_i$ | | R | |
| *Phylloscopus trochilus* | $C_i$ | C(?) | A | |
| *Phylloscopus borealis* | $C_i$ | | R | |
| *Muscicapa striata* | $C_i$ | C(?) | C | |
| *Ficedula hypoleuca* | $C_i$ | C(?) | C | |
| *Prunella modularis* | $C_i$/$H_g$ | | C | |
| *Anthus pratensis* | $C_i$ | | A | A |
| *Anthus trivialis* | $C_i$ | C(?) | U | |
| *Anthus cervinus* | $C_i$ | | | U |
| *Motacilla alba* | $C_i$ | | C | C |
| *Motacilla flava* | $C_i$ | | C | C |
| *Carduelis flammea* | $H_g$/$C_i$ | C(?) | A | |
| *Pyrrhula pyrrhula* | $H_g$ | R | | |
| *Pinicola enucleator* | O | R | C | |

Table 2 (continued)

| Ecosystem | | Relative abundance | | |
|---|---|---|---|---|
| Component | Function | Pine forest | Birch forest | Heath |
| Aves | | | | |
| *Fringilla montifringilla* | $C_i/H_g$ | U | A | |
| *Emberiza schoeniculus* | $C_i$ | | C | |
| *Calcarius lapponicus* | $C_i/H_g$ | | | A |
| *Plecrophenax nivalis* | O | | | C |
| Mammalia | | | | |
| *Sorex araneus* | $C_i$ | U(?) | C | U(?) |
| *Sorex minutus* | $C_i$ | C | C | U(?) |
| *Lepus timidus* | $H_h$ | C | C | |
| *Sciurus vulgaris* | $H_g/H$ | R–A | | |
| *Lemmus lemmus* | H | | R–A | R–A |
| *Clethrionomys rutilus* | H | R–C(?) | R–C | R–C |
| *Clethrionomys rufocanus* | H | | R–C | R–C |
| *Microtus oeconomus* | H | C | R–C | R–C |
| *Vulpes vulpes* | $C_v$ | | C | U(?) |
| *Alopex lagopus* | $C_v$ | C | R | R |
| *Mustella erminea* | $C_v$ | C(?) | C | |
| *Mustella rixosa* | $C_v$ | R | C | |
| *Gulo gulo* | $C_v$ | U | R | R |
| *Alces alces* | H | C | U | |
| *Rangifer tarandus* | $H_h$ | | A | A |

Lemmings *(Lemmus lemmus)* were found at the study sites only during the very high lemming peak in 1970–1971. *Microtus oeconomus* also occurs in birch forests although it prefers more open and moist areas.

The hare *(Lepus timidus)* is common in the area and, like the above, fluctuates violently in numbers.

Shrews (*Sorex araneus* and *S. minutus*) are common insectivores.

Among predatory mammals *Mustela erminea* and *M. rixosa* are the most abundant, but also the fox *(Vulpes vulpes)* is common, especially at lower elevation. The arctic fox *(Alopex lagopus)* is an occasional visitor to the area, but less regularly so than the wolverine *(Gulo gulo)*. See also Table 2.

Table 3 gives density estimates of birds by the line census method in the neighborhood of the birch site in 1972. According to the table, the willow warbler *(Phylloscopus trochilus)*, the brambling *(Fringilla montifringilla)* and the redpoll *(Carduelis flammea)* are the dominant species. The rather high number of pied flycatchers *(Ficedula hypoleuca)* at the site (Table 3) is the result of putting up nest-boxes to the area.

Among species not observed in the line surveys, the following ones may be regarded as typical of the birch and pine sites: the rough-legged buzzard *(Buteo lagopus)*, the hawk owl *(Surnia ulula)*, the willow tit *(Parus montanus)*, the Siberian tit *(Parus cinctus)* and the redwing *(Turdus iliacus)*, although the first two are not common (Table 2).

Table 3. Density estimates of birds based on line censuses performed between June 6 and June 20, 1972 in the neighborhood of the Kevo birch forest site. Area of the main field 73 hectares

| Species | Pairs $km^{-2}$ |
|---|---|
| *Lagopus lagopus* | 1.4 |
| *Anthus trivialis* | 2.7 |
| *Motacilla flava* | 2.7 |
| *Acrocephalus schoenobaenus* | 2.7 |
| *Phylloscopus trochilus* | 52 |
| *Ph. borealis* | 1.4 |
| *Ficedula hypoleuca* | 9.6 |
| *Oenanthe oenanthe* | 6.8 |
| *Phoenicurus phoenicurus* | 2.7 |
| *Luscinia svecica* | 1.4 |
| *Fringilla montifringilla* | 25 |
| *Carduelis flammea* | 16 |
| *Pinicola enucleator* | 1.4 |
| *Emberiza schoeniclus* | 2.7 |
| | $\sum$ 128 |

The bird fauna of the alpine site is not representative of alpine heaths in the surroundings because of its situation on the summit of a hill with a limited area of heath-like vegetation. Wheatears *(Oenanthe oenanthe)*, meadow pipits *(Anthus pratensis)* and lapland buntings *(Calcarius lapponicus)* are abundant in the neighborhood of the site (Table 2), but for example the snow bunting *(Plectrophenax nivalis)*, the ptarmigan *(Lagopus mutus)* and the dotterel *(Eudromias morinellus)*, typical of larger alpine heaths, do not breed at the site itself.

The common lizard *(Lacerta vivipara)* and the common frog *(Rana temporaria)* can be found, in small numbers, at least in birch and pine forest sites.

## Invertebrates

**The Soil and Ground Layer.** The microfaunal material was collected using a steel borer (area about 10.5 $cm^2$). The animals were extracted using dry Tullgren funnels with 40 W lamps for five days (Acarina and Collembola) or by wet Baermann funnels with 40 W lamps in 24 hrs (Nematoda).

The figures for microfauna are taken from the year 1973. The numbers of Acarina and Collembola in June, July, and August of 1973 from soil with a moss layer varying from 0 to 6 cm and a humus layer from 1 to 8 cm, are given in Table 4 as individuals $m^{-2} \pm$S.E. The numbers of Acarina in low alpine heath and in birch forest were clearly higher than in pine forest. The numbers of Collembola were highest in low alpine heath, second highest in birch forest, and lowest in pine forest. The lowest individual numbers of Collembola in each habitat were noted in June. The dominant group among Acarina was Cryptostigmata, the percentual proportions of which in birch forest, low alpine heath, and pine forest were 67%, 77%, and 72%, respectively (August 1973). The most abundant Col-

Table 4. The numbers of Acarina and Collembola $m^{-2}$ (with S.E.) at the IBP sites at Kevo, 1973

| | June | July | August |
|---|---|---|---|
| Birch forest | | | |
| Acarina | 109417±10256 ($n$=21) | 136712±22288 ($n$=30) | 210975±21095 ($n$=26) |
| Collembola | 15051± 2199 ($n$=21) | 23928± 3653 ($n$=28) | 39115± 4734 ($n$=26) |
| Pine forest | | | |
| Acarina | 116847±25719 ($n$=12) | 103958±12323 ($n$=17) | 114359±21904 ($n$=17) |
| Collembola | 18957± 4253 ($n$=12) | 26987± 4802 ($n$=17) | 13550± 2818 ($n$=17) |
| Low alpine heath | | | |
| Acarina | 212666±17064 ($n$= 6) | 102377±25742 ($n$=18) | 171461±22845 ($n$=18) |
| Collembola | 28037± 5864 ($n$= 6) | 38073± 7246 ($n$=18) | 52038± 9635 ($n$=18) |

Table 5. Live biomass (g) of Acarina and Collembola $m^{-2}$ at the IBP sites at Kevo, August 1973

| | Birch forest | Pine forest | Low alpine heath |
|---|---|---|---|
| | $\bar{x}$ S.E. | $\bar{x}$ S.E. | $\bar{x}$ S.E. |
| Acarina | 2.754±0.706 | 1.634±0.502 | 3.460±0.817 |
| Collembola | 0.403±0.152 | 0.232±0.079 | 0.612±0.133 |
| $n$ | 7 | 3 | 3 |

lembola family was Isotomidae. The live biomass of Acarina and Collembola in some samples in August, 1973, is given in Table 5. The highest biomasses in both groups were in low alpine heath and the lowest one in pine forest.

The number of large, active Nematoda (note the extracting method) was 115000 in birch forest, 42000 in pine forest, and in low alpine heath 33000 ind. $m^{-2}$.

The numbers of main studied microfaunal groups, Acarina and Collembola, were rather similar to data for the forest floor in Central Finnish Lapland (HUHTA et al., 1967).

The mesofaunal samples were taken using 25×25 cm quadrats and extracted by large dry Tullgren funnels (see HUHTA, 1972) with 40 W lamps for seven days.

The numbers of abundant mesofaunal groups from the year 1973 are given in Tables 6 and 7. The total individual numbers of mesofauna per $m^2$ were clearly higher in pine and birch forests than in low alpine heath. The highest densities of Araneae, Coleoptera adults and larvae, and Diptera adults were from birch forest, and of Hemiptera and Diptera larvae from pine forest. The numbers of all studied groups were higher than those in studies made by HUHTA et al. (1967) from Central Finnish Lapland; the same was true when the data were compared with those of FRANZ (1950) from subalpine coniferous forests. One, probably very important, reason could be the difference in extracting methods.

Table 6. Density of soil and ground layer mesofauna (numbers $m^{-2}$ with S.E.) at the birch forest site in 1973

| | June | July | August |
|---|---|---|---|
| Mollusca | 0 | 2.0 ± 1.4 | 1.3 ± 0.9 |
| Chilopoda | 0 | 0 | 7.3 ± 4.2 |
| Phalangida | 12.0 ± 9.3 | 1.0 ± 1.0 | 0.7 ± 0.7 |
| Araneae | 267.2 ± 67.9 | 228.0 ± 36.9 | 220.7 ± 23.1 |
| Linyphiidae and Theridiidae | 238.7 ± 64.3 | 200.0 ± 39.3 | 173.3 ± 21.2 |
| Insecta adults | 381.3 ± 42.0 | 315.0 ± 28.1 | 180.7 ± 21.0 |
| Coleoptera | 53.0 ± 12.6 | 59.0 ± 9.7 | 68.7 ± 11.3 |
| Diptera | 189.3 ± 32.8 | 78.0 ± 17.5 | 10.0 ± 4.8 |
| Hemiptera | 118.7 ± 35.7 | 109.0 ± 18.7 | 66.7 ± 11.9 |
| Insecta larvae | 314.7 ± 43.1 | 215.0 ± 42.7 | 432.0 ± 70.9 |
| Coleoptera | 213.3 ± 30.3 | 138.0 ± 33.5 | 134.7 ± 16.4 |
| Diptera | 101.3 ± 23.9 | 73.0 ± 17.4 | 286.0 ± 66.2 |
| Total mesofauna | 975.7 ± 124.8 | 760.0 ± 67.2 | 839.9 ± 70.2 |
| *n* | 12 | 16 | 24 |

Table 7. Density of soil and ground layer mesofauna (numbers-m² with S.E.) at low alpine and pine forest sites at Kevo in 1973

| | June | | July | | August | |
|---|---|---|---|---|---|---|
| | Alpine heath | Pine forest | Alpine heath | Pine forest | Alpine heath | Pine forest |
| Araneae | 29.3 ± 11.3 | 112.0 ± 47.3 | 100.0 ± 35.9 | 106.0 ± 32.3 | 115.7 ± 29.7 | 120.0 ± 11.8 |
| Insecta adults | 224.0 ± 51.9 | 704.0 ± 242.8 | 138.0 ± 35.8 | 498.8 ± 61.2 | 88.0 ± 27.6 | 176.0 ± 32.4 |
| Insecta larvae | 222.3 ± 53.9 | 376.0 ± 116.6 | 120.0 ± 37.4 | 316.5 ± 38.0 | 193.2 ± 63.7 | 409.6 ± 77.3 |
| Total mesofauna | 478.3 ± 93.3 | 1197.3 ± 340.1 | 366.3 ± 63.2 | 923.3 ± 104.5 | 400.1 ± 85.8 | 710.6 ± 80.9 |
| *n* | 6 | 6 | 8 | 8 | 10 | 10 |

Some groups of mesofauna are absent in northern Finnish Lapland (e.g. terrestrial Isopoda and Diplopoda) or are very rare (Pseudoscorpionida) and therefore do not appear in the material. Densities of Lumbricidae (*Dendrobaena octaedra*) were very low; the same was true of Phalangida (*Mitopus morio*, 4 ind. $m^{-2}$ in birch forest), Chilopoda (*Lithobius curtipes*, max. 7 ind. $m^{-2}$ in birch forest), and Mollusca.

The surface-active mesofauna was studied using pitfall traps with ethylene glycol over a number of years (KOPONEN, 1971). Great fluctuations in both absolute and relative numbers were observed between different years. The following figures are from the year 1973 (periods: June 5–July 5, July 5–August 5, August 5–September 5). Adult Coleoptera formed the most numerous group both in pine (29.3%) and birch forests (24.5%), and Phalangida in low alpine heath (23.7%). Other abundant groups were Diptera, Araneae, Hemiptera, and Hymenoptera (Table 8); in pine forest Psocoptera was also common and in low alpine heath Cicadina (Hom.).

The diversity of the sites studied can be shown using numbers of spider species, especially that of Linyphiidae. The number of spider species in the

Table 8. The percentual proportion of the most important mesofaunal groups in the pitfall trap material at IBP sites at Kevo, 1973

| Group | Total number | Birch forest (%) | Pine forest (%) | Low alpine heath (%) | Mean rank of group |
|---|---|---|---|---|---|
| Coleoptera ad. | 3511 | 24.5 | 29.3 | 15.3 | 2.0 |
| Diptera ad. | 2268 | 20.2 | 11.2 | 16.1 | 2.6 |
| Phalangida | 2298 | 18.9 | 8.1 | 23.7 | 3.3 |
| Araneae | 2566 | 15.4 | 22.0 | 13.9 | 3.6 |
| Hemiptera | 860 | 4.4 | 10.0 | 1.5 | 5.7 |
| Hymenoptera ad. | 1175 | 13.6 | 5.4 | 4.6 | 6.0 |
| Psocoptera | 503 | 0.4 | — | 8.4 | — |

trapped material was in birch forest 42 (28 Linyphids), in pine forest 39 (24 Linyphids), and in low alpine heath 23 (11 Linyphids).

Abundant mesofaunal species on the ground at Kevo, not listed above, were e.g. *Euconulus fulvus* (Moll.) and *Limotettix transversus* (Hom., Cicadina), the former in birch forest and the latter especially in pine forest and low alpine heath. Abundant Coleoptera species in birch forest were *Tachinus elongatus* and *Mannerheimia arctica*, in low alpine heath *Hypnoides rivularius* and *Notiophilus germinyi*, and in pine forest *Tachinus elongatus* and *Mycetoporus brunneus*. The most abundant species of spiders in the trapped materials, ranked according to numbers, were in birch forest *Pardosa eiseni*, *Hahnia ononidum*, and *Pelecopsis mengei*, in low alpine heath *Pardosa eiseni*, *P. tarsalis*, and *Oxyptila rauda*, and in pine forest *Pardosa eiseni*, *Hahnia ononidum*, and *Alopecosa aculeata*. See Koponen (1975) for more details of the spider population.

**The Field Layer.** The material was collected by netting. Each sample consisted of 100 net sweeps, and collecting was carried out only in fine weather. The net material comprised over 8500 invertebrate specimens collected during the years 1971–73. A more detailed description of the mesofauna inhabiting the field layer is given by Koponen and Ojala (1974).

The most numerous groups in birch forest were Diptera (35.3%), Homoptera, Hymenoptera adults, and Araneae; in low alpine heath Diptera (32.6%), Homoptera, Hymenoptera larvae, and Hymenoptera adults; and in pine forest Homoptera (42.1%), Diptera, and Hymenoptera adults (Tables 9–10). The most important groups by biomass were Symphyta (Hym.) larvae in birch forest and low alpine heath, and Cicadina (Hom.) in pine forest.

The most common species of the important herbivorous group Cicadina were in birch forest *Thamnotettix confinis* and *Colladonus torneellus;* in low alpine heath *Limotettix transversus* and *Diplocolenus limbatellus;* and in pine forest *L. transversus.*

Other numerous herbivorous species in birch forest were *Mecomma ambulans* (Hem.), *Cidaria caesiata* (Lep.), *Pristiophora mollis* (Hym., also common in low alpine heath), *Polydrosus ruficornis* (Col., also common in low alpine heath), and *Corymbites costalis* (Col.).

Table 9. The percentual proportion of some herbivorous groups of the total herbivore mesofauna in the field layer of birch forest. *n* based on numbers, *b* based on biomass

| | June | | | | July | | | | August | | | | September | |
|---|---|---|---|---|---|---|---|---|---|---|---|---|---|---|
| | 1972 | | 1973 | | 1972 | | 1973 | | 1972 | | 1973 | | 1972 | |
| | *n* | *b* | *n* | *b* | *n* | *b* | *n* | *b* | *n* | *b* | *n* | *b* | *n* | *b* |
| Homoptera | 44.7 | 11.5 | 42.0 | 7.6 | 77.4 | 37.8 | 74.2 | 15.5 | 95.0 | 79.4 | 89.4 | 48.5 | 82.9 | 58.2 |
| Psyllina | 3.8 | 0.2 | — | — | 45.2 | 7.6 | 49.9 | 4.0 | 36.6 | 10.6 | 63.5 | 14.4 | 7.1 | 2.6 |
| Cicadina | 41.0 | 11.3 | 42.0 | 7.6 | 32.2 | 30.2 | 22.4 | 11.3 | 57.8 | 68.7 | 22.4 | 32.9 | 75.7 | 55.5 |
| Hymenoptera larvae | 17.8 | 35.9 | 10.0 | 5.6 | 4.7 | 54.4 | 19.7 | 82.3 | 0.6 | 11.3 | 1.8 | 1.0 | 1.4 | 10.9 |
| Lepidoptera larvae | 18.6 | 44.0 | 16.0 | 46.1 | 1.2 | 0.7 | 1.0 | 1.1 | 1.2 | 0.8 | 3.5 | 7.9 | 10.0 | 11.5 |
| Total ind. | 349 | | 50 | | 407 | | 407 | | 161 | | 170 | | 70 | |

Table 10. The percentual proportion of some herbivorous groups of the total herbivore mesofauna in the field layer of low alpine heath and pine forest at Kevo. *n* based on numbers, *b* based on biomass

| | Low alpine heath | | | | Pine forest | |
|---|---|---|---|---|---|---|
| | 1972 | | 1973 | | 1973 | |
| | *n* | *b* | *n* | *b* | *n* | *b* |
| Homoptera | 82.5 | 23.4 | 45.2 | 3.5 | 90.0 | 67.2 |
| Cicadina | 81.2 | 23.2 | 44.6 | 3.5 | 89.4 | 66.8 |
| Hymenoptera larvae | 11.0 | 67.5 | 47.7 | 88.3 | — | — |
| Lepidoptera larvae | 5.2 | 9.0 | 1.7 | 3.0 | 1.9 | 29.6 |
| Total ind. | 154 | | 175 | | 160 | |

The most numerous species of spiders in birch forest were *Theridion ohlerti*, *Gonatium rubellum*, and *Theridion varians* and in pine forest, *T. ohlerti* and *Bolyphantes index*; in low alpine heath with low field layer no species was particularly numerous. The most abundant genus of predatory Coleoptera was *Malthodes*.

**The Tree Layer.** In studying fauna inhabiting the tree layer main attention was directed towards birch forest, the dominant forest type of northernmost Fennoscandia. The material was collected by netting. A branch of a birch was pushed quickly into a net, cut off, stored in a plastic sack, and handled in the lab. The material from the years 1971–73 comprised approx. 8800 mesofaunal animals. The most numerous invertebrate groups from birch foliage at Kevo were as follows:

| | |
|---|---|
| Homoptera | 5504 ind. |
| Hymenoptera | 1138 ind. |
| Diptera adults | 830 ind. |
| Araneae | 352 ind. |
| Coleoptera | 312 ind. |
| Lepidoptera | 284 ind. |

Two abundant, gall-inhabiting groups are excluded from the above list: Cecidomyiidae larvae (Dipt.) and Eriphyidae mites (Arachn., Acarina).

Abundant herbivorous species of the birch were e.g. *Psylla hartigi* (Hom.), *Dineura viridiorsata* (Hym.), *Coeliodes* spp. (Col.), *Eriocrania* spp. (Lep.), and *Massalongia rubra* (Dipt.). A more detailed description of birch herbivores is given by HAUKIOJA and KOPONEN (1975) and by KOPONEN (1973).

The most abundant predator groups in birch foliage were Araneae and parasiting Hymenoptera. The most abundant spiders were *Araneus proximus* and *Philodromus cespitum*. Numerous groups not feeding on birch leaves or birch herbivores were Diptera adults and Psocoptera.

In low alpine heath with very scattered birch bush formation, the abundance of *Eriocrania* spp. larvae (cf. KOPONEN, 1974) and that of *Polydrosus ruficornis* was clearly higher than in the birch forest site.

In the poor material from pines, frequent herbivorous groups were Homoptera, Hemiptera, Hymenoptera larvae (e.g. *Diprion* spp.), and Coleoptera (many Curculionidae and Scolytidae species). Abundant predator groups were Araneae (*Bolyphantes index* dominating), Hemerobidae (Neur.), and parasiting Hymenoptera.

**Night-Flying Fauna.** The night-flying insect fauna of the birch forest was investigated using a 500 W blended lamp with a trap, based on JALAS' model (1960). The trapping period was from June 13 to September 12, 1973, including about 60 days of continuous light. The lamp was turned on at 21.00 hr and off at 05.00 hr.

About 79% of the total material belonged to the order Diptera, and the bulk of Diptera to the suborder Nematocera (92.6%). Approximately 10% of the material were Lepidoptera, and other groups frequently encountered were Hymenoptera, Homoptera, Trichoptera, and Coleoptera. Numerous species were e.g. *Chloroperla borealis* (Plec.), *Thamnottetix confinis* (Hom.), *Malthodes guttifer* and *Podistra pilosa* (Col.). Abundant Lepidoptera species in the early summer were *Ancylis myrtillana, Paraswammerdamia conspersella*, and *Argyroploce hyperboreana*, in mid-summer *Cidaria caesiata* and *Eana osseana*, and in late summer and autumn *Oporinia autumnata*.

Minimum numbers of the total insect fauna were recorded from the end of July to the beginning of August. When this minimum time was compared with temperature figures, no correlation was observed. The dates of the minimum corresponds, however, to the end of everlasting days in late summer in the area (July 24). Because this phenomenon was also observed with other insect catches using light traps in the Kevo area, the change in light conditions may have some inhibiting effects on the activity of insects or an their reactions to attractive light sources.

## References

FRANZ, H.: Bodenzoologie als Grundlage der Bodenpflege. Berlin: Akademie-Verlag 1950.

HAUKIOJA, E., KOPONEN, S.: Birch herbivores and herbivory at Kevo, northern Finland. In: WIELGOLASKI, F. E. (Ed.): Fennoscandian tundra ecosystems, Part 2: Animals and systems analysis, pp. 181–188. Berlin-Heidelberg-New York: Springer 1975.

HUHTA, V: Efficiency of different dry-funnel techniques in extracting Arthropoda from raw humus forest soil. Ann. Zool. Fennici **9**, 42–48 (1972).

HUHTA, V., KARPPINEN, E., NURMINEN, M., VALPAS, A.: Effect of silvicultural practices upon arthropod, annelid and nematode populations in coniferous forest soil. Ann. Zool. Fennici **4**, 87–143 (1967).

JALAS, I.: Eine leichtgebaute, leichttransportable Lichtreuse zum Fangen von Schmetterlingen. Ann. Ent. Fennici **26**, 44–50 (1960).

KÄRENLAMPI, L., KAUHANEN, H.: A direct gradient analysis of the vegetation of the surroundings of the Kevo Subarctic Station. Rep. Kevo Subarctic Res. Stat. **9**, 82–98 (1972).

KOPONEN, S.: On the abundance relations of mesofaunal groups in the ground layer of three subarctic habitats. Rep. Kevo Subarctic Res. Stat. **7**, 57–59 (1971).

KOPONEN, S.: Herbivorous invertebrates of the mountain birch at Kevo, Finnish Lapland. Rep. Kevo Subarctic Res. Stat. **10**, 20–28 (1973).

KOPONEN, S.: On the occurence and ecology of *Eriocrania spp. (Lep., Eriocraniidae)* and other mining insects of the birch in northernmost Fennoscandia in 1973. Rep. Kevo Subarctic Res. Stat. **11**, 52–64 (1974).

KOPONEN, S.: Spider populations in a subalpine birch forest. In: WIELGOLASKI, F. E. (Ed.): Fennoscandian tundra ecosystems. Part 2: Animals and systems analysis, pp. 66–72. Berlin-Heidelberg-New York: Springer 1975.

KOPONEN, S., OJALA, H.: On the mesofauna of the field layer of three subarctic habitats. Rep. Kevo Subarctic Res. Stat. **11**, 65–71 (1974).

SONESSON, M., WIELGOLASKI, F. E., KALLIO, P.: Description of Fennoscandian tundra ecosystems. In: WIELGOLASKI, F. E. (Ed.): Fennoscandian tundra ecosystems. Part I: Plants and microorganisms, pp. 3–28. Berlin-Heidelberg-New York: Springer 1975.

# Faunal Structure of Hardangervidda, Norway

T. SOLHØY, E. ØSTBYE, H. KAURI, A. HAGEN, L. LIEN, and H.-J. SKAR

The outline of vertebrates is mostly based on year-round investigations from 1959 onwards in the Finse area, which is situated in the north-western part of the Hardangervidda plateau (ØSTBYE et al., 1972a; SONESSON et al., 1975). The material is supplemented with data from investigations and scattered observations from the whole mountain plateau (references cited in ØSTBYE et al., 1972c; ØSTBYE et al., 1975).

Most of the accessible year-round invertebrate data are from three sites in the Stigstuv area in the middle of the mountain plateau (SONESSON et al., 1975) and from some investigations in the Finse area. The invertebrate structure on Hardangervidda will be exemplified by the Stigstuv data. Due to the large complex of different plant communities in this mountain area, it is at present impossible to give a general description of the invertebrate fauna for the whole of Hardangervidda.

The Section on invertebrates has been prepared by T. SOLHØY and H. KAURI, while E. ØSTBYE is responsible for the vertebrate Section.

## Vertebrates

The vertebrate fauna of the Hardangervidda mountain plateau encompasses both species with a distribution restricted to mountain areas, and species with a wider occurrence in other biomes, such as alpine birch forests, coniferous forests,

Table 1. The number of vertebrate faunal elements at Hardangervidda

| Systematic group | Total number of species registered | Breeders | Number of migrants | Visitors |
|---|---|---|---|---|
| Pisces | 6[a] | 4 | — | — |
| Amphibia | 1 | 1 | — | — |
| Reptilia | 1 | 1 | — | — |
| Aves | 114 | 61 | 31 | 22 |
| Mammalia | 26 | 17 | — | 9 |
| Total | 148 | 84 | 31 | 31 |

[a] Two species introduced by man not spawning in the area.

grassland and coastal areas. In addition to the species which regularly inhabit and breed in these alpine environments, many pass through these areas, either on seasonal migrations or on occasional visits. Most members of the bird fauna which breed in the mountains are migrants to lower elevations of Fennoscandia or to more southern latitudes in winter. At Hardangervidda 148 species of vertebrates have been registered altogether. Of these only 84 can be regarded as regular breeders, while 31 must be characterized as migrants and 31 as visitors; there are also 2 fish species introduced which do not spawn in the area. The number of species in each of these categories is shown organized into systematic groups in Table 1.

**The Breeding Fauna.** Within an ecosystem context the breeding fauna must be regarded as a more important component than migrants and visitors. Many birds and mammals in arctic, alpine and boreal regions have a pronounced cyclic occurrence. In Fennoscandia a cyclic periodicity of 3–4 years between peaks in population density is the normal picture. This is characteristic for ptarmigans, birds of prey and owls, shrews, small rodents, small mustelids and foxes (for references see LIEN et al., 1974). Newer investigations have indicated that passerines, and even waders show, to a certain extent, a parallel cyclic occurrence (LIEN et al., 1974, 1975).

The number of individuals of the different species varies greatly during a cyclic period. For instance the density of small rodents may range from less than one animal per hectare during a year with a low population density to approximately 300 per hectare during a peak year. The number of carnivores, which is rather high during a year with a large population of small rodents, is drastically reduced in "low" years. Many species either do not breed in such a year, e.g. arctic fox; or they may be totally absent from the area, e.g. the snowy owl. The number of species usually breeding on Hardangervidda, together with indications of their functioning as components of a mountain ecosystem, and their relative abundance, is given in Table 2. Their function is expressed in terms of their trophic status. When reading this Table one should keep in mind that it is most often inaccurate to place an animal in a straight trophic level, as it may be, for example,

Table 2. The vertebrate species usually breeding at Hardangervidda. H – herbivore; $H/C_i$ – the adult herbivorous, the juvenile carnivorous taking invertebrates; $H_h$ – grazers' browsers; C – carnivore; $C_{i,v}$ – carnivore taking both invertebrates and vertebrates; $C_v/C_i$ – adult carnivore taking vertebrates, juvenile carnivore taking invertebrates; $C_i$, (O) – mainly carnivore taking invertebrates, there is however some plant material in its food; O – omnivore. A – abundant, C – common, U – uncommon, R – rare

| Ecosystem Component | Function | Relative abundance | Ecosystem Component | Function | Relative abundance |
|---|---|---|---|---|---|
| Pisces | | | Amphibia | | |
| *Salmo trutta* | $C_{i,v}$ | A | *Rana temporaria* | $C_i$ | R |
| *Salvelinus alpinus* | $C_{i,v}$ | A | | | |
| *Coregonus lavaretus* | $C_i$ | R | Reptilia | | |
| *Phoxinus phoxinus* | $C_i$ (O) | R | *Vipera berus* | $C_{i,v}$ | R |

Table 2 (continued)

| Ecosystem Component | Function | Relative abundance | Ecosystem Component | Function | Relative abundance |
|---|---|---|---|---|---|
| Aves | | | *Pica pica* | O | R[b] |
| *Gavia arctica* | $C_v/C_i$ | U | *Parus major* | O | R[b] |
| *Anas platyrhynchos* | H, (O) | U | *Cinclus cinclus* | $C_i$, (O) | U |
| *A. crecca* | H, (O) | C | *Turdus pilaris* | O | C |
| *Aythya marila* | $C_i$ | C | *T. philomelos* | O | R |
| *Melanitta fusca* | $C_i$ | C | *T. iliacus* | O | R |
| *M. nigra* | $C_i$ | C | *T. torquatus* | O | U |
| *Mergus merganser* | $C_v$ | U | *T. merula* | O | R |
| *Aquila chrysaëtos* | $C_v$ | R | *Oenanthe oenanthe* | $C_i$ | A |
| *Buteo lagopus* | $C_v$ | A[a] | *Luscinia suecica* | $C_i$, (O) | U |
| *Falco peregrinus* | $C_v$ | R | *Phylloscopus trochilus* | $C_i$ | U |
| *F. rusticolus* | $C_v$ | R | *Anthus pratensis* | $C_i$ | A |
| *F. columbarius* | $C_v$ | U[a] | *Motacilla alba* | $C_i$ | C |
| *F. tinnunculus* | $C_v$ | C[a] | *Sturnus vulgaris* | O | C[b] |
| *Lagopus lagopus* | $H/C_i$ | A[a] | *Carduelis flavirostris* | H, (O) | C |
| *L. mutus* | $H/C_i$ | A[a] | *C. flammea* | H, (O) | U |
| *Grus grus* | O | U | *Emberiza schoeniclus* | $C_i$, (O) | U |
| *Vanellus vanellus* | $C_i$ | U | *Calcarius lapponicus* | O | C |
| *Charadrius hiaticula* | $C_i$ | C | *Plectrophenax nivalis* | H, (O) | A |
| *Pluvialis apricaria* | $C_i$ | A | *Passer domesticus* | O | R[b] |
| *Eudromias morinellus* | $C_i$, (O) | A | Mammalia | | |
| *Gallinago gallinago* | $C_i$, (O) | U | *Sorex araneus* | $C_i$ | C |
| *G. media* | $C_i$, (O) | R | *S. minutus* | $C_i$ | U |
| *Numenius phaeopus* | O | R | *Neomys fodiens* | $C_i$ | U |
| *Tringa glareola* | $C_i$ | C | *Lepus timidus* | $H_h$ | A |
| *T. totanus* | $C_i$ | C | *Castor fiber* | $H_h$ | U |
| *Actitis hypoleucos* | $C_i$ | C | *Lemmus lemmus* | H, (O) | A[a] |
| *Calidris maritima* | $C_i$ | C | *Clethrionomys glareolus* | H | C[a] |
| *C. temminckii* | $C_i$ | C | *C. rufocanus* | H | A[a] |
| *C. alpina* | $C_i$ | C | *Microtus agrestis* | H | U[a] |
| *Limicola falcinellus* | $C_i$ | R | *M. oeconomus* | H | A[a] |
| *Philomachus pugnax* | $C_i$, (O) | U | *Arvicola terrestris* | H | R |
| *Phalaropus lobatus* | $C_i$ | R | *Vulpes vulpes* | O | A |
| *Stercorarius longicaudus* | O, ($C_v$) | R | *Alopex lagopus* | O | U[a,c] |
| *Larus canus* | O | C | *Mustela vison* | $C_v$ | U |
| *Cuculus canorus* | $C_i$/O | C | *M. erminea* | $C_v$ | C[a] |
| *Nyctea scandiaca* | $C_v$ | R[a] | *M. nivalis* | $C_v$ | C[a,d] |
| *Asio flammeus* | $C_v$ | C[a] | *Gulo gulo* | $C_v$ | R[e] |
| *Alauda arvensis* | O | R | *Lynx lynx* | $C_v$ | R[e] |
| *Eremophila alpestris* | $C_i$, (O) | C | *Alces alces* | $H_h$ | U[e] |
| *Delichon urbica* | $C_i$ | U[b] | *Rangifer tarandus* | $H_h$ | A |
| *Corvus corax* | O | A | *Capreolus capreolus* | $H_h$ | R[e] |

[a] Cyclic.
[b] Partly synantropic.
[c] Population in slight increase.
[d] *M. nivalis* or *M. rixosa*, systematic position not clear.
[e] Probably not breeding within the mountain plateau.

for part of the year carnivorous, and the rest of the year omnivorous. Some animals, such as ptarmigans, may have a carnivore diet as juveniles, and a herbivore one as adults.

Some species, here mentioned as synantropic, can only exist in connection with human settlements, e.g. magpie and house sparrow.

**Migrators and Visitors.** Most of the bird migrators pass over the Hardangervidda area on their seasonal migrations in spring and autumn to and from their breeding grounds in other biomes. The pattern of their migrational movement, e.g. with respect to direction, is not fully known at present. It seems, however, as though the main migrational directions are east ↔ west or north ↔ south, to a large extent following such topographical lead-lines as valleys, watercourses etc. Some birds, e.g. the short-eared owl *(Asio flammeus)* and the long-eared owl *(Asio otus)*, seem to have a certain multiannual migrational pattern as they move over larger areas of Fennoscandia and Eurasia in a cyclic manner (MYSTERUD, 1970). Observations from Hardangervidda strongly support this hypothesis.

In later decades large carnivores like wolf and bear seem to have been exterminated from the area, but wolverine and lynx occasionally pass through the mountain plateau. Many animals, e.g. moose and ring-ouzel, which usually breed in the alpine birch forests or in the upper part of the coniferous forests, enter the borders of the mountain plateau quite regularly in late summer, obviously for feeding purposes.

**Regional Distribution of Species.** The Hardangervidda area belongs in a zoogeographical context to the Norwegian arctoalpine region, even if it differs quite markedly from other mountain areas in both mid and northern Norway. Despite many intrinsic differences, all mountain areas seem to have so many features and typical elements in common that they can be treated as one region.

Many species have their main distribution range on the Eurasian continent, with a Fennoscandian outlet that has its southern distribution limit at the southern borders of Hardangervidda, e.g. the arctic fox and the snowy owl (SIIVONEN, 1968; HAFTORN, 1971). Some species extend their distribution range from the south, e.g. roe deer (SIIVONEN, 1968).

In the latter decades some species have extended their distribution range, for example the moose has expanded its territory from the coniferous forests at the eastern border of Hardangervidda, over the mountain plateau, to the forests on the west coast of Norway. Other species have also extended their distribution area, e.g. the common gull, redwing, skylark and American mink.

The arctic fox, which was formerly widely distributed in mountain areas and along the northern coast of Norway, was drastically reduced in number and distribution at the beginning of this century. The reduction went so far that after 1930 the species, when it became protected by law, was almost regarded as extinct in southern Norway. However, this species seems to have shown a slight increase in number at Hardangervidda over the last 10–15 years, and we hope that it may be conserved for the future. Our mountains would be much poorer without this true arctic and mountain element.

## Invertebrates

The Scandinavian alpine tundras are characterized by relatively low diversity. There is a decreasing number of species from the continental eastern and northeastern tundra regions to the atlantic-influenced alpine tundra in Scandinavia. In the Kevo area, Inari Lapland in Finland (69°45′ N, 27° E) a total number of 318 ichneumonid species has been found (JUSSILA, 1965) against only 271 on Hardangervidda (JUSSILA, 1973). 62 species of Collembola are recorded in the tundra of Western Taimyr (ANANJEVA, 1973) but 38 species on Hardangervidda (FJELLBERG, 1975). The number of species of spiders in the alpine region of Saana Mountains in North Finland (69° N, 21° E) is 204 according to PALMGREN (1965), 116 in the alpine region on the mountains of Torneträsk, North Sweden (68°15′ N, 20° E) (HOLM, 1950) and 102 species on Hardangervidda (KAURI, unpublished). A decreasing number of species of stoneflies in Norway from east and northeast to the west is reported by LILLEHAMMER (1974).

The distribution on Hardangervidda of the following taxa is fairly well documented: Coleoptera (FJELLBERG, 1972). Ichneumonidae (JUSSILA, 1973), Psocoptera (MEINANDER, 1974), Tachydrominae (Diptera, Empididae) (CHVALA, 1974), Hemiptera (OSSIANNILSSON, 1974), Lumbricidae (STØP-BOWITZ, 1974), Aranaeae (KAURI, unpubl.), Gastropoda (SOLHØY, unpubl.), while Tipulidae (HOFSVANG, 1975) is studied in the Finse area and so are some other groups (e.g. HÅGVAR and ØSTBYE, 1975), particularly the Coleoptera species *Melasoma collaris* (HÅGVAR, 1975).

It is not always easy to judge whether a single individual of a species of the more vagile taxa found, i.e. winged insects and spiders, is an accidental visitor or has established itself permanently on Hardangervidda. There is a continual addition of species from the lowland, especially with termal winds (see GREVE, 1969).

Arthropods of small body size, *viz* Collembola, Acari, are successful under arctic-alpine conditions, dominating both in the edaphon and in the communities of the vegetation layer. A limited number of species appear in great abundance. Much of the synchronization of the emergence of imagines, egg laying, and larval development is dependent on climatic factors and a short summer period of 2.5–3 months. Quite a number of arthropods have a prolonged life cycle. Many opportunistic species, *viz* in Aphididae and Thysanoptera, appear in irregular cycles.

A negative characteristic feature of the invertebrate components is the definite absence of the groups Isopoda, Diplopoda, Chilopoda and Formicidae, important in the decomposition cycle in the surrounding lowland. The scantiness of Lumbricidae and Gastropoda (3 and 8 species, respectively) are also pronounced compared with the adjacent lowland of Hardanger, which has 10 species of lumbricids and about 55 species of gastropods (STØP-BOWITZ, 1969, 1974; SOLHØY, unpubl.).

Parallel to the decreasing number of species from lowland to the higher elevations, a decrease of the invertebrate biomass in the communities occurs. This may be illustrated by an example of oligotrophic lakes on the western slope of Hardanger mountains (KAURI and RADDUM, in press). At a mountain level of 1300 m the total average biomass $m^{-2}$ equals only a fourth of that of the most productive lakes of the lowland of Western Norway (Fig. 1). The few deviations

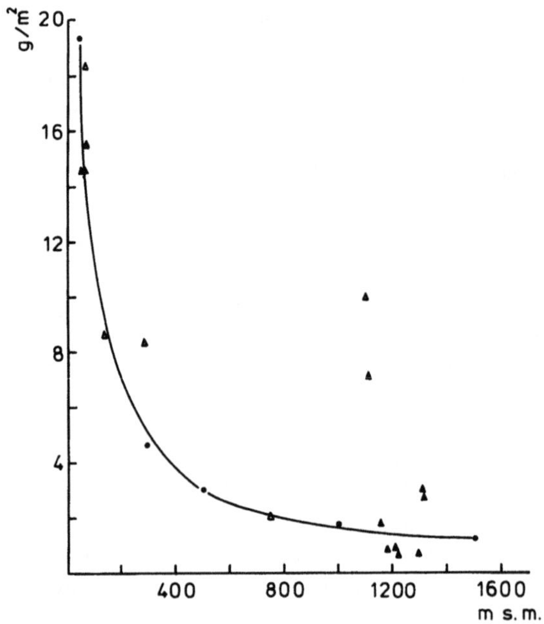


Fig. 1. Average biomass (g wet weight $m^{-2}$) of oligotrophic lakes plotted against altitude in western Norway

from the main regression are caused by the increase in dissolved nutrients, *viz.* Ca, and other biogenetic elements.

The composition of the invertebrate fauna at the three Stigstuv IBP sites, oligotrophic lichen heath (1220 m), relatively eutrophic dry (1275 m) and wet (1320 m) meadows (SONESSON et al., 1975), seems to be mostly governed by the soil-moisture conditions, and snow depth during winter time, but also to a lesser extent by soil temperature and soil properties. The lichen heath is characterized by a shallow snow cover (20–70 cm), while the dry and wet meadow have more than 100 cm of snow. The lichen heath is easily subjected to drought and so is the dry meadow, while the wet meadow has a high humidity. In spite of the differences in snow depth, the duration of the snow cover of the sites does not differ greatly, lasting from the beginning or middle of October to late May or the first week of June. Data on soil properties are given by HINNERI et al. (1975) and on climate by SKARTVEIT et al. (1975).

The lichen heath has a xeric, cold-tolerant fauna. On the dry meadow the fauna hibernate at soil temperatures from −1° C to −3° C, but are comparatively easily subjected to drought conditions. The site, however, varies to a certain extent from xeric microhabitats to wetter ones. On the wet meadow the winter temperatures below snow are seldom lower than −2° C, and drought periods never occur. The fauna here consist of hygric species.

The ratio Collembola/Acari can be used as a characteristic of the sites (Table 3). Collembolas are most abundant in the wet habitat, Acari however in the mesic and xeric sites. From Table 3 it can be seen that this applies also to arctic tundra sites in Spitsbergen, U.S.S.R. and U.S.A. This

Table 3. The ratio Collembola to Acari at the Stigstuv sites compared with some arctic areas

| Habitats \ Territories | Hardangervidda, Norway. Average of the years 1969, 1970 and 1971 | Ny Ålesund, Svalbard BENGTSON et al., 1974 | Point Barrow, U.S.A. (S. MACLEAN jr. pers. comm.) | Taimyr, U.S.S.R. (CHERNOV et al., 1972) |
|---|---|---|---|---|
| Wet moss tundra | | 11.02 | | |
| Wet meadow | 1.26 | | 4.90<br>4.20<br>3.77 | |
| Trough | | | | 6.71 |
| Mesic meadow | | | 1.04 | |
| Grassland | | 1.08 | | |
| Dry meadow | 0.64 | | | |
| Lichen tundra | | 1.58<br>0.98 | | |
| Raised polygon | | | 1.02 | |
| Lichen heath | 0.45 | | | |
| Ridge | | | | 0.93 |

is, however, not the case at the Kevo sites (HAUKIOJA and KOPONEN, 1975), where Acari is much more numerous at all the sites. According to JANETSCHEK (1948) this may indicate more mature soil conditions which correspond to a more advanced level of community development. Similar conclusions are expressed by MANI (1962). CHERNOV (1972) found that Collembola predominate in the tundra zone, but Oribatei in more southern zones and in the taiga.

The characteristic invertebrate species of several important groups are given in Table 4. A preliminary invertebrate species list has been given for the Stigstuv sites by KAURI (1971) and for the Finse area by ØSTBYE and HÅGVAR (1972b).

Table 4. Some characteristic invertebrate species at the Stigstuv sites. 0 denotes absence from the site; n.i. – not investigated

| Taxon \ Site | Lichen heath | Dry meadow | Wet meadow |
|---|---|---|---|
| Enchytraeidae | n.i. | *Bryodrilus ehlersi*<br>*Cognettia glandulosa*<br>*C. sphagnetorum*<br>*Fredericia bulboides*<br>*Henlea nasuta* | *Cernosvitoviella sp.*<br>*Cognettia glandulosa*<br>*C. sphagnetorum*<br>*Henlea nasuta* |
| Lumbricidae | 0 | *Dendrobaena norvegica*<br>*D. octaedra* | *Dendrobaena octaedra* |
| Gastropoda | 0 | *Euconulus fulvus*<br>*Vertigo arctica*<br>*Vitrina pellucida* | *Vitrina pellucida* |
| Araneae | *Conigerella borealis*<br>*Oxyptila rauda*<br>*Trichopterna mengei*<br>*Trochosa terricola* | *Arctosa alpigena*<br>*Caledonia evansi*<br>*Erigone atra*<br>*Hilaira frigida*<br>*Pardosa palustris*<br>*Tiso aestivus* | *Erigone psychrophila*<br>*E. atra*<br>*Hilaira nubigena*<br>*Leptorhoptrum robustum*<br>*Robertus scoticus* |

Table 4 (continued)

| Taxon \ Site | Lichen heath | Dry meadow | Wet meadow |
|---|---|---|---|
| Opiliones | *Mitopus morio* | *Mitopus morio* | *Mitopus morio* |
| Oribatei | *Calyptozetes sarekensis*<br>*Carabodes labyrinthicus*<br>*Chamobates cuspidatiformis*<br>*Liochthonius lapponicus*<br>*Oppia ornata*<br>*Tectocepheus velatus* | *Camisia lapponica*<br>*Carabodes marginatus*<br>*C. minusculus*<br>*Edwardzetes edwardsi*<br>*Eupelops plicatus*<br>*Hemileius initialis*<br>*Liebstadia similis*<br>*Liochthonuis sp.*<br>*Oribatula tibialis*<br>*Tectocepheus velatus* | *Eupelops plicatus*<br>*Malaconothrus egregius*<br>*Mucronothrus nasalis*<br>*Liochthonius sp.*<br>*Oppia translamellata*<br>*Oromurcia bicuspidata*<br>*Tectocepheus velatus* |
| Collembola | *Anurida forsslundi*<br>*Anurophorus binoculatus*<br>*Folsomia brevicauda*<br>*Karlsteinia norvegica*<br>*Xenylla maritima*<br>*Bourletiella pistillum*<br>*Tetracanthella wahlgreni* | *Folsomia quadrioculata*<br>*Tetracanthella britannica*<br>*Tullbergia krausbaueri*<br>*Isotomodella pusilla* | *Frisea mirabilis*<br>*Hypogastura denticulata*<br>*Isotoma olivacea*<br>*Isotomiella minor*<br>*Sminthurinus aureus*<br>*Tetracanthella britannica*<br>*Tullbergia krausbaueri* |
| Heteroptera | *Nysius groenlandicus* | 0 | 0 |
| Aphidoidea | 0 | *Acyrtosiphon malvae* | *Acyrtosiphon malvae* |
| Auchenorrhyncia | *Psammotettix frigidus* | *Psammotettix frigidus* | 0 |
| Coccoidea | *Arctorthecia cataphracta* | *Arctorthecia cataphracta* | 0 |
| Psylloidea | 0 | *Triozoa femoralis* | 0 |
| Thysanoptera | 0 | *Anapotrips stylifer*<br>*Aptinotrips secticornis* | *Anapotrips stylifer*<br>*Aptinotrips secticornis* |
| Neuroptera | *Helicoconis lutea* | *Helicoconis lutea* | 0 |
| Mecoptera | *Boreus hyemalis* | *Boreus hyemalis* | *Boreus hyemalis* |
| Psocoptera | *Reuterella helvimacula* | 0 | 0 |
| Carabidae | *Amara alpina*<br>*Carabus problematicus*<br>*Cymindis vaporarium*<br>*Miscodera arctica* | *Amara praetermissa*<br>*Notiophilus aquaticus*<br>*Patrobus septentrionis*<br>*Calathus melanocephalus* | *Elaphrus lapponicus*<br>*Notiophilus aquaticus*<br>*Patrobus septentrionis*<br>*P. assimilis*<br>*Pelophila borealis* |
| Staphylinidae | *Mycetoporus nigrans*<br>*Ocyusa nivicola*<br>*Bryoporus rugipennis* | *Anthophagus alpinus*<br>*Arpedium brunnescens*<br>*Atheta munsteri*<br>*Mycetoporus erichsonanus*<br>*Olophrum boreale*<br>*Quedius boops*<br>*Boreaphihus henningianus* | *Arpedium tenue*<br>*Atheta arctica*<br>*A. gramminicola*<br>*A. islandica*<br>*A. subplana*<br>*Cylletron nivale*<br>*Stenus carbonarius*<br>*Quedeus boopoides* |
| Curculionidae | *Otiorrhyncus dubius* | *Otiorrhyncus dubius*<br>*Apion frumentarium* | *Otiorrhyncus dubius*<br>*Gryphus equiseti* |
| Scarabaeidae | *Aphodius lapponum* | *Aphodius lapponum* | 0 |
| Hydrophilidae | 0 | *Hydrophorus glacialis* | *Hydrophorus glacialis* |
| Byrrhidae | *Byrrhus fasciatus* | *Byrrhus pilula* | *Byrrhus pilula* |

Most of the invertebrate activity occurs in the litter layer and the upper 1–3 cm of the soil (SOLHØY, 1972). However, during drought periods a substantial part of at least Acari, Collembola and Enchytraeidae populations on the dry meadow and lichen heath move a few cm deeper; even then, however, only a few percent penetrate lower than 6 cm, probably due to diminishing pore space.

The abundance and dry-weight biomass of some of the important taxa are given in Table 5. The highest number of Enchytraeidae and Coleoptera are found on the wet meadow, while Araneae, Collembola, Hemiptera and Lepidoptera larvae are most abundant on the dry meadow. This is, however, not always connected with the highest biomass due to differences in individual weights of the different species.

The composition of the microfaunal groups investigated, i.e. Oribatei, Collembola and Enchytraeidae is given in special papers in this volume (SOLHØY, 1975a, 1975b; FJELLBERG, 1975), and the characteristic species at the sites are also shown in Table 4. Some of the mesofaunal groups are also treated in separate papers, i.e. Thysanoptera and Aphididae (KJELLSEN, 1975; TAMBS-LYCHE, 1975). Data are given here on some of the other important groups.

The Tables on species' dominance are based exclusively on quantitative samples (KAURI et al., 1969; SOLHØY, 1972) for 2–4 years. However, the number of species (Table 6) is based on this material together with pitfall trapping, sweeping, handpicking, etc.

**Araneae.** The Araneae form a conspicuous part of the high-mountain ecosystems and are without exception carnivorous.

The dominant species on the lichen heath is *Conigerella borealis*, on the dry meadow *Tiso aestivus* and on the wet meadow *Erigone psychrophila* and *Erigone atra* (Table 7). Highest diversity of spiders is found on the dry meadow and the total number of species at the Stigstuv sites is about 40. In material from about 170 terrestrial biotopes (FJELLBERG, 1972; KAURI, unpubl.), a total of 102 species occurs above the subalpine birch forest. The most common species collected were *Bolyphantes index*, *Hilaira frigida*, *Leptorhoptrum robustum* and the two *Pardosa* species, *amentata* and *palustris*.

**Coleoptera.** Among the carabid beetles *Notiophilus aquaticus* is the dominant species both on the dry and wet meadow (Table 8), *Amara praetermissa* is also a characteristic species at the dry site while *Patrobus septentrionis*, *P. assimilis* and *Elaphrus lapponicus* are fairly abundant at the wet meadow site. *Amara alpina* seems to be the most common species on the lichen heath, but the material is scanty.

*Quedius boops* is the dominant species of Staphylinidae on the dry meadow. The highest diversity of these beetles is found on the wet meadow with three species of nearly equal dominance and three others fairly dominant (Table 8). *Ocyusa nivicola* and *Mycetophorus nigrans* are most abundant on the lichen heath.

The other families of Coleoptera found at the sites are represented by few species of which some are given in Table 4. The Chrysomelid beetle *Melasoma collaris* is abundant in snowbeds at Finse (HÅGVAR, 1975) but is absent from the Stigstuv sites.

Table 5. Mean seasonal and minimum-maximum abundance and

| | 1969 | |
|---|---|---|
| | Number m$^{-2}$ | Biomass mg dw m$^{-2}$ |
| Lichen health | | |
| Araneae mean (min-max) | 25 (10–40) | 2.0 (0.8–3.2) |
| Acari mean (min-max) | 55400 (37300–73400) | 184.1 (123.9–243.9) |
| Collembola mean (min-max) | 27800 (24100–31500) | 68.8 (59.7–78.0) |
| Hemiptera mean (min-max) | 80 (76–83) | 1.4 (1.4–1.5) |
| Coleoptera adults mean (min-max) | 8 (0–16) | |
| Lepidoptera larvae mean (min-max) | 6 (4–7) | 6.5 (4.0–7.6) |
| Dry meadow | | |
| Enchytraeidae mean (min-max) | | |
| Araneae mean (min-max) | 46 (38–56) | 3.7 (3.0–4.5) |
| Acari mean (min-max) | 101400 (33400–159100) | 337.1 (110.2–525.0) |
| Collembola mean (min-max) | 68200 (12100–121700) | 168.4 (29.9–300.6) |
| Hemiptera mean (min-max) | 446 (280–584) | 7.8 (4.9–10.2) |
| Coleoptera adults mean (min-max) | 3 (2–11) | |
| Lepidoptera larvae mean (min-max) | 12 (3–25) | 13.0 (3.2–27.0) |
| Wet meadow | | |
| Enchytraeidae mean (min-max) | | |
| Araneae mean (min-max) | 62 (44–82) | 3.7 (2.6–4.9) |
| Acari mean (min-max) | 32100 (11300–55000) | 153.1 (53.1–258.5) |
| Collembola mean (min-max) | 39400 (10400–63300) | 214.0 (56.5–341.8) |
| Hemiptera mean (min-max) | 28 (6–62) | 1.4 (0.3–3.1) |
| Coleoptera adults mean (min-max) | 27 (7–52) | |
| Lepidoptera larvae mean (min-max) | 1 (0–2) | 0.2 (0.0–0.4) |

Table 6. Species number of some invertebrate groups at Stigstuv

| | Lichen heath | Dry meadow | Wet meadow |
|---|---|---|---|
| Oligochaeta | | | |
| Enchytraeidae | n. i. | 9 | 10 |
| Lumbricidae | 0 | 3 | 1 |
| Mollusca | | | |
| Gastropoda | 0 | 5 | 2 |
| Arachnoidea | | | |
| Araneae | 16 | 23 | 12 |
| Opiliones | 1 | 1 | 1 |
| Acari | | | |
| Oribatei | 17 | 30 | 18 |
| Labiata | | | |
| Collembola | 20 | 24 | 28 |
| Heteroptera | 1 | 1 | 0 |
| Homoptera: Aphidoidea | | 6 | 6 |
| Homoptera: Auchenorrhyncha | 2 | 1 | 0 |
| Homoptera: Psylloidea | 0 | 4 | 2 |
| Thysanoptera | 1 | 4 | 5 |
| Neuroptera | 1 | 1 | 0 |
| Mecoptera | 1 | 2 | 1 |
| Psocoptera | 2 | 2 | 3 |
| Coleoptera | 13 | 31 | 25 |
| Hymenoptera | | | |
| Ichneumonidae | 26 | 21 | 14 |
| Ceraphronidae | 2 | 7 | 0 |

standing crop of some invertebrate taxa at the sites at Stigstuv

| 1970 | | 1971 | |
|---|---|---|---|
| Number $m^{-2}$ | Biomass mg dw $m^{-2}$ | Number $m^{-2}$ | Biomass mg dw $m^{-2}$ |
| 43 (16–69) | 3.4 (1.3–5.5) | 32 (5–65) | 2.5 (0.4–5.2) |
| 88 200 (56 100–140 800) | 293.1 (186.4–467.9) | 125 500 (59 100–185 600) | 417.0 (196.4–616.7) |
| 36 300 (19 200–75 100) | 90.3 (47.5–185.9) | 54 700 (43 900–73 000) | 135.4 (108.7–180.7) |
| 106 (44–190) | 1.9 (0.8–3.4) | 161 (45–336) | 2.9 (0.8–6.0) |
| 4 (1–6) | | 4 (0–8) | |
| 11 (6–18) | 12.1 (6.5–19.4) | 9 (4–16) | 9.7 (4.3–17.3) |
| | | | |
| 8 800 (5060–11 310) | 313.0 (179.6–401.5) | 20 000 (14.400–36 800) | 308.0 (221.8–566.7) |
| 54 (32–88) | 4.3 (2.5–7.0) | 46 (22–94) | 4.4 (2.0–8.5) |
| 112 100 (46 300–164 100) | 372.5 (152.8–541.5) | 118 700 (22 400–276 400) | 866.9 (163.5–2017.7) |
| 79 600 (46 700–123 200) | 197.2 (112.0–406.6) | 64 800 (15 100–105 700) | 214.2 (49.8–348.8) |
| 405 (160–636) | 7.1 (2.8–11.1) | 2 333 (234–4 882) | 13.3 (1.3–27.8) |
| 6 (4–12) | | 8 (4–10) | |
| 10 (4–16) | 10.8 (4.3–17.3) | 7 (6–8) | 9.8 (8.4–11.2) |
| | | | |
| 29 700 (11 700–39 000) | 857.0 (337.0–1123.2) | 39 000 (30 400–48 400) | 891.0 (693.1–1103.5) |
| 52 (12–128) | 9.5 (2.2–23.3) | 35 (13–69) | 5.5 (2.0–10.8) |
| 33 600 (18 300–42 600) | 160.2 (87.1–202.8) | 44 900 (20 800–106 500) | 233.5 108.2–553.8) |
| 52 800 (18 000–71 700 | 287.2 (97.7–389.3) | 44 200 (28 500–72 100) | 229.3 (147.6–373.5) |
| 19 (3–46) | 1.0 (0.2–2.4) | 11 (4–25) | 0.5 (0.2–1.1) |
| 13 (9–16) | | 8 (4–12) | |
| 1 (0–3) | 0.2 (0.0–0.6) | 2 (0–5) | 0.5 (0.0–1.3) |

**Diptera.** This is a very conspicuous group in the high-mountain ecosystem of Hardangervidda. About 25 families are recorded from the sites. On the dry meadow Sciaridae, Bibionidae, and Cecidomyidae are the most commonly occurring families in the quantitative samples, while Chironomidae, Borboridae and Chloropidae dominate on the wet meadow (Table 9). No data exist for the lichen heath, and only scanty data on the larval densities at the sites. SOLHØY (unpubl.) recorded 300–800 ind. $m^{-2}$ of Chironomidae larvae at the wet meadow, and HOFSVANG (unpubl.) about 60 ind. $m^{-2}$ of *Tipula subnodicornis* at the same site. At Finse HOFSVANG (1975) recorded 20–50 $m^{-2}$ of *Tipula excisa.*

**Hymenoptera Parasitica.** Only the family Ichneumonidae is determined to species level. The relative abundance of the different families found hitherto at the dry and wet meadow sites, can however be given (Table 10). The dominating family on the dry meadow (Eulophidae) accounts for about 25% of the species, but is absent on the wet meadow. The same is true for the Ceraphronidae. The families Myrmaridae and Ichneumonidae are common at both sites (JUSSILA, 1973). 47 species of Ichneumonidae are known from the sites, but in very small numbers, so that no dominance diagram can be given. To get an idea of which are the more common species, the species found at Stigstuv are ranked in Table 11 according to frequency of occurrence in the 170 terrestrial biotopes investigated 1967–1972 (FJELLBERG, 1972). The three dominant species seem to be *Aclastus minutus, Phygadeuon leviventris* and *Stenomacrus affinis.* Only 16 of the 47 species, however, occurred in 5 or more of the biotopes investigated.

Table 7. The relative abundance of Araneae at the Stigstuv sites

| Lichen heath ($n=261$) | | Dry meadow ($n=885$) | | Wet meadow ($n=552$) | |
|---|---|---|---|---|---|
| *Conigerella borealis* | 74.7 | *Tiso aestivus* | 77.5 | *Erigone psychrophila* | 57.4 |
| *Meioneta gulosa* | 9.6 | *Erigone atra* | 7.2 | *Erigone atra* | 33.7 |
| *Tiso aestivus* | 9.2 | *Oedothorax retusus* | 5.4 | *Drepanotylus uncatus* | 4.2 |
| *Cornicularia karpinskii* | 5.0 | *Metapobactus prominulus* | 4.1 | *Hilaira nubigena* | 1.4 |
| *Oxyptila rauda* | 0.8 | *Rhaebothorax morulus* | 3.2 | *Ceratinella brevipes* | 1.3 |
| *Hilaira frigida* | 0.4 | *Oxyptila rauda* | 0.6 | *Cornicularia cuspidata* | 1.1 |
| *Rhaebothorax morulus* | 0.4 | *Ceratinella brevipes* | 0.5 | *Agyneta cauta* | 0.2 |
| | | *Cornicularia karpinskii* | 0.5 | *Rhaebothorax sphagnicola* | 0.2 |
| | | *Trichopterna mengei* | 0.5 | *Robertus scoticus* | 0.2 |
| | | *Rhaebothorax monticola* | 0.2 | *Tiso aestivus* | 0.2 |
| | | *Gonatium rubens* | 0.1 | *Trichopterna mengei* | 0.2 |
| | | *Latithorax faustus* | 0.1 | | |
| | | *Meioneta rurestris* | 0.1 | | |
| | | *Pardosa palustris* | 0.1 | | |

Table 8. The relative abundance of Carabidae and Staphylinidae at the Stigstuv sites

| Dry meadow | | Wet meadow | | Lichen hearth | |
|---|---|---|---|---|---|
| Carabidae ($n=48$) | | Carabidae ($n=58$) | | Carabidae ($n=8$) | |
| *Nothiophilus aquaticus* | 62.5 | *Nothiophilus aquaticus* | 43.1 | *Amara alpina* | 62.5 |
| *Amara praetermissa* | 35.4 | *Patrobus septentrionis* | 24.1 | *Miscodera arctica* | 25.0 |
| *Calathus melanocephalus* | 2.1 | *P. assimilis* | 19.0 | *Amara quenseli* | 12.5 |
| | | *Elaphrus lapponicus* | 10.3 | | |
| | | *Pelophila borealis* | 1.7 | | |
| | | *Pterostichus oblongpunctatus* | 1.7 | | |
| Staphylinidae ($n=99$) | | Staphylinidae ($n=325$) | | Staphylinidae ($n=137$) | |
| *Quedius boops* | 58.6 | *Atheta arctica* | 18.5 | *Ocyusa nivicola* | 53.3 |
| *Mycetophorus erichsonanus* | 12.1 | *Arpedium tenue* | 17.5 | *Mycetoporus nigrans* | 24.8 |
| *Atheta arctica* | 7.1 | *Stenus carbonarius* | 16.9 | *Bryoporus rugipennis* | 10.2 |
| *Ocyusa nivicola* | 5.1 | *Arpedium brunnescens* | 12.3 | *Atheta atramentaria* | 2.9 |
| *Oxypoda annularis* | 4.0 | *Quedius boopoides* | 10.5 | *Coryphium angusticolle* | 2.9 |
| *Anthophagus alpinus* | 3.0 | *Atheta graminicola* | 7.4 | *Aploderus caelatus* | 1.5 |
| *Atheta alpestris* | 3.0 | *A. alpestris* | 2.8 | *Atheta laevicauda* | 0.7 |
| *Arpedium brachypterium* | 2.0 | *A. arctica* | 2.5 | *Arpedium brachypterium* | 0.7 |
| *Boreaphilus henningianus* | 2.0 | *Arpedium brachypterium* | 2.2 | *A. tenue* | 0.7 |
| *Othius melanocephalus* | 2.0 | *Olophrum consimile* | 2.2 | *Atheta cinnamoptera* | 0.7 |
| *Olophrum boreale* | 1.0 | *Atheta islandica* | 1.8 | *A. dwinensis* | 0.7 |
| | | *Cylletron nivale* | 1.8 | *Boreaphilus henningianus* | 0.7 |
| | | *Boreaphilus henningianus* | 1.5 | | |
| | | *Oxypoda annularis* | 0.9 | | |
| | | *Atheta atramentaria* | 0.6 | | |
| | | *Arpedium quadrum* | 0.3 | | |
| | | *Quedius fulvicollis* | 0.3 | | |
| | | *Oxypoda funebris* | 0.3 | | |

Table 9. The relative abundance of Diptera families at the Stigstuv meadow sites

| Dry meadow ($n = 504$) | | Wet meadow ($n = 225$) | |
|---|---|---|---|
| Sciaridae | 54.6 | Chironomidae | 24.4 |
| Bibionidae | 23.8 | Borboridae | 20.4 |
| Cecidomyidae | 11.3 | Chloropidae | 13.3 |
| Borboridae | 3.0 | Muscidae | 8.4 |
| Muscidae | 1.8 | Sciaridae | 7.6 |
| Phoridae | 1.0 | Cecidomyidae | 6.2 |
| Agromyzidae | 0.6 | Empididae | 6.2 |
| Chironomidae | 0.6 | Agromyzidae | 5.8 |
| Chloropidae | 0.6 | Anthomyiidae | 1.8 |
| Dolichopodidae | 0.6 | Anthomyzidae | 1.8 |
| Simulidae | 0.6 | Ceratopogonidae | 1.3 |
| Anthomyiidae | 0.4 | Simulidae | 1.3 |
| Empididae | 0.4 | Cordiluridae | 0.4 |
| Ceratopogonidae | 0.2 | Dolichopodidae | 0.4 |
| Ephydridae | 0.2 | Mycetophilidae | 0.4 |
| Mycetophilidae | 0.2 | | |
| Odinidae | 0.2 | | |

Table 10. The relative abundance of Hymenoptera Parasitica families at the Stigstuv meadow sites

| Dry meadow ($n = 198$) | | Wet meadow ($n = 92$) | |
|---|---|---|---|
| Eulophidae | 24.7 | Myrmaridae | 42.4 |
| Ceraphronidae | 19.7 | Ichneumonidae | 19.5 |
| Myrmaridae | 15.2 | Scelionidae | 14.1 |
| Ichneumonidae | 11.6 | Platygasteridae | 6.5 |
| Pteromalidae | 10.6 | Brachonidae | 5.4 |
| Platygasteridae | 8.1 | Eulophidae | 4.3 |
| Scelionidae | 3.0 | Pteromalidae | 3.3 |
| Brachonidae | 2.5 | Cynipidae | 1.1 |
| Encyrtidae | 2.5 | Encyrtidae | 1.1 |
| Cynipidae | 1.0 | Proctotrupidae | 1.1 |
| Diapriidae | 1.0 | Trichogrammatidae | 1.1 |

**Other Groups.** Five species of Lumbricidae are found above 1000 m (STØP-BOWITZ, 1974). *Lumbricus rubellus, Dendrobaena octaedra,* and *D. norvegica* are recorded from the dry meadow, and *D. octaedra* from the wet meadow. No density estimates are available, but the values seem to be low, probably less than 20 ind. $m^{-2}$.

*Vitrina pellucida, Euconulus fulvus* and *Vertigo arctica* are the most common gastropod species of the eleven species found above 1000 m on Hardangervidda. In willow thickets near the dry meadow, they account for 63%, 21%, and 14% respectively of the total (SOLHØY, unpubl.). They were, however, found at very low densities on the meadow sites and were definitely absent from the lichen heath.

Table 11. The most common Ichneumonidae (Hymenoptera Parasitica) species found at Stigstuv ranked according to the number of localities in which they are found on the western parts of Hardangervidda

| | |
|---|---|
| *Aclastus minutus* | 29 |
| *Phygadeuon leviventris* | 25 |
| *Stenomacrus affinis* | 25 |
| *Aclastus gracilis* | 15 |
| *Atractodes picipes* | 14 |
| *Phygadeuon brachyurus* | 11 |
| *Probles xanthopus* | 11 |
| *Pantorhaestes albipes* | 10 |
| *Atractodes croceicornis* | 9 |
| *Stenomacrus silvaticus* | 9 |
| *Eusterinx oligomera* | 6 |
| *Gelis tonsus* | 6 |
| *Atractodes acuminator* | 5 |
| *Barycnemis karpurus* | 5 |
| *Gelis rufipes* | 5 |
| *Mastrus carri* | 5 |
| 31 species | $\leqq 3$ |

*Mitopus morio alpinus* is the only opilionid found and is a characteristic species of all the sites, density 0.5–2.0 ind. $m^{-2}$.

Prostigmata, Mesostigmata, and Oribatei are the mite groups found at the sites. On the lichen heath, preliminary figures indicate that they account numerically for 36%, 1%, and 63% of the total, respectively. When the biomass is taken into account, the figures are about 9%, 1%, and 90%, and seem to remain the same also on the dry meadow. On the wet meadow, however, the Prostigmata seem to account for a higher proportion of both the number and biomass, due to abundant occurrence of the species *Eustigmaeus (Ledermulleria)* sp.

The Hemiptera fauna of the sites is characterized by comparatively few species. Both greatest diversity and abundance are found on the dry meadow due to the high diversity of vascular plants. Here the dominant leafhopper is *Psammotettix frigidus* (OSSIANNILSSON, 1974).

Scale bugs, dominated by *Arctorthecia cataphracta*, are conspicuous elements of the lichen heath and the dry meadow. For Hemiptera, total densities fluctuate much from year to year, but in a peak year more than 2000 ind. $m^{-2}$ may be found.

The number for Lepidoptera larvae is found to be about 10 ind. $m^{-2}$ on the dry meadow and lichen heath and only 1 ind. $m^{-2}$ on the wet meadow. Most of the material is undetermined, but a common species on the dry meadow is *Zygaena exulans*.

No data exist for the groups Protozoa, Nematoda, Rotifera, Crustacea (i.e. Harpacticidae and Ostracoda), Tardigrada and/or the total number of soil-living insect larvae. The total invertebrate biomass of the sites is estimated at approximately 30–60 kg ha (fresh weight). It is, however, evidently highest on the wet meadow and lowest on the lichen heath.

## References

ANAJEVA, S. J.: Collembola of Western Taimyr. In: TIKHOMIROV, B. A., MATVEYEVA, N. V. (Eds.): Biogeocenoses of Taimyr Tundra and their productivity. Vol. 2, 152–165. Leningrad: Publ. House „NAUKA" 1973.

BENGTSON, S. A., FJELLBERG, A., SOLHØY, T.: Abundance of tundra arthropods in Spitsbergen. Ent. Scand. **5**, 137–142 (1974).

CHERNOV, Y. I.: Animal population structure in the subzone of typical tundras of the western Taimyr. In: WIELGOLASKI, F. E., ROSSWALL, T. (Eds.): Biological productivity of tundra, Tundra Biome, pp. 63–79. Oslo-Stockholm: IBP Tundra Biome Steering Committee 1972.

CHVALA, M.: Tachydrominae (Diptera, Empididae). Fauna of Hardangervidda, **4**, 9–12 (1974).

FJELLBERG, A.: Coleoptera from Hardangervidda. Fauna of Hardangervidda **1**, 1–74 (1972).

FJELLBERG, A.: Organization and dynamics of Collembola populations on Hardangervidda. In: Wielgolaski, F. E. (Ed.): Fennoscandian tundra ecosystems. Part 2: Animals and systems analysis, pp. 73–79. Berlin-Heidelberg-New York: Springer 1975.

GREVE, L.: An aerial-drift of Neuroptera from Hardangervidda, Western Norway. Årb. Univ. Bergen. Mat.-Nat. Ser. **2**, 1–15 (1969).

HAFTORN, S.: Norges fugler. Oslo-Bergen-Tromsø: Universitetsforlaget 1971.

HÅGVAR, S.: Studies on the ecology of *Melasoma collaris* L. (Col. Chrysomelidae) in alpine habitats at Finse, south Norway. Norw. J. Ent. **22**, 31–47 (1975).

HÅGVAR, S., ØSTBYE, E.: Occurrence and role of different invertebrate groups in alpine *Salix herbacea* snowbeds at Hardangervidda. In: WIELGOLASKI, F. E. (Ed.): Fennoscandian tundra ecosystems. Part 2: Animals and systems analysis, pp. 88–93. Berlin-Heidelberg-New York: Springer 1975.

HAUKIOJA, E., KOPONEN, S.: Faunal structure of investigated areas at Kevo, Finland. In: WIELGOLASKI, F. E. (Ed.): Fennoscandian tundra ecosystems. Part 2: Animals and systems analysis, pp. 19–28. Berlin-Heidelberg-New York 1975.

HINNERI, S., SONESSON, M., VEUM, A. K.: Soils of Fennoscandian IBP tundra ecosystems. In: WIELGOLASKI, F. E. (Ed.): Fennoscandian tundra ecosystems. Part I: Plants and microorganisms, pp. 31–40. Berlin-Heidelberg-New York: Springer 1975.

HOFSVANG, T.: Life cycle and energy budget of *Tipula excisa* Schum. (Diptera, Tipulidae). In: WIELGOLASKI, F. E. (Ed.): Fennoscandian tundra ecosystems. Part 2: Animals and systems analysis, pp. 145–150. Berlin-Heidelberg-New York: Springer 1975.

HOLM, Å.: Studien über die Spinnenfauna des Torneträskgebietes. Zool. Bidrag, Uppsala **29**, 103–213 (1950).

JANETSCHEK, H.: Tierische Succession auf Hochalpinen Neuland. Schlern-Schriften **67**, 1–215 (1948).

JUSSILA, R.: The Ichneumonidae of the Kevojoki area in Inari Lapland (Finland). Ann. Univ. Turku A, **11:34**, 1–186 (1965).

JUSSILA, R.: Ichneumonidae from Hardangervidda. Fauna of Hardangervidda **2**, 1–50 (1973).

KAURI, H.: List of animal species at Stigstuv area, Hardangervidda, Norway. Evertebrata I. Univ. of Bergen: Zool. Mus. 1971.

KAURI, H., MOLDUNG, T. J., SOLHØY, T.: Turnbull and Nicholls "Quick Trap" for acquiring standing crop of evertebrates in high mountain grassland communities. Norsk Ent. Tidsskr. **16**, 133–136 (1969).

KAURI, H., RADDUM, G. G.: Decreasing gradient of the average biomass of the oligotrophic lakes in Western Norway (in press).

KJELLSEN, E. K.: Dynamics of Thysanoptera populations on Hardangervidda. In: WIELGOLASKI, F. E. (Ed.): Fennoscandian tundra ecosystems. Part 2: Animals and systems analysis, pp. 80–83. Berlin-Heidelberg-New York: Springer 1975.

LIEN, L., ØSTBYE, E., HAGEN, A., SKAR, H.-J., SVALASTOG, D.: Density variations of bird populations in high moutain habitats, Hardangervidda. In: WIELGOLASKI, F. E. (Ed.): Fennoscandian tundra ecosystems. Part 2: Animals and systems analysis, pp. 105–110. Berlin-Heidelberg-New York: Springer 1975.

LIEN, L., ØSTBYE, E., HOGSTAD, O., HAANDE, K. M., HAANDE, P. S., HAGEN, A., SKAR, H.-J., SKARTVEIT, A., SVALASTOG, D.: Bird surveys in the high mountain habitats of Finse and Stigstuv, Hardangervidda, South Norway, 1967–72. Norw. J. Zool. **22**, 1–14 (1974).

LILLEHAMMER, A.: Norwegian stoneflies. II. Distribution and relationship to the environment. Norw. J. Ent. **21**, 195–250 (1974).

MANI, M. S.: Introduction to high altitude entomology. London: Methuen 1962.

MEINANDER, M.: Psocoptera. Fauna of Hardangervidda **3**, 1–9 (1974).

MYSTERUD, I.: Hypothesis concerning characteristics and causes of population movements in Tengmalm's Owl (*Aegolius funereus* (L.)). Nytt Mag. Zool. **18**, 49–74 (1970).

OSSIANNILSSON, F.: Hemiptera (Heteroptera, Auchenorrhyncha and Psylloidea). Fauna of Hardangervidda **5**, 13–35 (1974).

ØSTBYE, E., BERG, A., BLEHR, O., ESPELAND, M., GAARE, E., HAGEN, A., HESJEDAL, O., HÅGVAR, S., KJELVIK, S., LIEN, L., MYSTERUD, I., SANDHAUG, A., SKAR, H.-J., SKARTVEIT, A., SKRE, O., SKOGLAND, T., SOLHØY, T., STENSETH, N. C., WIELGOLASKI, F. E.: Hardangervidda. In: ROSSWALL, T., HEAL, O. W. (Eds.): Structure and Function of Tundra Ecosystems. Ecol. Bull. **20**, pp. 225–264. Stockholm: Swed. Nat. Sci. Res. Council 1975.

ØSTBYE, E., HAGEN, A., HOGSTAD, O., LIEN, L., MYSTERUD, I., SKAR, H.-J., SVALASTOG, D.: List of vertebrates of the Finse area, Hardangervidda, South Norway. Preliminary list. Rapp. Høyfjellsøkol. Forsk. Stn., Finse, Norge 1972 **3**, 1–17 (1972a).

ØSTBYE, E., HÅGVAR, S.: List of terrestrial invertebrates of the Finse area, Hardangervidda, South Norway. Preliminary list. Rapp. Høyfjellsøkol. Forsk. Stn. Finse, Norge 1972 **2**, 1–21 (1972b).

ØSTBYE, E., MYSTERUD, I., LØKKEN, S.: Bibliografi over naturvitenskapelig litteratur fra Hardangervidda med tilgrensende fjellområder. Foreløbig utgave. Rapp. Høyfjellsøkol. Forsk. Stn. Finse, Norge 1972 **1**, 1–66 (1972c).

PALMGREN, P. Die Spinnenfauna der Gegend von Kilpisjärvi in Lapland. Acta Zool. Fennica **110**, 1–70 (1965).

SIIVONEN, L.: Nordeuropas däggdjur. Stockholm: P. A. Nordstedt och Söners förlag 1968.

SKARTVEIT, A., RYDEN, B. E., KÄRENLAMPI, L.: Climate and hydrology of some Fennoscandian tundra ecosystems. In: WIELGOLASKI, F. E. (Ed.): Fennoscandian tundra ecosystems. Part I: Plants and microorganisms, pp. 41–53. Berlin-Heidelberg-New York: Springer 1975.

SOLHØY, T.: Quantitative invertebrate studies in mountain communities at Hardangervidda, South Norway. I. Norsk Ent. Tidsskr. **19**, 99–108 (1972).

SOLHØY, T.: Dynamics of Enchytraeidae populations on Hardangervidda. In: WIELGOLASKI, F. E. (Ed.): Fennoscandian tundra ecosystems. Part 2: Animals and system analysis, pp. 55–59. Berlin-Heidelberg-New York: Springer 1975a.

SOLHØY, T.: Dynamics of Oribatei populations on Hardangervidda. In: WIELGOLASKI, F. E. (Ed.): Fennoscandian tundra ecosystems. Part 2: Animals and systems analysis, pp. 60–65. Berlin-Heidelberg-New York: Springer 1975b.

SONESSON, M., WIELGOLASKI, F. E., KALLIO, P.: Description of Fennoscandian tundra ecosystems. In: WIELGOLASKI, F. E. (Ed.): Fennoscandian tundra ecosystems. Part 1: Plants and microorganisms, pp. 3–28. Berlin-Heidelberg-New York: Springer 1975.

STØP-BOWITZ, C.: A contribution to our knowledge of the systematics and zoogeography of Norwegian earthworms (Annelida Oligochaeta Lumbricidae). Nytt Mag. Zool. **17**, 169–280 (1969).

STØP-BOWITZ, C.: Earthworms (Annelida Oligochaeta Lumbricidae). Fauna of Hardangervidda **6**, 37–46 (1974).

TAMBS-LYCHE, H.: Dynamics of Aphididae populations on Hardangervidda. In: WIELGOLASKI, F. E. (Ed.): Fennoscandian tundra ecosystems. Part 2: Animals and systems analysis. pp. 84–87. Berlin-Heidelberg-New York: Springer 1975.

# Faunal Structure of the IBP Tundra Site and Its Surroundings, Abisko, Sweden

S. Jonsson

Faunistically the Torneträsk area is characterized by heterogeneity in species composition and abundance depending on steep gradients in precipitation and radiation; for a climatic review, see Skartveit et al. (1975). In the western part the precipitation exceeds 1000 mm $yr^{-1}$; in the rain shadow, 20 km to the east, it is about 300 mm $yr^{-1}$. In the same way the solar radiation varies with shifting degrees of cloudiness. Thus a macrorelief in faunal distribution and composition has been developed, varying from a hygrophil to a more xerophil fauna. Nevertheless, this fact is apparently overshadowed by the microrelief which often gives a variation within a few meters e.g. from a shore biotope to a rocky ground.

## Interchange with the Surroundings

On the Swedish IBP Tundra site, the Stordalen mire, described by Sonesson et al. (1975), diurnal and seasonal migration of animals from the neighboring birch forests, lakes, small streams, and rocky grounds considerably influences the mire fauna. Thus, predators and omnivores of the site, such as spiders and ants, receive their food supply mainly in the form of insects from the vicinity. Prodigious quantities of insects belonging to Nematocera, Trichoptera, Ephemeroptera, and Plecoptera hatch from the surrounding waters. Most of the approximately 50 species of Trichoptera in the Abisko region live in the lakes and ponds, only a few species in streaming water, the abundant *Apatania stigmatella* Zett. in both rivers and lakes. *Oporinia autumnata* Bkh. and other Lepidoptera come from the mountain birch forests. Like the water-insects, the butterflies contribute to the nutrient pool of the mire, and during years with mass outbreaks they can also establish populations on its dwarf shrubs, e.g. on *Betula nana* L. and *Vaccinium uliginosum* L.

However, covering a relatively small area, the most intensive interchange of the mire with the surroundings is occasioned by the vertebrates. This is particularly valid for many of the species recorded at the Stordalen mire (Table 1). While not being a normal ingredient of the site, several of the other vertebrates listed for the whole Torneträsk region may, however, in outbreak periods also exert a regulatory effect.

Table 1. Vertebrate species of the Torneträsk region. H – herbivore; $H_g$ – granivore (seeds); $H_h$ – grazer, browser, etc.; C – carnivore; $C_i$ – "insectivore" (invertebrate feeder); $C_v$ – predator; O – omnivore. A – abundant; C – common; U – uncommon; R – rare

| Ecosystem Component | Function | Relative abundance |
|---|---|---|
| Pisces | | |
| *Salmo trutta* | $C_i$ | C[c] |
| *Salvelinus alpinus* | $C_i$ | A[c] |
| *Coregonus pidschian* | $C_i/C_v$ | A[c] |
| *Thymallus thymallus* | $C_i/C_v$ | C |
| *Esox lucius* | $C_v$ | U |
| *Phoxinus phoxinus* | $C_i/C_v$ | C |
| *Gasterosteus pungitius* | $C_i$ | C |
| *Lota lota* | $C_v/C_i$ | C |
| *Perca fluviatilis* | $C_i/C_v$ | C |
| *Cottus poecilopus* | $C_i/C_v$ | U |
| Amphibia | | |
| *Rana t. temporaria* | $C_i$ | C |
| Reptilia | | |
| *Lacerta vivipara* | $C_i$ | U |
| Aves | | |
| *Gavia a. arctica* | $C_v$ | U[a] |
| *Gavia stellata* | $C_v$ | C[a] |
| *Podiceps a. auritus* | $C_v/C_i$ | U[a] |
| *Anas p. platyrhynchos* | $H_g/H_h$ | U[a] |
| *Anas c. crecca* | $H/C_i$ | A[a] |
| *Anas penelope* | $H_h$ | C[a] |
| *Anas acuta* | $H_h$ | U[a] |
| *Aythya m. marila* | $C_i$ | U[a] |
| *Aythya fuligula* | $C_i$ | C[a] |
| *Clangula hyemalis* | $C_i$ | U[a] |
| *Melanitta f. fusca* | $C_i$ | C[a] |
| *Melanitta n. nigra* | $C_i/H$ | C |
| *Bucephala c. clangula* | $C_i$ | U |
| *Mergus s. serrator* | $C_v$ | C[a] |
| *Mergus m. merganser* | $C_v$ | R |
| *Anser erythropus* | $H_h$ | R |
| *Anser f. fabalis* | $H_h$ | R |
| *Cygnus cygnus* | $H_h$ | R |
| *Aquila c. chrysaëtos* | $C_v$ | R |
| *Buteo l. lagopus* | $C_v$ | C[a] |
| *Accipiter n. nisus* | $C_v$ | R[b] |
| *Haliaeetus albicilla* | $C_v$ | R |
| *Pandion h. haliaëtus* | $C_v$ | R |
| *Falco p. peregrinus* | $C_v$ | C |
| *Falco r. rusticolus* | $C_v$ | R[a] |
| *Falco columbarius aesalon* | $C_v$ | C |
| *Falco t. tinnunculus* | $C_v/C_i$ | C[a] |
| *Lagopus l. lagopus* | $H_g/H_h$ | U–A[a] |
| *Lagopus m. mutus* | $H_g/H_h$ | R–C |
| *Lyrurus t. tetrix* | $H_g/H_h$ | R |
| *Tetrao u. urogallus* | $H_h$ | R |
| *Grus g. grus* | O | U |
| *Vanellus vanellus* | $C_i$ | R[b] |
| *Charadrius hiaticula tundrae* | $C_i$ | C[a] |
| *Eudromias morinellus* | $C_i$ | C |
| *Pluvialis apricaria altifrons* | $C_i/H$ | C[a] |
| *Gallinago g. gallinago* | $C_i$ | C[a] |
| *Gallinago media* | $C_i$ | R[a] |
| *Lymnocryptes minimus* | $C_i$ | R[a] |
| *Scolopax rusticola* | $C_i$ | R |
| *Numenius a. arquata* | $C_i/H_g$ | R[a,b] |
| *Numenius p. phaeopus* | $C_i/H_g$ | C[a] |
| *Tringa ochropus* | $C_i$ | R[a,b] |
| *Tringa glareola* | $C_i$ | C[a] |
| *Tringa hypoleucos* | $C_i$ | A[a] |
| *Tringa t. totanus* | $C_i$ | C[a] |
| *Tringa erythropus* | $C_i$ | R |
| *Tringa nebularia* | $C_i$ | C[a] |
| *Calidris maritima* | $C_i$ | R |
| *Calidris temminckii* | $C_i$ | C[a] |
| *Calidris a. alpina* | $C_i$ | R |
| *Philomachus pugnax* | $C_i/H_g$ | U[a] |
| *Phalaropus lobatus* | $C_i$ | A[a] |
| *Stercorarius longicaudus* | $C_v$ | R–C[a] |
| *Larus f. fuscus* | $C_v/C_i$ | R[a] |
| *Larus a. argentatus* | C | U[a] |
| *Larus c. canus* | C | C[a] |
| *Larus ridibundus* | $C_i$ | U |
| *Sterna paradisaea* | $C_i$ | R[a] |
| *Cuculus canorus* | $C_i$ | C[a] |
| *Bubo b. bubo* | $C_v$ | R |
| *Nyctea scandiaca* | $C_v$ | R–C |
| *Surnia u. ulula* | $C_v$ | R–C |
| *Asio o. otus* | $C_v$ | R[b] |
| *Asio f. flammeus* | $C_v$ | R–C |
| *Aegolius f. funerus* | $C_v$ | R[b] |
| *Apus a. apus* | $C_i$ | R[a] |
| *Dendrocopos m. minor* | $C_i$ | U |
| *Picoides t. tridactylus* | $C_i$ | R |
| *Dryocopus m. martius* | $C_i$ | R[b] |
| *Alauda a. arvensis* | $H_g/C_i$ | R[a] |
| *Eremophila alpestris flava* | $H_g/C_i$ | U |
| *Delichon u. urbica* | $C_i$ | C[c] |
| *Riparia r. riparia* | $C_i$ | R[a] |
| *Corvus c. corax* | O | C[a,c] |
| *Corvus corone cornix* | O | C[a,c] |
| *Pica pica fennorum* | O | C[a,c] |
| *Perisoreus i. infaustus* | O | R |

Table 1 (continued)

| Ecosystem Component | Function | Relative abundance |
|---|---|---|
| *Parus m. major* | $H_g/C_i$ | C[c] |
| *Parus cinctus lapponicus* | $H_g/C_i$ | U[a] |
| *Parus montanus borealis* | $H_g/C_i$ | C[a] |
| *Cinclus c. cinclus* | $C_i$ | C |
| *Troglodytes t.troglodytes* | $C_i$ | R[b] |
| *Turdus pilaris* | $H_g/C_i$ | A[a] |
| *Turdus ph. philomelos* | $H_g/C_i$ | U |
| *Turdus i. iliacus* | $H_g/C_i$ | A[a] |
| *Turdus t. torquatus* | $H_g/C_i$ | U |
| *Oenanthe oe. oenanthe* | $C_i$ | C[a] |
| *Saxicola rubetra* | $C_i$ | R |
| *Phoenicurus ph. phoenicurus* | $C_i/H_g$ | C |
| *Luscinia s. svecica* | $C_i$ | C[a] |
| *Acrocephalus schoenobaenus* | $C_i$ | C[a] |
| *Sylvia b. borin* | $C_i/H_g$ | C |
| *Phylloscopus trochilus acredula* | $C_i$ | A[a] |
| *Phylloscopus collybita abietinus* | $C_i$ | R |
| *Phylloscopus borealis talovka* | $C_i$ | R |
| *Muscicapa s. striata* | $C_i$ | R[a] |
| *Ficedula h. hypoleuca* | $C_i$ | C |
| *Prunella m. modularis* | $H_g/C_i$ | C[a] |
| *Anthus p. pratensis* | $C_i$ | A[a] |
| *Anthus t. trivialis* | $C_i$ | U |
| *Anthus cervinus* | $C_i$ | U[a] |
| *Motacilla a. alba* | $C_i$ | C[a,c] |
| *Motacilla flava thunbergi* | $C_i$ | C[a] |
| *Bombycilla g. garrulus* | $H/C_i$ | R |
| *Lanius e. excubitor* | C | R |
| *Sturnus v. vulgaris* | $C_i/H_g$ | C[a,c] |
| *Carduelis f. flammea* | $H_g/C_i$ | A[a] |
| *Carduelis hornemanni exilipes* | $H_g/C_i$ | R[a] |
| *Pinicola e. enucleator* | $H_g$ | R |
| *Fringilla c. coelebs* | $H_g/C_i$ | R[a] |
| *Fringilla montifringilla* | $H_g/C_i$ | A[a] |
| *Emberiza c. citrinella* | $C_i/H_g$ | R[b] |
| *Emberiza s. schoeniclus* | $H_g/C_i$ | C[a] |
| *Calcarius l. lapponicus* | $H_g/C_i$ | A[a] |
| *Plectrophenax n. nivalis* | $H_g/C_i$ | C |
| *Passer d. domesticus* | $H_g/C_i$ | U[c] |
| Mammalia | | |
| *Neomys fodiens* | C | R |
| *Sorex araneus* | $C_i$ | C |
| *Sorex caecutiens* | $C_i$ | R |
| *Sorex minutus* | $C_i$ | R |
| *Lepus timidus* | $H_h$ | C[a] |
| *Sciurus vulgaris* | O | C[c] |
| *Lemmus lemmus* | $H_h/O$ | A[a] |
| *Clethrionomys rutilus* | $H_h/O$ | C[a] |
| *Clethrionomys rufocanus* | $H_g/H_h$ | A[a] |
| *Microtus agrestis* | $H_h$ | C |
| *Microtus oeconomus* | $H_h$ | A |
| *Arvicola terrestris* | $H_h$ | R |
| *Canis l. lupus* | $C_v$ | R |
| *Alopex lagopus* | $C_v$ | R |
| *Vulpes vulpes* | $C_v/H_g$ | C[a] |
| *Ursus a. arctos* | O | R |
| *Mustela erminea* | $C_v$ | C[a] |
| *Mustela rixosa* | $C_v$ | R |
| *Martes martes* | $C_v$ | R |
| *Gulo gulo* | $C_v$ | R |
| *Lutra lutra* | $C_v$ | U |
| *Felis lynx* | $C_v$ | R |
| *Alces a. alces* | H | C[a] |
| *Rangifer t. tarandus* | H | A[a,c] |

[a] Species recorded at the Stordalen mire.
[b] Rare and infrequent bird species occasionally breeding within the Torneträsk area.
[c] Completely or partly synanthropic.

In addition to the vertebrates listed in Table 1, 47 bird species have been recorded as rare or infrequent visitors in the area, e.g. hen harrier (*Circus c. cyaneus* L.), oystercatcher (*Haematopus o. ostralegus* L.), great blackbacked gull (*Larus marinus* L.), great grey owl (*Strix n. nebulosa* Thunb.), swallow (*Hirundo r. rustica* L.), and jay (*Garrulus g. glandarius* L.). Altogether 200 vertebrate species are observed in the Torneträsk area, of which 153 are relatively regular, but only 147 species breed.

The mire acts as a nesting place for waders and ducks, which mainly draw their nourishment from the surrounding lakes, ponds, and small streams. Without removing any considerable quantity of the plant material, the birds substantially influence primary production of the mire by adding nutrients in the form of feces, while at the same time being prey for warm-blooded predators of the mire, e.g. the weasels.

Vertebrates nesting and lairing in the surroundings and frequently foraging on the mire account for similar contributions to the site. To this category belongs willow grouse (*Lagopus l. lagopus* L.), hare (*Lepus timidus* L.), and rodents (especially *Clethrionomys rufocanus* Sund., *C. rutilus* Pall., and *Lemmus lemmus* L.). The lemming, endemic to the Scandinavian mountain chain and the Kola Peninsula, characterizes the Torneträsk area. Feeding mainly on grass and moss species, the lemmings prefer moist biotopes such as mires during the summer season, moving to dry fell fields in the winter (CURRY-LINDAHL, 1963). Reindeer (*Rangifer t. tarandus* L.) and also moose (*Alces. a. alces* L.), which often stay in the neighboring mountain birch forests, may affect the vegetation of the mire, more, however, by trampling and leaving feces than by grazing.

## Migration

The annual migration of birds and reindeer, depending upon changes in climate and food supply, is a characteristic of the vertebrates of the region. Herded by the Lapps, the reindeer alternate between the winter feeding area in the coniferous forest region and the summer pasturage high up in the mountains during the vegetation period. During fall and especially in spring, when many of the mires in the area are free of snow, these communities are important for reindeer foraging.

Most of the 56 annual migratory bird species of the mire stay within the area between 80 and 120 days $yr^{-1}$ (JONSSON, 1975). Only 8 out of the 64 observed bird species of the Stordalen mire are non-migratory, 5 of these resident bird species are abundant or common to the area, i.e. redpoll (*Carduelis f. flammea* L.), willow tit (*Parus montanus borealis* S.-L.), magpie (*Pica pica fennorum* Lönnberg), raven (*Corvus c. corax* L.), and willow grouse.

By varying their choice of food, e.g. by adding seeds and other parts of the vegetation to the insect diet in late summer and fall, many migrant passerines are able to extend their stay in the region.

## Adaptation

The animals originally arrived in northern Fennoscandia from three different directions; arctic species invading the area from the north and west, probably to a large extent from the unfrozen shores of Norway during the last Würm stadial, Siberian taiga species from the east and a European faunal type from the south. Belonging to these three faunal types, the variety of animal species within the region is comparatively abundant (SIIVONEN, 1974).

The postglacial colonization and relocalization of plant and animal species is still going on in the region, depending on topography and changes in food and climate. This may explain the absence of many soil-inhabiting species.

Due to the relatively favorable climate caused by the Gulf Stream and to the above-mentioned immigration history of the area, this northern latitude fauna includes species from different light- and temperature-climates. Many passerines extend their working day and have a shorter nestling period in the arctic than at lower latitudes owing to the "length of daylight" (Armstrong, 1954).

Representing a gradient from a maritime to a continental climate within a short distance, the whole Torneträsk area allows an extended distribution range for maritime species with a broad amplitude such as great blackbacked gull, lesser blackbacked gull (*Larus f. fuscus* L.), oystercatcher, and lapwing (*Vanellus vanellus* L.).

## Invertebrate Fauna

The invertebrate fauna in the Torneträsk area may be exemplified by some well-known groups. Of the 46 species recorded of Rhopalocera and Hesperoidea (Nordström, 1955) and the 21 species of Sphinges, Bombycimorpha, etc. (Nordström et al., 1961) *Argynnis frigga* Thunb., *Pyrgus centaureae* Rbr., *Eriogaster arbusculae* Frr., and *Hyphoraia lapponica* Thunb. feed principally on cloudberry (*Rubus chamaemorus* L.). *Argynnis sifanica aquilonaris* Stich., *Vacciniina optilete* Kn., *Callophrys rubi* L., *Colias palaeno* L., and *Zygaena exulans* Hochw. feed mainly on *Vaccinium* spp., and *Erebia disa* Thunb., *Oeneis norna* Thunb., *Coenonympha tullia* Müll., and *Fumea casta* Pall. on grass and *Carex* spp. of the swampy grounds. The larval period of some butterfly species, e.g. *Hyphoraia lapponica* Thunb. (Sandberg, 1884), lasts for two years or more. Also the pupal stage, e.g. of some *Anarta* and *Eriogaster* spp., may last for more than one year (Gullander, 1963, 1971).

On the ombrotrophic elevated areas of mires of the Stordalen type the dominant beetle species are *Boreaphilus henningianus* Sahlb. and *Dyschirius norvegicus* Munst., while on the minerotrophic depressions the dominants are *Stenus niveus* Fauv., *S. pubescens* Steph., *S. hyperboreus* J. Sahlb., *Lathrobium punctatum* Zett., *Myllaena gracilicornis* Fairm., *Pterostichus diligens* Strm., *Gymnusa brevicollis* Payk., and *Cyphon variabilis* Thunb. (Brundin, 1934).

Of the 173 species of spiders in the Torneträsk area *Rhaebothorax sphagnicola* Holm, *Pardosa lapponica* Thorell, *P. hyperborea* Thorell, and *P. atrata* Thorell are the dominant species on *Sphagnum* bogs at Stordalen (Holm, 1950). Altogether 127 oribatid species were found by Dalenius (1960) in the Torneträsk region, of which 65 species at bogs and fens, where *Oppia translamellata* Willm. were the most constant and abundant. No species of Chilopoda, Diplopoda, and Pseudoscorpiones are to be found in the area (Holm, 1963).

From the Stordalen mire Svensson (1972) reported the following invertebrates as abundant: (Homoptera) *Psylla betulae-nanae* Oss., *Cicadula intermedia* Boh., *Delphacodes discolor* Boh., (Coleoptera)*Patrobus septentrionis* Dej., *Anthobium lapponicum* Mnh., *Podistra pilosa* Payk., (Hymenoptera) *Formica gagatoides*

Ruzsky, *Vespula norvegica* F., (Diptera) *Chrysops nigripes* Zett., *Anachrostichus lucidus* Zett., *Pararamphomyia obscura* Zett., *Hydrophorus alpinus* Wahlb., *Lispocephala erythrocera* Desv., *Spilogona contractifrons* Zett., *S. dorsata* Zett., *Okeniella caudata* Zett., (Araneae) *Pardosa atrata* Thorell, and *P. lapponica* Thorell.

AGRELL (1941) found 68 species of Collembola within the Torneträsk region of which 32 species in swampy grounds. Gamasid mites and larval stages of several species of Diptera and Coleoptera act as soil-living predators, Collembola species often being their prey.

The culicids, appearing in very dense populations, are the most important blood-suckers on the vertebrates within the area and form at the same time an important part of the food of the passerines. Parasitic nematodes and insects are common on the invertebrates.

The omnivorous dorylaimoid nematodes appear to make up 40% of the nematode biomass of about 0.2 g d.w. and plant-feeding nematodes may make up about 30% of the total nematode fauna (LAGERLÖF et al., 1975). Although few in numbers, the lumbricids, e.g. *Eiseniella tetraedra* Sav. and *Dendrobaena octaedra* Sav. (JULIN, 1949), play an important role through their function and energetically through relatively large biomass. The enchytraeids seem to have very low population densities in the mire (LUNDKVIST, 1973). Up to now *Cognettia lapponica* Nurm. is the only Enchytraeidae species found in the mire, but several species within the group are found in the region (NURMINEN, 1965).

Considering its northern situation, the Torneträsk area has an unusually rich molluscfauna with 23 species. Palsa bogs within the region are often inhabited by molluscs (JAECKEL, 1961). However, the Stordalen mire is characterized by low pH and calcium values and sift samples give malacologically negative results (NILSSON, 1968).

## Density Fluctuations

Remarkable fluctuations occur in the Torneträsk region primarily in the population densities of *Oporinia autumnata* Bkh. and *Operophthera brumata* L. (TENOW, 1972), and of lemming. The herbivore consumption results in a fragmentation of the plant material, increasing the litter surface for attacks of saprovorous organisms.

Due to the specific food supply, the predatory fauna varies both quantitatively and qualitatively. Within the Torneträsk area, covering about 2500 km², seven species of owls can be recorded mainly as immigrants during peak years, compared with normally only two fairly numerous species. In lemming years the short-eared owl (*Asio f. flammeus* Pont.) is more abundant, often being a prey to the peregrine (*Falco p. peregrinus* Tunst.) (BERG, 1913). The diet of the snowy owl (*Nyctea scandiaca* L.) consists mainly of lemming during the breeding season in years when lemming populations are high, which results in successful breeding (ANDERSSON and PERSSON, 1971).

The population of the gyr falcon (*Falco r. rusticolus* L.) varies with the supply of *Lagopus* spp. The importance of willow grouse on the Stordalen mire is negligible in terms of fowl production. Nevertheless, by removing buds and seeds they influence the primary production probably both quantitatively and through changing the relative frequencies of the plant species.

The population density of some birds of prey in Swedish Lapland is here given as a rough mean for recent years between top and bottom years in numbers $km^{-2}$: golden eagle (*Aquila c. chrysaëtos* L.) 0.002, roughlegged buzzard (*Buteo l. lagopus* Pont.) 0.15, white-tailed eagle (*Haliaeetus albicilla* L.) 0.0002, osprey (*Pandion h. haliaëtus* L.) 0.003, peregrine falcon 0.0001, gyr falcon 0.0003, and eagle owl (*Bubo b. bubo* L.) 0.0004.

The density of most abundant passerines in the Abisko area is approximately $km^{-2}$ (ENEMAR, 1963): fieldfare (*Turdus pilaris* L.) 100, redwing (*Turdus i. iliacus* L.) 60, red start (*Phoenicurus ph. phoenicurus* L.) 40, willow warbler (*Phylloscopus trochilus acredula* L.) 240, pied flycatcher (*Ficedula h. hypoleuca* Pallas) 40, meadow pipit (*Anthus p. pratensis* L.) 200, red poll 50, and brambling (*Fringilla montifringilla* L.) 120.

Some birds may have two clutches $yr^{-1}$ in the district, e.g. redpoll, hedge sparrow (*Prunella m. modularis* L.), fieldfare, and redwing. Though the number of hatched nestlings fairly well corresponds to the number of eggs, usually only a small proportion will be fledged due to the available food-supply.

In a very cold spring in the region, recordings from 14 nests of Lapland bunting (*Calcarius l. lapponicus* L.) gave the relatively high number of 5 eggs per nest (ROWELL, 1957). The population density on *Sphagnum* heath areas was estimated at 480 $km^{-2}$. However, Lapland bunting seems to be on the decrease since the beginning of this century.

Ample food supply within the Torneträsk region as well as overproduction of birds in other faunal regions may sometimes cause occasional appearance of invasion birds, e.g. jay, nutcracker (*Nucifraga caryocatactes macrorhynchos* Brehm), coal tit (*Parus a. ater* L.), and parrot crossbill (*Loxia pytyopsittacus* Borkh.).

## The Impact of Man

For centuries the fauna of the Torneträsk region has been influenced by primitive hunting and fishing, reindeer breeding, and setting out salmonoid fishes. During this century, the fauna has been heavily affected by continual hunting with modern weapons, disturbance of animals by tourism and scooter traffic, changes of biotopes through railway construction and permanent settlements, etc.

Man's impact is mainly registered by a reduction in predatory animals. The populations of wolf (*Canis l. lupus* L.), wolverine (*Gulo gulo* L.), and arctic fox (*Alopex lagopus* L.) have decreased to $<1\%$ in 70 years and are now close to extinction (AHLÉN, 1975).

A principle cause for the decrease of birds in recent years is the use of biocides, to which the migrants are especially exposed during their stay in more southern regions. This may be one reason for the reduction in predatory birds, e.g. golden eagle, white-tailed eagle, peregrine, gyr falcon, and eagle owl, observed also in the Torneträsk region through the last decades. Other reasons are illegal hunting and egg collecting.

The decimation of predators has caused a displacement in the balance between species, permitting a colonization of other species such as the moose, which has invaded the district over the last 40 years. Other species, which have recently expanded their distribution area to the Torneträsk region, are red fox (*Vulpes vulpes* L.), hooded crow (*Corvus corone cornix* L.), starling (*Sturnus v. vulgaris* L.), spotted flycatcher (*Muscicapa s. striata* Pallas), and hedge sparrow, the latter two coming from the Norwegian coast.

## References


AGRELL, I.: Zur Ökologie der Collembolen. Opusc. Ent., Suppl. **3**, 1–236 (1941).

AHLÉN, I.: Hotade ryggradsdjur (exkl. fiskar) i Sverige. Sveriges Natur **66**, 126–129 (1975).

ANDERSSON, N. Å., PERSSON, B.: Något om fjällugglans *Nyctea scandiaca* näringsval i Lappland. Vår Fågelvärld **30**, 227–231 (1971).

ARMSTRONG, E. A.: The behaviour of birds in continuous daylight. Ibis **96**, 1–30 (1954).

BERG, B.: Pilgrimsfalken och lemmelvandringarna. Fauna och Flora **8**, 218–224 (1913).

BRUNDIN, L.: Die Coleopteren des Torneträskgebietes. Lund: C. Bloms tryckeri 1934.

CURRY-LINDAHL, K.: Lapplands ryggradsdjur. In: CURRY-LINDAHL, K. (Ed.): Natur i Lappland, Part 1, pp. 290–314. Uppsala: Almqvist & Wiksells 1963.

DALENIUS, P.: Studies on the Oribatei (Acari) of the Torneträsk territory in Swedish Lapland. I. A list of the habitats, and the composition of their oribatid fauna. Oikos **11**, 80–124 (1960).

ENEMAR, A.: The density of birds in the subalpine birch forest of the Abisko area Swedish Lapland, in 1961. K. Fysiogr. Sällsk. Handl. **73**, 1–21 (1963).

GULLANDER, B.: Nordens svärmare och spinnare. Stockholm: P. A. Norstedt & Söner 1963.

GULLANDER, B.: Nordens nattflyn. Stockholm: P. A. Norstedt & Söner 1971.

HOLM, Å.: Studien über die Spinnenfauna des Torneträskgebietes. Zool. Bidr. Uppsala **29**, 103–213 (1950).

HOLM, Å.: Spindlar i Lappland. In: CURRY-LINDAHL, K. (Ed.): Natur i Lappland. Part 1, pp. 415–422. Uppsala: Almqvist & Wiksells 1963.

JAECKEL, S. G. A.: Zur Molluskenfauna des nördlichen Schwedisch-Lappland. Schriften Naturw. Ver. Schleswig-Holstein **32**, 21–46 (1961).

JONSSON, S.: Vertebrate fauna of the Torneträsk area. IBP Swedish Tundra Biome Project Tech. Rep. **5**, 1–20 (1975).

JULIN, E.: De svenska daggmaskarterna. Arkiv Zool. **42**, 1–58 (1949).

LAGERLÖF, J., MAGNUSSON, C., ROSSWALL, T.: Investigation of nematodes at the Stordalen site. IBP Swedish Tundra Biome Project Tech. Rep. **18**, 1–32 (1975).

LUNDKVIST, H.: Occurrence of Enchytraeidae at Stordalen. IBP Swedish Tundra Biome Project Tech. Rep. **16**, 29–34 (1973).

NILSSON, A.: De på land levande molluskerna inom Abisko nationalpark. Redog. f. undersökn. utförda under åren 1966–1967. Landskrona: Mimeographed 1968.

NORDSTRÖM, F.: De fennoskandiska dagfjärilarnas utbredning. Lund: K. Fysiogr. Sällsk. Handl. **66**, 1–177 (1955).

NORDSTRÖM, F., OPHEIM, M., SOTAVALTA, O.; De fennoskandiska svärmarnas och spinnarnas utbredning. Lund: K. Fysiogr. Sällsk. Handl. **72**, 1–93 (1961).

NURMINEN, M.: Enchytraeids (Oligochaeta) from northern Norway and western Lapland. Ann. Zool. Fenn. **2**, 11–15 (1965).

ROWELL, C. H. F.: The breeding of the lapland bunting in Swedish Lapland. Bird Study **4**, 33–50 (1957).

SANDBERG, G.: Fortsatte iagttagelser over arktiske sommerfugles metamorphoser. Ent. Tidskr. **5**, 139–144 (1884).

SIIVONEN, L.: Changes in the vertebrate fauna of the North Calotte area. In: BYLUND, E. (Ed.): Ecological problems of the circumpolar area, pp. 49–57. Luleå: Alltryck AB 1974.

SKARTVEIT, A., RYDÉN, B. E., KÄRENLAMPI, L.: Climate and hydrology of some Fennoscandian tundra ecosystems. In: WIELGOLASKI, F. E. (Ed.): Fennoscandian tundra ecosystems. Part 1: Plants and microorganisms, pp. 41–53. Berlin-Heidelberg-New York: Springer 1975.

SONESSON, M., WIELGOLASKI, F. E., KALLIO, P.: Description of Fennoscandian tundra ecosystems. In: WIELGOLASKI, F. E. (Ed.): Fennoscandian tundra ecosystems. Part 1: Plants and microorganisms, pp. 3–28. Berlin-Heidelberg-New York: Springer 1975.

SVENSSON, B. G.: An inventory of the insect fauna at Stordalen mire 1970. IBP Swedish Tundra Biome Project Tech. Rep. **3**, 1–22 (1972).

TENÓW, O.: The outbreaks of *Oporinia autumnata* Bkh. and *Operophthera* spp. (Lep., Geometridae) in the Scandinavian mountain chain and northern Finland 1862–1968. Zool. Bidr. Uppsala, Suppl. **2**, 1–107 (1972).

# 1.2. Organization and Dynamics of Populations

## Dynamics of Enchytraeidae Populations on Hardangervidda

T. SOLHØY

The dynamics of two Enchytraeidae populations at Stigstuv were studied over a period of 3 years. The sites selected were the dry and wet meadow (SONESSON et al., 1975). For various reasons, the group had to be mostly treated collectively during this investigation.

The distribution and species composition of Enchytraeidae in Fennoscandia and Spitzbergen have been fairly well documented by BACKLUND (1946, 1947), NIELSEN and CHRISTENSEN (1959, 1961, 1963), NURMINEN (1967b) and ABRAHAMSEN (1968). Ecological studies have been performed in various coniferous forests in Norway (ABRAHAMSEN, 1972) and Finland (NURMINEN, 1967a). No studies from Fennoscandian alpine and arctic communities have so far been published.

### Methods

The sampling procedure for enchytraeids is identical to that used for soil-dwelling microarthropods (SOLHØY, 1972). However, when the soil was frozen, squares of about 20 × 20 × 6 cm were cut out and brought back to the laboratory. They were allowed to thaw at + 5° C for about 24 hrs, and 2–4 soil cores were then taken from each square.

The sample units were extracted in modified O'Connor funnels and the specimens were counted immediately to minimize fragmentation.

### Results and Discussion

13 species have hitherto been found at the two sites (Table 1). The dominant species on the wet meadow are *Cognettia glandulosa* (Mich.), *C. sphagnetorum* (Vejd.) and *Cernosvitoviella* sp. The *Cognettia* species are also dominant on the dry meadow together with *Bryodrilus ehlersi* Ude, *Henlea nasuta* (Eisen) and *Fredericia bulboides* N. and Chr.

70–95% of the population occurred in the uppermost 3 cm of the soil at both sites (Fig. 1). Exceptions to this were only found during drought periods on the dry meadow during summer and also at both sites in late fall, probably due to frost. A downward migration was clearly apparent in these cases.

Table 1. Dominance of the Enchytraeidae recorded at the two sites

| Species | Wet meadow | Dry meadow |
|---|---|---|
| *Cognettia glandulosa* (Mich.) | 41.2 | 11.2 |
| *C. sphagnetorum* (Vejd.) | 22.1 | 21.3 |
| *Bryodrilus ehlersi* Ude | 1.7 | 19.9 |
| *Mesenchytraeus armatus* (Lev.) | 2.5 | — |
| *M. flavus* (Lev.) | 1.2 | — |
| *Cernosvitoviella* sp. | 20.1 | — |
| *Henlea nasuta* (Eisen) | 7.9 | 28.1 |
| *H. perpusilla* Friend, augm. Cern. | 2.0 | 3.4 |
| *Hemifredericia parva* N. & Chr. | — | 2.2 |
| *Fredericia bulboides* N. & Chr. | — | 11.2 |
| *F. bulbosa* (Rosa) | 0.3 | 1.1 |
| *F. ratzeli* (Eisen) | 1.0 | — |
| *Enchytraeus buchholzi* Vejd. | — | 2.2 |

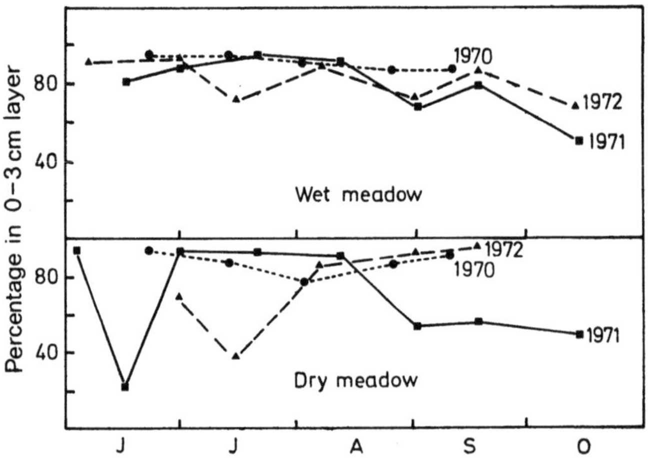


Fig. 1. Seasonal variation in the number of individuals in the uppermost 3 cm of the soil on the dry and wet meadows expressed as percentage of total number found

The seasonal trends in population densities during the vegetation period are shown in Fig. 2. At both sites population minima were found at the beginning of the vegetation period (first week of June) after winter mortality. This winter mortality is well documented on the wet meadow (Fig. 3) where it is clear that the population resisted the environmental stress until the middle of March. Although the soil temperature did not change at that time, the population diminished markedly within a few weeks and remained at this minimum till the snow had melted.

The apparent differences in seasonal trends on the dry meadow seem to be explicable in terms of drought effects. This is consistent with the results of NIELSEN (1955a, b) from Denmark and of O'CONNOR (1957) from England. Severe drought periods occurred at the dry meadow site at various times in both 1968 and 1969. This fact, together with ordinary winter mortality, must account for the

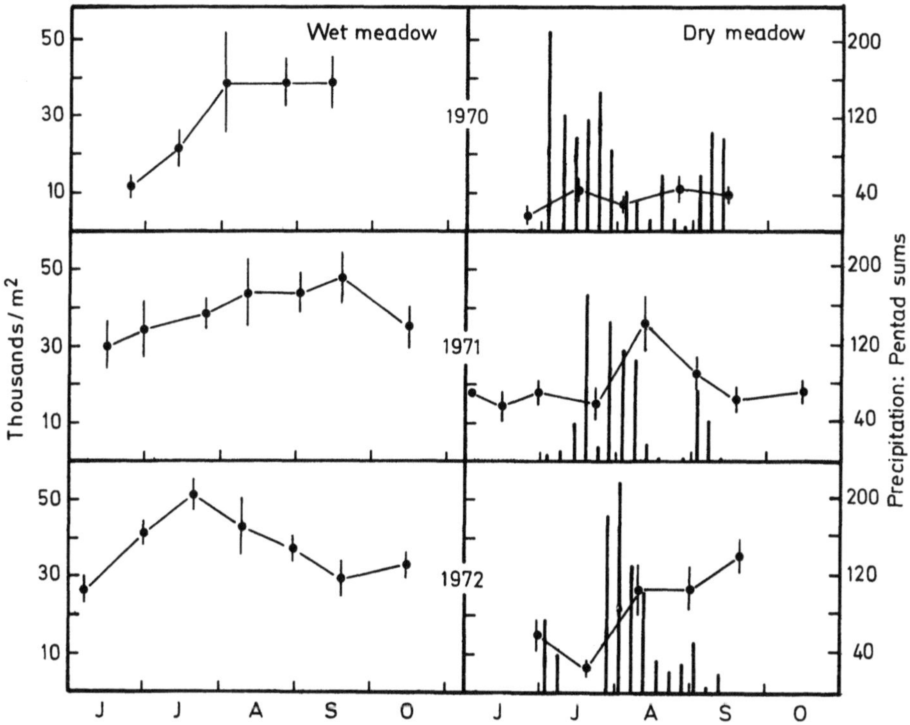


Fig. 2. The abundance (± S.E.) of Enchytraeidae at the two Stigstuv sites during the vegetation period in three successive years. Vertical bars show pentad sums of precipitation

low population density observed in 1970. The pentad-sums of precipitation (SKARTVEIT, 1974) showed no pronounced dry spells during this season (Fig. 2), but nevertheless the density did not change significantly till the onset of winter. In the following two years a marked increase in abundance occurred between July and August. In 1971 this high density was followed by a decrease in numbers due to drought, while a slight increase occurred towards winter in 1972. In that year, a dry spell in July also caused a temporary drop in density. Thus the results clearly indicate that the dry meadow population is vulnerable to changes in moisture.

On the wet meadow, where the soil moisture is apparently not a limiting factor, the population increased towards autumn maxima in both 1970 and 1971. The decrease in abundance after the peak in Juli 1972, however, has not been found to be correlated to any observed abiotic factor.

The series of samples below the snow cover on the wet meadow 1971–1973 indicated that substantial growth of the population also occurred early in the period of snow cover (Fig. 3). In contrast to the dry meadow, ground frost did not develop before the middle of January, except for a short period in October before the snow fell. The soil temperature (at 2 cm depth) was just above freezing point (+ 0.4°–+ 0.1° C). The increase in enchytraeid-population at this low temperature is surprising, but the same tendency has been reported from northern Finland by NURMINEN (1967a) and by PEACHY (1963) from *Nardus* grassland at Moor

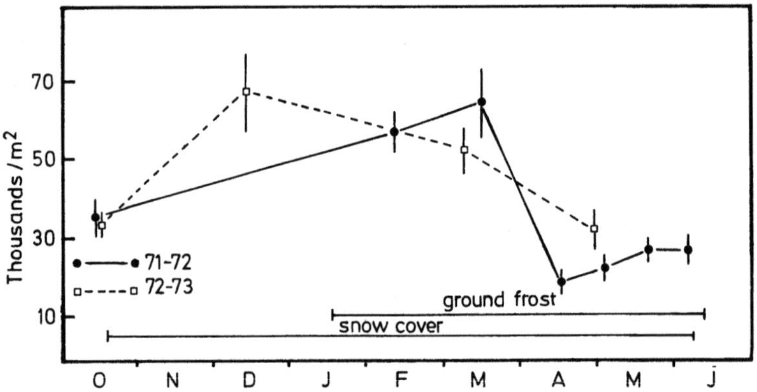


Fig. 3. The abundance of Enchytraeidae (±S.E.) on the wet meadow during the winters 1971/1972 and 1972/1973. The duration of snow cover and ground frost are also indicated

Table 2. The abundance of Enchytraeidae in various Fennoscandian coniferous forest soils (from ABRAHAMSEN, 1972) compared with the corresponding values from the Stigstuv sites

| Site | Sample depth cm | Nos. $m^{-2}$ in thousands | | | Reference/country |
|---|---|---|---|---|---|
| | | min. | mean | max. | |
| Spruce loc. 1 | ca. 5 | 1.8 | 16.2 | 53.6 | NURMINEN, 1967a; Finland |
| Spruce | 6 | 7.6 | 43.0 | 85.5 | ABRAHAMSEN, 1969; Norway |
| Spruce | 6 | 12.7 | 34.7 | 73.9 | ABRAHAMSEN, 1972; Norway |
| Wet meadow | 6 | 12.2 | 38.0 | 69.0 | Present paper |
| Pine loc. 7 | ca. 5 | 0.8 | 10.7 | 36.8 | NURMINEN, 1967a; Finland |
| Pine | ca. 5 | 3.2 | 34.7 | 54.7 | ABRAHAMSEN, 1970; Norway |
| Pine | 6 | 0.8 | 22.9 | 47.8 | ABRAHAMSEN, 1972; Norway |
| Dry meadow | 6 | 5.0 | 17.7 | 36.8 | Present paper |

House, England. Further evidence of growth of the wet meadow population in early winter is given by the fact that the mean dry weight per individual was found to increase from about 30 μg to 40 μg at this time in both years.

The mean population density on the wet meadow during the three years was 38,000 ind. $m^{-2}$, and the corresponding value on the dry meadow was 18,000 ind. $m^{-2}$. The wet meadow abundance corresponds fairly well with values from Fennoscandian spruce forests and the dry meadow abundance with those from Fennoscandian pine forests (Table 2). As pine forest soils in general are more vulnerable to drought than spruce forest soils during dry spells, this can to a certain extent explain the difference in abundance between the two forest types (ABRAHAMSEN, 1972). The same conclusion applies to the difference in abundance found between the dry and wet meadows at Stigstuv.

*Acknowledgement.* The species composition of some samples has kindly been determined by Dr. G. ABRAHAMSEN, at the Norwegian Forest Research Institute, and his valuable help is hereby acknowledged.

## References


ABRAHAMSEN, G.: Records of Enchytraeidae (Oligochaeta) in Norway. Meddr Norske Skogfors Ves. **25**, 209–230 (1968).

ABRAHAMSEN, G.: Sampling design in studies on population densities in Enchytraeidae (Oligochaeta). Oikos **20**, 54–66 (1969).

ABRAHAMSEN, G.: Skoggjødsling og jordbunnsfaunaen. Tidsskr. Skogbr. 78, 296–303 (1970).

ABRAHAMSEN, G.: Ecological study of Enchytraeidae (Oligochaeta) in Norwegian coniferous forest soils. Pedobiologia **12** 26–82 (1972).

BACKLUND, H. O.: Swedish Enchytraeidae. I. Kungl. Fysiogr. Sällsk. Handl. N.F. **57**, 1–22 (1946).

BACKLUND, H. O.: Swedish Enchytraeidae. II. Kungl. Fysiogr. Sällsk. Handl. N.F. **58**, 1–31 (1947).

NIELSEN, C. O.: Studies on Enchytraeidae. 2. Field studies. Natura Jutl. **4–5**, 1–58 (1955a).

NIELSEN, C. O.: Studies on Enchytraeidae. 5. Factors causing seasonal fluctuations in numbers. Oikos **6**, 153–169 (1955b).

NIELSEN, C. O., CHRISTENSEN, B.: The Enchytraeidae, critical revision and taxonomy of European species. Natura Jutl. **8–9**, 1–160 (1959).

NIELSEN, C. O., CHRISTENSEN, B.: The Enchytraeidae, critical revision and taxonomy of European species. Suppl. 1 Natura Jutl. **10**, 1–23 (1961).

NIELSEN, C. O., CHRISTENSEN, B.: The Enchytraeidae, critical revision and taxonomy of European species. Suppl. 2. Natura Jutl. **10**, 1–19 (1963).

NURMINEN, M.: Ecology of enchytraeids (Oligochaeta) in Finnish coniferous forest soil. Ann. Zool. Fenn. **4**, 147–157 (1967a).

NURMINEN, M.: Faunistic notes on North-European enchytraeids (Oligochaeta). Ann. Zool. Fenn. **4**, 567–587 (1967b).

O'CONNOR, F. B.: An ecological study of the enchytraeid worm population of a coniferous forest soil. Oikos **8**, 161–199 (1957).

PEACHY, J. E.: Studies on the Enchytraeidae (Oligochaeta) of moorland soil. Pedobiologia **2**, 81–95 (1963).

SKARTVEIT, A.: Netto strålingsbalanse og energibalanse ved jordyta. University of Bergen: Unpublished Thesis 1974.

SOLHØY, T.: Quantitative invertebrate studies in mountain communities at Hardangervidda, South Norway. I. Norsk Ent. Tidsskr. **19**, 99–108 (1972).

SONESSON, M., WIELGOLASKI, F. E., KALLIO, P.: Description of Fennoscandian tundra ecosystems. In: WIELGOLASKI, F. E. (Ed.): Fennoscandian tundra ecosystems. Part 1: Plants and microorganisms, pp. 3–28. Berlin-Heidelberg-New York: Springer 1975.

# Dynamics of Oribatei Populations on Hardangervidda

T. SOLHØY

The abundance of soil- and litter-dwelling mites at three sites at Stigstuv, Hardangervidda, South Norway (1220–1320 m) described by SONESSON et al. (1975) has been sampled regularly for some years in the snow-free period (June–Oct.). Sampling has also been carried out below the snow-cover from January to June 1972. On an average, Oribatei seem to account for about 50% of the total mite abundance and 70–90% of mite biomass.

Owing to taxonomic and technical problems only part of the material has been worked upon. This contribution deals with some aspects of the dynamics of the group in 1970 at two of the sites studied, the lichen heath and the wet meadow.

## Methods

Procedures for sampling and extracting mites above and below ground have been described by KAURI et al. (1969) and SOLHØY (1972). Specific biomass values (of adults) were calculated from the average individual weight, obtained usually from 20–30 specimens. Sex was determined and the presence of mature eggs in the females detected using standard clearing methods. The larval stage was recognized by the occurrence of three pairs of legs, and the nymphal stages were distinguished by the number of genital suckers (WALLWORK, 1969).

## Results and Discussion

Due to insufficient knowledge of the immature stages (of which some are undescribed), attention had to be focused on the adults with the exception of the two most common species *Oromurcia bicuspidata* and *Tectocepheus velatus* (where the immature stages were also studied).

Fluctuations in the number and biomass of adult Oribatei on the lichen heath are given in Table 1. Immature stages of Brachychthoniidae are also included summarily due to the difficulty of separating the different stages. Two peaks of biomass were encountered, the first in June, about four weeks after the ground frost had disappeared, and the second in August/Spetember. All the common

Table 1. Number of adults (in thousands) and their biomass in mg (dry weight) $m^{-2}$ of species of Oribatei on the lichen heath in 1970. Other species in addition to those named in the text: *Eobrachychthonicus oudemansi* v.d.H., *Liochthonius lapponicus* (Träg.), *Camisia biurus* (C.L.K.), *C. horrida* (Herm.), *Paulonothrus longisetotus* (Willm.), *Epidamaeus tatricus* (Kulz.), *Porobelba spinosa* (Sell.), *Oppia bicarinata* (Paoli), *O. nova* (Oudms.), *O. ornata* (Oudms.), *O. subpectinata* (Oudms.), *Oribatula tibialis* (Nic.)

| Species | 20.5 | | 12.6 | | 24.6 | | 14.7 | | 4.8 | | 27.8 | | 12.9 | | mean | |
|---|---|---|---|---|---|---|---|---|---|---|---|---|---|---|---|---|
| | *n* | dw | *n* | dw | *n* | dw | *n* | dw | *n* | dw | *n* | dw | *n* | dw | *n* | dw |
| Brachychthoniidae | 7.90 | 0.8 | 13.10 | 1.3 | 17.09 | 1.7 | 7.91 | 0.8 | 8.33 | 0.8 | 26.53 | 2.7 | 23.60 | 2.4 | 14.92 | 1.5 |
| *Phthiracarus piger* (Scop.) | 0.10 | 6.6 | 0.01 | 0.7 | 0.06 | 3.9 | 0.07 | 4.6 | 0.03 | 2.0 | 0.11 | 7.2 | 0.11 | 7.2 | 0.07 | 4.6 |
| *Carabodes labyrinthicus* (Mich.) | 3.20 | 24.4 | 4.92 | 37.4 | 5.91 | 44.9 | 1.83 | 13.9 | 1.91 | 14.5 | 3.04 | 23.1 | 3.31 | 25.2 | 3.45 | 26.2 |
| *Tectocepheus velatus* (Mich.) | 10.48 | 17.8 | 4.64 | 7.9 | 4.45 | 7.6 | 4.61 | 7.8 | 2.21 | 3.8 | 11.58 | 19.7 | 10.23 | 17.4 | 6.86 | 11.7 |
| *Chamobates borealis* (Träg.) | 0.54 | 4.1 | 0.41 | 3.1 | 0.11 | 0.8 | 0.12 | 0.9 | 0.14 | 1.1 | 0.43 | 3.3 | 0.04 | 0.3 | 0.26 | 1.9 |
| *Calyptozetes sarekensis* (Träg.) | 0.52 | 5.8 | 0.92 | 10.3 | 0.82 | 9.2 | 0.59 | 6.6 | 0.32 | 3.6 | 1.21 | 13.6 | 0.63 | 7.1 | 0.72 | 8.0 |
| Other species | 0.01 | 0.3 | 0.45 | 5.7 | 0.42 | 1.7 | 0.01 | 0.3 | 0.26 | 2.4 | 0.21 | 0.4 | 0.31 | 3.7 | 0.24 | 2.1 |
| Total | 22.75 | 59.7 | 24.45 | 66.4 | 28.86 | 69.5 | 15.14 | 34.9 | 13.20 | 28.2 | 43.11 | 70.0 | 38.23 | 63.3 | 26.52 | 56.0 |

Table 2. Number of adults (in thousands) and their biomass in mg (dry weight) $m^{-2}$ of species of Oribatei on the wet meadow in 1970. × = less than 10 specimens or 0.1 mg dry weight $m^{-2}$.
Other species in addition to those named in the Text: *Liochthonius* sp., *Mucronothrus nasalis* Willm., *Trimalaconothrus* sp., *Camisia lapponica* (Träg.), *Nothrus borussicus* (Selln.), *Platynothrus peltifer* (C.L.K.), *P. punctatus* (L.Koch), *Fuscozetes fuscipes* (C.L.K.), *Melanozetes mollicomus* (C.L.K.), *Trichoribates setiger* (Träg.)

| Species | 12.6. | | 22.6. | | 15.7. | | 3.8. | | 26.8. | | 11.9. | | mean | |
|---|---|---|---|---|---|---|---|---|---|---|---|---|---|---|
| | *n* | dw | *n* | dw | *n* | dw | *n* | dw | *n* | dw | *n* | dw | *n* | dw |
| *Malaconothrus egregius* Berl. | 0.05 | 0.1 | 0.01 | × | 0.23 | 0.5 | 0.31 | 0.6 | 0.25 | 0.5 | 0.20 | 0.4 | 0.18 | 0.4 |
| *Tectocepheus velatus* (Mich.) | 0.15 | 0.3 | 0.04 | 0.1 | 0.52 | 0.9 | 1.46 | 2.5 | 1.16 | 1.2 | 1.73 | 2.9 | 0.84 | 1.3 |
| *Oppia translamellata* (Willm.) | 1.60 | 1.1 | 0.50 | 0.4 | 3.34 | 2.3 | 3.06 | 2.1 | 2.43 | 1.7 | 1.09 | 0.8 | 2.00 | 1.4 |
| *Oribatula tibialis* (Nic.) | 0.05 | 0.4 | 0.01 | 0.1 | 0.08 | 0.6 | 0.02 | 0.2 | 0.02 | 0.2 | 0.13 | 1.0 | 0.05 | 0.4 |
| *Liebstadia similis* (Mich.) | 0.20 | 1.6 | 0.05 | 0.4 | 0.22 | 1.8 | 0.03 | 0.2 | 0.02 | 0.2 | 0.13 | 1.0 | 0.11 | 0.9 |
| *Edwardzetes edwardsi* (Nic.) | 0.25 | 6.7 | 0.03 | 0.8 | 0.25 | 6.7 | 0.40 | 10.7 | 0.10 | 2.7 | 0.28 | 7.5 | 0.22 | 5.9 |
| *Oromurcia bicuspidata* S.Thor | 0.35 | 7.7 | 1.06 | 23.2 | 1.60 | 35.0 | 1.63 | 35.7 | 1.09 | 23.9 | 2.53 | 55.4 | 1.38 | 30.2 |
| *Eupelops plicatus* (C.L.K.) | 0.30 | 5.0 | 0.51 | 8.5 | 0.37 | 6.1 | 0.67 | 11.1 | 0.50 | 8.3 | 0.69 | 11.5 | 0.51 | 8.4 |
| Other species | 0.10 | 2.9 | × | × | 0.06 | 1.3 | 0.55 | 0.6 | 0.35 | 0.6 | 0.16 | 0.8 | 0.20 | 1.0 |
| Total | 3.05 | 25.8 | 2.21 | 33.5 | 6.67 | 55.6 | 8.13 | 63.7 | 5.92 | 39.3 | 6.94 | 81.3 | 5.49 | 49.9 |

species showed the same trend. A reasonable explanation is that in this dry lichen heath biotope, moisture conditions are most favorable just after the snow melts and in late summer when there is higher precipitation. A similar pattern of two peaks has been found in dry biotopes in Austria (LOUB and HAYBACH, 1967) and in Denmark (WEBB, 1970). The low abundance in midsummer cannot, however, be explained solely by high mortality. Some estivation of a part of the population seems to occur which results in less efficient extraction, a view strongly supported by the data given by WEBB (1970). Of the 17 species found on the lichen heath *Carabodes labyrinthicus*, *Tectocepheus velatus* and *Calyptozetes sarekensis* comprised 82% of the mean seasonal biomass.

18 species were found on the wet meadow (Table 2). *Edwardzetes edwardsi*, *Oromurcia bicuspidata*, *Eupelops plicatus* formed 89% of the mean seasonal biomass. The seasonal fluctuations in number and biomass of adults (Table 2) are less synchronous than those found on the lichen heath. However, total biomass and the biomass of the dominant species *O. bicuspidata* were at a minimum just after the snow had disappeared in June and at a maximum in September. A similar pattern seems to occur on wet meadows at Point Barrow, Alaska (S. MACLEAN, pers. comm.). It is also to a certain extent similar to the results obtained for enchytraeids (SOLHØY, 1975) and collembola (FJELLBERG, 1975). The melting of a thick cover of snow (140 cm) caused a constant flooding of the site for five weeks from the beginning of May. Along with ordinary winter mortality, this event, though it occurs each spring, may account for the low numbers in June.

Females of the species studied were recorded with mature eggs during the whole sampling period (Fig. 1). Males and females of *O. bicuspidata* and *C. labyrinthicus* were found to occur in equal numbers, while *T. velatus* was found to be thelytokous. Comparatively few of the females of *C. labyrinthicus* and *T. velatus* (12% and 26% respectively) were found to have overwintered with eggs on the lichen heath. 90% of *O. bicuspidata* and *T. velatus* on the wet meadow contained mature eggs about two weeks after the snow had disappeared. Analysis of the

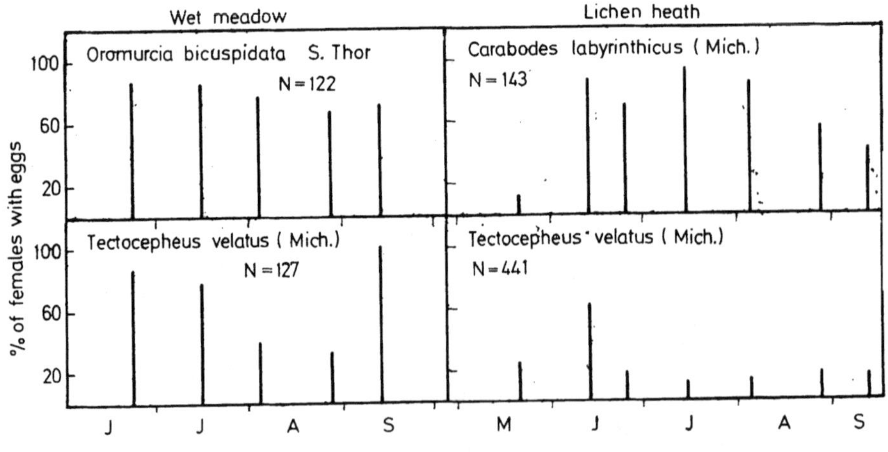


Fig. 1. Percentage of females with mature eggs in three species of Oribatei during the snow-free period in 1970

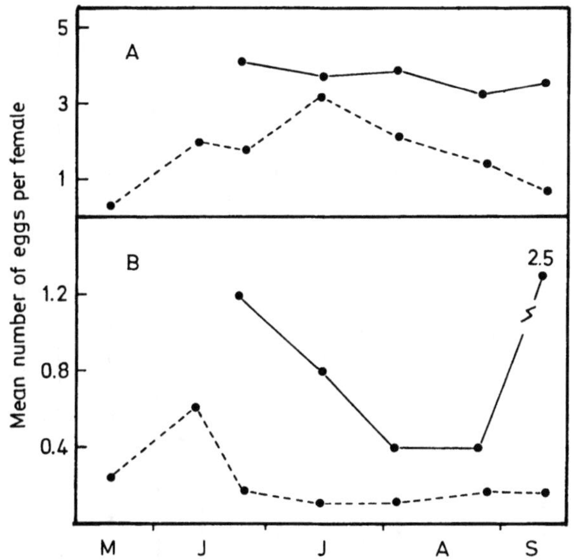


Fig. 2A and B. Mean number of mature eggs per female in three species of Oribatei during the snow-free period in 1970. (A) *Oromurcia bicuspidata* from the wet meadow (whole line) and *Carabodes labyrinthicus* from the lichen heath (stippled line). (B) *Tectocepheus velatus* from the wet meadow (whole line) and *T. velatus* from the lichen heath (stippled line)

females of *O. bicuspidata* from winter samples 1972 also showed that 90% contained mature eggs.

In 1970, no definite period of maximum egg-laying was found in *O. bicuspidata* and *C. labyrinthicus* (Figs. 1 and 2). *T. velatus* seemed to have one period of maximum egg-laying in June on the lichen heath, and in June–July on the wet meadow.

All female specimens of *T. velatus* on the wet meadow also contained mature eggs in September, but it is uncertain whether there was a second period of intensive egg-laying at that time, or if the eggs were deposited the next year. If they were deposited in September, they certainly did not hatch before the next season, due to the low prevailing temperatures.

Larvae, the three nymphal stages, and adults occurred simultaneously during the snow-free period. No significant seasonal pattern could be found in the development of the two species studied (Fig. 3). The adults have been omitted from the calculations, as the extraction of adults is far more efficient than for the juveniles (FORSSLUND, 1948; MACFADYEN, 1953). It is supposed by the author, however, that the various stages of juveniles are extracted with equal efficiency.

Figure 3 suggests that the eggs of *T. velatus* laid in June on the lichen heath could have developed into protonymphae by the beginning of September. In a Danish soil HAARLÖV (1960) found that *T. velatus* produced two generations a year, and at Moor House in England, BLOCK (1966) recorded one generation a year. At the Stigstuv sites, where the soil temperature at 2 cm depth is at or below zero for eight months of the year and the mean summer temperature is 8–10° C

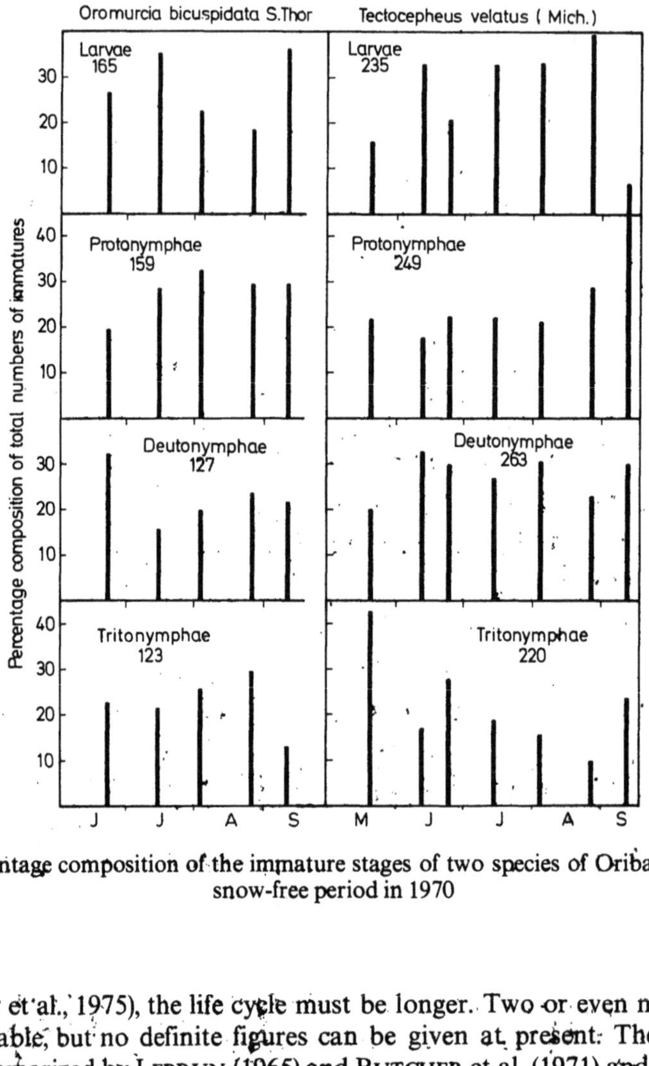


Fig. 3. Percentage composition of the immature stages of two species of Oribatei during the snow-free period in 1970

(SKÅRTVEIT et al., 1975), the life cycle must be longer. Two or even more years is most probable, but no definite figures can be given at present. The laboratory studies summarized by LEBRUN (1965) and BUTCHER et al. (1971) and the study of *Hermannia gibba* (C. L. K.) by BÄUMLER (1970) may support this view. At 10° C the development time of *H. gibba* from egg to adult was about 500 days. Development is usually quicker in small species than in bigger ones (BUTCHER et al., 1971), but even in the case of the small species, *T. velatus*, MURPHY and JALIL (1964) found a postembryonic development time of 102 days at 25° C.

The duration of the adult stage is not known, but BLOCK (1965) has estimated more than one year for some species at Moor House, England. Adults of *C. labyrinthicus* have been kept alive at room temperature for about one year by the author.

*Acknowledgements.* The author is strongly indebted to Dr. P. DALENIUS, Hedemora, Sweden, Professor C. PÉREZ-INNIGO, Madrid, Spain and the late Professor K.-H. FORSSLUND, Stockholm, Sweden for their help with identification problems.

## References

BÄUMLER, W.: Zur Morphologie, Biologie und Ökologie von *Hermannia gibba* (C. L. Koch) (Acarina: Oribatei) unter Berücksichtigung einiger Begleitarten. II. Z. Angew. Entomol. **66**, 337–362 (1970).

BLOCK, W. C.: The life histories of *Platynothrus peltifer* (Koch, 1839) and *Damaeus clavipes* (Hermann, 1804) (Acarina: Cryptostigmata) in soils of Pennine moorland. Acarologia **7**, 735–743 (1965).

BLOCK, W. C.: Seasonal fluctuations and distribution of mite populations in moorland soils, with a note on biomass. J. Anim. Ecol. **35**, 487–503 (1966).

BUTCHER, J. W., SNIDER, R., SNIDER, J. R.: Bioecology of edaphic Collembola and Acarina. In: SMITH, R. F., MITTLER, T. E. (Eds.): Annual review of entomology. **16**, 249–288 (1971).

FJELLBERG, A.: Organisation and dynamics of Collembola populations on Hardangervidda. In: WIELGOLASKI, F. E. (Ed.): Fennoscandian tundra ecosystems. Part 2: Animals and systems analysis, pp. 73–79. Berlin-Heidelberg-New York: Springer 1975.

FORSSLUND, K.-H.: Något om innsamlingsmetodiken vid markfaunaundersökningar. Medd. Skogforsk. Inst. Stockh. **37**, 1–22 (1948).

HAARLØV, N.: Microarthropods from Danish soils. Oikos, Suppl. **3**, 1–176 (1960).

KAURI, H., MOLDUNG, T., SOLHØY, T.: Turnbull and Nicholls "Quick Trap" for acquiring standing crop of evertebrates in high mountain grassland communities. Norsk Ent. Tidsskr. **16**, 113–136 (1969).

LEBRUN, P.: Contribution a l'étude écologique des Oribatides de la litière dans une fôret de Moyenne-Belgique. Mem. Inst. Sci. Nat. Belg. **153**, 1–96 (1965).

LOUB, W., HAYBACH, G.: Jahreszyklische Beobachtungen der Mikroflora und Mikrofauna von Böden im südlichen Wienerwald. Rev. Ecol. Biol. Sol. **4**, 59–80 (1967).

MACFADYEN, A.: Notes on methods for the extraction of small soil arthropods. J. Anim. Ecol. **22**, 65–77 (1953).

MURPHY, P. W., JALIL, M.: Some observations on the genus *Tectocepheus*. C. R. I. Congr. Int. Acarol. (1963). Acarologia **6** (h. s.), 566–591 (1964).

SKARTVEIT, A., RYDÉN, B. E., KÄRENLAMPI, L.: Climate and hydrology of some Fennoscandian tundra ecosystems. In: WIELGOLASKI, F. F. (Ed.): Fennoscandian tundra ecosystems. Part 1: Plants and microorganisms, pp. 41–53. Berlin-Heidelberg-New York: Springer (1975).

SOLHØY, T.: Quantitative invertebrate studies in mountain communities at Hardangervidda, South Norway. I. Norsk Ent. Tidsskr. **19**, 99–108 (1972).

SOLHØY, T.: Dynamics of Enchytraeidae populations on Hardangervidda. In: WIELGOLASKI, F. E. (Ed.): Fennoscandian tundra ecosystems. Part 2: Animals and systems analysis, pp. 55–59. Berlin-Heidelberg-New York: Springer 1975.

SONESSON, M., WIELGOLASKI, F. E., KALLIO, P.: Description of Fennoscandian tundra ecosystems. In: WIELGOLASKI, F. E. (Ed.): Fennoscandian tundra ecosystems. Part 1: Plants and microorganisms, pp. 3–28. Berlin-Heidelberg-New York: Springer 1975.

WALLWORK, J. A.: Some basic principles underlying the classification and identification of cryptostigmatid mites. In: SHEALS, J. G. (Ed.): The soil ecosystem. Systematics Assosiation Publications, **8**, 155–169 (1969).

WEBB, N. R.: Population metabolism of *Nothrus silvestris* Nicolet (Acari). Oikos **21**, 155–159 (1970).

# Spider Populations in a Subalpine Birch Forest

S. KOPONEN

## Introduction and Study Area

Spiders (Araneae) are an important predator group in most terrestrial ecosystems. The great number of spiders in IBP sites at Kevo has earlier been shown to exist both in the ground (KOPONEN, 1971) and in vegetation layers (KOPONEN and OJALA, 1974). The spiders inhabiting the ground layer, especially the family Lycosidae, constitute a large proportion of the surface-active mesofauna in many habitats, especially when the pitfall-trapping method is used. The aim of the present paper is to deal with the spiders of the ground layer of a simple habitat in subarctic conditions, with special attention in Lycosidae populations.

The study area is situated near the Kevo Subarctic Research Station (69° 45′N., 27° E.) in northernmost Finland. The site lies on a slightly north-exposed slope of the hill Puksalskaidi, approximately 190 m above sea level and 120 m above Lake Kevojärvi. The area is a rather dry, scattered mountain birch forest. There are *Betula tortuosa* bushes (2–3 m tall); the short and patchy field layer is dominated by *Empetrum hermaphroditum* and the ground layer by *Cladonia* spp. and a few mosses.

## Methods and Material

The material was collected using pitfall traps with ethylene glycol. One hundred traps were placed in ten lines, each trap approx. 5 m from the next. Traps were emptied every day, about 17.00 h; the peak of activity of Lycosidae at Kevo is from 14.00 h to 17.00 h (KOPONEN, 1972). The trapping period was from June 6 to September 11, 1973, including a period of 60 days with continuous daylight. The material comprised 2877 specimens and 47 species of spiders.

## Results

**Family and Species Composition.** The present material can be divided into nine families (Table 1, systematics and nomenclature according to LEHTINEN, 1967). The vast majority of species (55.3%) belonged to the family Linyphiidae (incl. Erigoninae), and the family Lycosidae greatly dominated in number of specimens (67.1%). Other important families were Gnaphosidae, Thomisidae,

Table 1. Numbers of species and specimens in different families

| Family | Number of species | % | Number of specimens | % |
|---|---|---|---|---|
| Lycosidae | 5 | 10.6 | 1931 | 67.1 |
| Linyphiidae | 26 | 55.3 | 334 | 11.6 |
| Gnaphosidae | 6 | 12.8 | 266 | 9.2 |
| Thomisidae | 4 | 8.5 | 196 | 6.8 |
| Hahniidae | 1 | 2.1 | 71 | 2.5 |
| Zoridae | 1 | 2.1 | 55 | 1.9 |
| Dictynidae | 2 | 4.3 | 13 | 0.5 |
| Salticidae | 1 | 2.1 | 9 | 0.3 |
| Araneidae | 1 | 2.1 | 2 | 0.1 |
| Total | 47 | 100 | 2877 | 100 |

Hahniidae, and Zoridae; in the two last-named only one species. Abundance relations of families were rather similar to results found in other studies of dry mountain birch zone in northernmost Fennoscandia (KOPONEN, unpubl.).

The great number of very active wolf spiders (Lycosidae) is typical of pitfall-trap materials. This was true also in the present material, the four most abundant species belonged to Lycosidae: *Pardosa hyperborea* 753 ind. (26.2%), *P. tarsalis* 553 ind., *P. eiseni* 344 ind., and *Alopecosa aculeata* 137 adults and approx. 100 juveniles. Other numerous species, ranked according to numbers, were *Xysticus obscurus*, *Hahnia ononidum*, *Macrargus multesimus* (the most abundant Linyphid species), *Zora nemoralis palmgreni*, *Haplodrassus soerenseni*, *Gnaphosa lapponum*, *Micaria alpina*, *Gonatium rubens*, *Meioneta gulosa*, and *Tiso aestivus*. Other typical, but less numerous, species of Finnish Fell-Lapland were e.g. *Gnaphosa intermedia*, *Arctella lapponica*, *Oxyptila rauda*, *Ceratinella oculatissima*, *Meioneta similis*, *Saloca strandi*, and *Tibioplus diversus*.

The present fauna was, in general lines, rather similar to that of the Kilpisjärvi area, Finnish West-Lapland (PALMGREN, 1965). However, great differences were noted in abundance relations of species; this was caused by our different sampling methods. Some common species seemed to occur at considerably lower altitudes in Kevo than in Kilpisjärvi, e.g. *Tiso aestivus* and *Meioneta gulosa*.

In Fig. 1 a kind of "dominance-diversity" curve, is shown for the material with curves for geometric series and for a non-overlapping, random niche system (see WHITTAKER, 1965; ODUM, 1971, p. 236). The curves are not based on the true dominance relation of species, but on the activity of species. This is caused by the trapping method. Number of individuals is considered as an expression for activity. The curve for spiders was found to be intermediate between the geometric curve which may occur in very rigorous conditions and the curve of a non-overlapping random "niche-part" system with little or no competition. Intermediate types of curves seem to be typical of most natural communities (WHITTAKER, 1965). However, competition, at least between closely related species with similar habits and niche types (e.g. *Pardosa* spp.; some Linyphids), must be remarkable, especially in the present study habitat with simple vegetation structure and few "niche-segments".

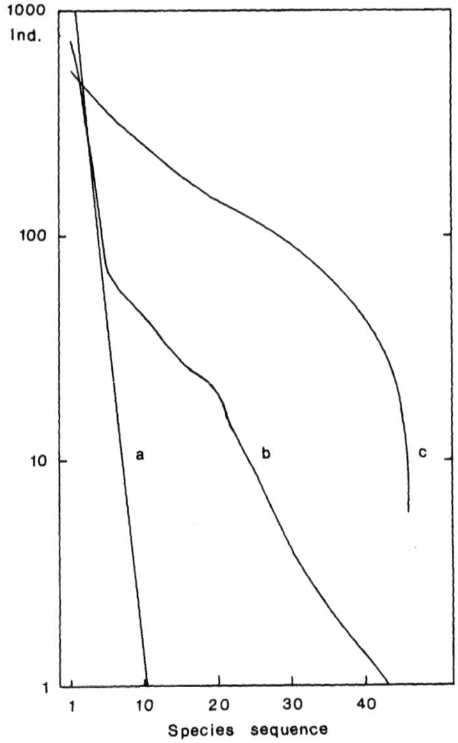


Fig. 1. "Dominance-diversity" curve for 47 spider species caught (*b*), for geometric series (*a*) and for non-overlapping random niche hypothesis (*c*). Explanations in text

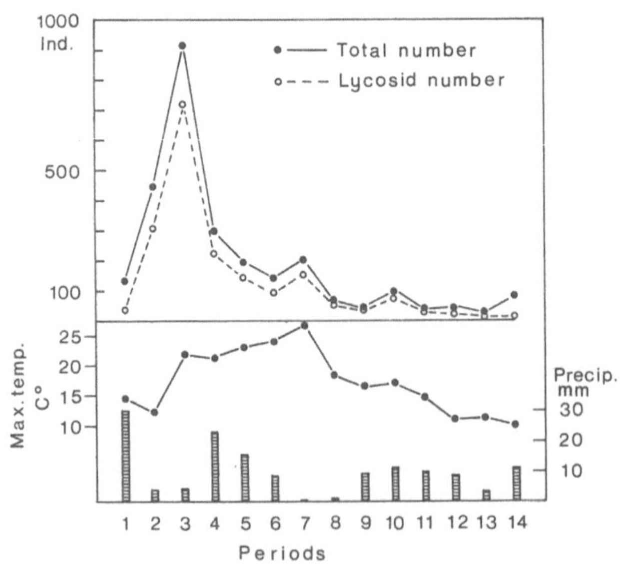


Fig. 2. Individual number, mean maximum temperature, and precipitation for sampling periods (weeks)

The phenology of families was, in general, clear. Lycosidae had a great peak of occurrence in the third sampling week (Fig. 2), Thomisidae in early summer, Gnaphosidae, Hahniidae, and Zoridae in early and mid-summer, and Linyphiidae in spring and autumn. The peak of occurrence in the third week was common e.g. to all Lycosidae species, *Xysticus obscurus*, *Haplodrassus soerenseni*, *Hahnia ononidum*, and *Agyneta conigera*. However, it must be emphasized that the third week was the first warm period in 1973. *Gnaphosa lapponum* and *Micaria alpina* had their peak in the 5–6th weeks, and *Macrargus multesimus* in the last one. When these phenological data are compared with those of PALMGREN (1965) from West-Lapland, the figures are rather similar.

**Spider Numbers and Weather Factors.** Daily pitfall catches of spiders and weather station measurements were compared, using multiple regression analyses (KOPONEN et al., 1975). In Gnaphosidae and Lycosidae, solar radiation and maximum temperature were the best explaining factors (among several often highly intercorrelated weather factors), both very positively significant ($p<0.001$). In Linyphiidae minimum temperature was negatively significant ($p<0.01$). The highest correlation coefficients for a family was 0.639 for Gnaphosidae (solar radiation) and for a species 0.577 for *Pardosa tarsalis* (radiation) and *Micaria alpina* (hours of sunshine). Higher weekly maximum temperatures indicated increased numbers of spiders in the same week apart from the third week, especially of Lycosidae (Fig. 2, weeks 7 and 10).

**Structure of Lycosidae Populations.** The vast majority of the fauna trapped belonged to Lycosidae. Rather dense populations of three *Pardosa* and one *Alopecosa* species (*A. aculeata* and the rarer *A. pinetorum*) were living in this simple and small-area habitat. The activity peak of Lycosidae was very clear, about 66% of the total Lycosidae material was caught during weeks 2–4, and all species had their activity or occurrence peak in the third week (Table 2).

Some phenological differences among the occurrence of *Pardosa* species were observed. The period of adult females in late summer was 3 weeks shorter in *P. tarsalis* than in other species. The peak of males of *P. hyperborea* and *P. tarsalis* was on June 19 and the last male was caught on July 14 and July 15, respectively. The number of "male-days" was 40 for both species. The peak of males of *P. eiseni* was later, June 23, and the last male caught on July 8, number of "male-days" only 34.

The first egg cocoon and pullus was observed as follows:

| | *P. hyperborea* | *P. tarsalis* | *P. eiseni* | *A. aculeata* |
|---|---|---|---|---|
| First cocoon | June 26 | June 24 | July 1 | July 8 (the only one) |
| First pullus | July 17 | July 17 | July 20 | July 24 |

The number of egg cocoons and the percentual proportion of *Pardosa* females with cocoons of the total female material from weeks 3–14 are shown in Table 3. The first peak was observed in the fifth week and the second one in the tenth week in all species. The proportion of females with cocoon was clearly higher in the least common *Pardosa* species, *P. eiseni* (53.7%), and rather similar in *P. hyperbo-*

Table 2. Catches of *Pardosa* species per one-week period, males/females/juveniles–total

| Species | 1<br>6–12 June | 2<br>13–19 June | 3<br>20–26 June | 4<br>27 June–3 July |
|---|---|---|---|---|
| *P. hyperborea* | 7/2/9–18 | 124/18/4–146 | 311/15/2–328 | 80/10/3–93 |
| *P. tarsalis* | 1/0/2–3 | 86/21/3–110 | 151/31/2–184 | 14/21/8–43 |
| *P. eiseni* | 4/0/1–5 | 5/2/2–9 | 99/5/2–106 | 34–25/6–65 |
| Species | 5<br>4–10 Juli | 6<br>11–17 July | 7<br>18–24 July | 8<br>25–31 July |
| *P. hyperborea* | 27/10/6–43 | 2/7/3–12 | 0/42/12–54 | 0/15/1–16 |
| *P. tarsalis* | 3/18/14–35 | 1/18/24–43 | 0/48/23–71 | 0/3/15–18 |
| *P. eiseni* | 4/25/24–53 | 0/12/12–24 | 0/15/ 2–17 | 0/7/0–7 |
| Species | 9<br>1–7 August | 10<br>8–14 August | 11<br>15–21 August | 12<br>22–28 August |
| *P. hyperborea* | 0/7/1–8 | 0/9/1–10 | 0/8/2–10 | 0/6/1–7 |
| *P. tarsalis* | 0/1/8–9 | 0/4/10–14 | 0/2/4–6 | 0/0/5–5 |
| *P. eiseni* | 0/6/3–9 | 0/28/4–32 | 0/6/0–6 | 0/3/4–7 |
| Species | 13<br>29 August–4 Sept. | 14<br>5–11 September | | Total<br>6 June–11 Sept. |
| *P. hyperborea* | 0/4/0–4 | 0/3/1–4 | | 551/156/46–753 |
| *P. tarsalis* | 0/0/3–3 | 0/0/9–9 | | 256/167/130–553 |
| *P. eiseni* | 0/1/2–3 | 0/1/0–1 | | 146/136/62–344 |

Table 3. Numbers and percentual proportions of *Pardosa* females with egg cocoons. For dates of period numbers, see Table 2

Periods

| Species | 3 | | 4 | | 5 | | 6 | | 7 | |
|---|---|---|---|---|---|---|---|---|---|---|
| | *N* | % | *N* | % | *N* | % | *N* | % | *N* | % |
| *P. eiseni* | 0 | 0.0 | 18 | 72.0 | 24 | 96.0 | 11 | 91.7 | 3 | 20.0 |
| *P. hyperborea* | 1 | 6.7 | 7 | 70.0 | 10 | 100.0 | 4 | 57.1 | 0 | 0.0 |
| *P. tarsalis* | 2 | 6.5 | 11 | 52.4 | 16 | 88.9 | 9 | 50.0 | 2 | 4.2 |
| Species | 8 | | 9 | | 10 | | 11 | | 12 | |
| *P. eiseni* | 0 | 0.0 | 0 | 0.0 | 9 | 32.1 | 4 | 66.7 | 2 | 66.7 |
| *P. hyperborea* | 2 | 13.3 | 1 | 14.3 | 7 | 77.8 | 6 | 75.0 | 6 | 100.0 |
| *P. tarsalis* | 1 | 33.3 | 0 | 0.0 | 4 | 100.0 | 1 | 50.0 | – | – |
| Species | 13 | | 14 | | | | | | 3–14 | |
| *P. eiseni* | 0 | 0.0 | 1 | 100.0 | | | | | 72 | 53.7 |
| *P. hyperborea* | 3 | 75.0 | 3 | 100.0 | | | | | 50 | 36.8 |
| *P. tarsalis* | – | – | – | – | | | | | 46 | 31.5 |

*rea* and *P. tarsalis*. Total female numbers of each species were of the same class of magnitude: *P. tarsalis* 167, *P. hyperborea* 156, *P. eiseni* 136.

A small counting of egg numbers per cocoon was made. All studied cocoons were from the early or mid-summer and therefore the first cocoon of each of the specimens. The egg numbers per cocoon (with S. E.) were as follows:

| | | |
|---|---|---|
| *P. hyperborea* | $24.1 \pm 1.84$ | $(N=15)$, |
| *P. tarsalis* | $43.0 \pm 2.30$ | $(N=10)$, |
| *P. eiseni* | $61.5 \pm 3.09$ | $(N=14)$. |

The numbers of eggs in *P. tarsalis* and *P. eiseni* were similar to earlier reports from Sweden and Central Europe (HOLM, 1947; KESSLER, 1973). The egg number of *P. hyperborea* at Kevo was lower than in Sweden (HOLM, 1947: mean 33 eggs).

The number of eggs per cocoon depends, in general, on the individual size of each species (and specimen). This phenomenon has been shown earlier in the Lycosidae family by e.g. PETERSEN (1950) and KESSLER (1973). To test this point, random samples of females were weighed and other samples were measured. Live biomasses and the length and width of carapax of *Pardosa* females are shown in Table 4. Individuals for biomass measurements were caught using pitfall traps without preservation liquid. The number of eggs per cocoon correlated well with the weight and carapax size of each species.

The male to female ratio in *Pardosa* spp. correlated well with the total abundance of the species. Sex ratios and the proportion of juveniles were as follows:

| | Sex ratio (♂♂:♀♀) | % of juveniles |
|---|---|---|
| *P. hyperborea* | 3.53 | 6.1 |
| *P. tarsalis* | 1.53 | 23.5 |
| *P. eiseni* | 1.07 | 18.0 |
| *A. aculeata* | 1.63 | (45.6, all *Alopecosa* juv.) |

The sex ratio of spiders has often been discussed (see e.g. KESSLER-GESCHIERE, 1970). Many authors have pointed out that high male proportions were observed in areas where population density was low, i.e. in the marginal area of a population. However, in the present material *P. hyperborea* had a very high proportion of males. This situation was caused by the different ecological requirements and activity patterns of females and juveniles (note low juvenile proportion) to those of highly active males existing over a shorter period (cf. KOPONEN et al., 1975). In *P. eiseni*, the sex ratio was near 1, the proportion of females with egg cocoons high

Table 4. Length and width of carapex (in mm, with S.E.) and live biomass (in mg, with S.E.) of *Pardosa* females at Kevo

| Species | Length and | Width of carapax | *N* | Live biomass | *N* |
|---|---|---|---|---|---|
| *P. hyperborea* | $2.5 \pm 0.04$ | $1.9 \pm 0.03$ | 20 | $14.7 \pm 0.65$ | 26 |
| *P. tarsalis* | $2.7 \pm 0.04$ | $2.1 \pm 0.03$ | 20 | $22.4 \pm 1.07$ | 18 |
| *P. eiseni* | $3.0 \pm 0.05$ | $2.4 \pm 0.03$ | 20 | $41.3 \pm 4.81$ | 10 |

(54%) and the proportion of juveniles rather high (18%). Thus *P. eiseni* seemed to be near its optimal environmental requirements, although it was the most uncommon *Pardosa* species. However, no clear deductions can be made as pitfall trapping based purely on activity is the only sampling method used in the present work. According to large unpublished pitfall-trap and sieving material from Kevo, *P. eiseni* is the most eurytopic species occurring in considerable numbers in all habitats studied, the maximum abundance was observed in dry pine forest. *P. hyperborea* is also rather eurytopic; the optimum habitats seemed to be the present open birch forest and bogs. *P. tarsalis* favors open areas, e.g. fell heaths where it lives together with *P. lapponica*, a species not occurring in the present study sites. *Alopecosa aculeata* is more eurytopic than *A. pinetorum*. The optimum habitat of *A. aculeata* seemed to be dry pine and birch forests, and that of the other species, dense and moister birch forests.

*Acknowledgements.* The field work carried out by L. ISO-IIVARI is gratefully acknowledged.

## References

HOLM, A.: Svensk spindelfauna Araneae, Fam. 8–10. Stockholm: Entomologiska föreningen 1947.

KESSLER, A.: A comparative study of the production of eggs in eight *Pardosa* species in the field (Araneida, Lycosidae) Tijdschr. Entomol. **116**, 23–41 (1973).

KESSLER-GESCHIERE, A.: Sex ratio in spiders inhabiting the field layer. Bull. Mus. Nation. Hist. Nat. Paris **41**, Suppl. 1, 102–107 (1970).

KOPONEN, S.: On the abundance relations of mesofaunal groups in the ground layer of three subarctic habitats. Rep. Kevo Subarctic Res. Stat. **7**, 57–59 (1971).

KOPONEN, S.: On the spiders of the ground layer of a pine forest in Finnish Lapland, with notes on their diurnal activity. Rep. Kevo Subarctic Res. Stat. **9**, 32–34 (1972).

KOPONEN, S., HAUKIOJA, E., ISO-IIVARI, L.: Comparison of spider catches and weather in subarctic conditions. Proc. VI Intern. Arachnol. Congr., Amsterdam 1974, pp. 77–81. 1975.

KOPONEN, S., OJALA, H.: On the mesofauna of the field layer of three subarctic habitats. Rep. Kevo Subarctic Res. Stat. **11**, 65–71 (1974).

LEHTINEN, P. T.: Classification of the Cribellate spiders and some allied families, with notes on the evolution of the suborder Araneomorpha. Ann. Zool. Fennici **4**, 199–468 (1967).

ODUM, E. P.: Fundamentals of ecology. Philadelphia-London-Toronto: W. B. Saunders Co. 3d Ed. 1971.

PALMGREN, P.: Die Spinnenfauna der Gegend von Kilpisjärvi in Lappland. Acta Zool. Fennica **110**, 1–70 (1965).

PETERSEN, B.: The relation between size of mother and number of eggs and young in some spiders and its significance for the evolution of size. Experientia **6**, 96–101 (1950).

WHITTAKER, R. H.: Dominance and diversity in land plant communities. Science **147**, 250–260 (1965).

# Organization and Dynamics of Collembola Populations on Hardangervidda

A. FJELLBERG

## Introduction and Methods

The present study is part of the Norwegian International Biological Programme (IBP) concerning productivity of tundra ecosystems in alpine regions in central southern Norway. Three different habitats were examined: a wet meadow, a dry meadow and a dry lichen heath, as described by SONESSON et al. (1975), and SOLHØY (1972).

The fauna in vegetation and litter was sampled with a suction apparatus (quick trap) and extracted in Tullgren funnels, while soil fauna was sampled with a core sampler and extracted in Macfadyen high gradient apparatus. The sampling program, sampling methods and extraction techniques are described in detail by KAURI et al. (1969), and SOLHØY (1972).

## Results

The faunistic similarity between two biotopes may be expressed by the SØRENSEN (1948) index. Applying this to the Stigstuv Collembola data, the following similarities between wet meadow (W), dry meadow (D) and lichen heath (L) are obtained: $S_{L-D}=0.48$, $S_{L-W}=0.21$ and $S_{D-W}=1.41$ (excluding species with sporadic occurrence). The similarity is thus greatest between dry meadow and wet meadow and smallest between lichen heath and wet meadow.

Mean values for the period are summarized in Tables 1 and 2. The abundance is highest on the dry meadow, fluctuating between 45000 and 123000 ($\bar{X}=78000$) ind. $m^{-2}$, with the maximum in July. The wet meadow comes next with 17000—71000 ($\bar{X}=51000$) ind. $m^{-2}$, reaching a maximum in August. The lichen heath has the smallest numbers with 17000–75000 ($\bar{X}=35000$) ind. $m^{-2}$, reaching a maximum in August/September.

Generally the abundance is low in spring and increases rapidly in summer, due to the hatching of eggs. During the autumn there is a slow decrease. This trend is exemplified by the total abundance of Collembola at the wet meadow (Fig. 1) and by the abundance of *Folsomia brevicauda* on the lichen heath (Table 1).

Table 1. Mean abundance (N m$^{-2}$ · 10$^{3}$) of the most dominant Collembola species on each sampling date in 1970

| Species | 20.V. | 12.VI. | 23.VI. | 15.VII. | 3.VIII. | 26.VIII. | 10. IX. |
|---|---|---|---|---|---|---|---|
| Wet meadow | | | | | | | |
| *Hypogastrura denticulata* | | 3.5 | 2.3 | 4.7 | 9.1 | 6.8 | 13.3 |
| *Friesea mirabilis* | | 2.9 | 1.2 | 2.9 | 7.6 | 8.3 | 9.2 |
| *Tetracanthella britannica* | | 3.6 | 0.8 | 12.4 | 15.6 | 11.0 | 13.7 |
| *Folsomia quadrioculata* | | 4.4 | 2.1 | 12.5 | 10.1 | 5.2 | 16.5 |
| *Isotomiella minor* | | 1.6 | 3.3 | 20.4 | 13.6 | 2.5 | 5.3 |
| *Isotoma notabilis* | | 4.9 | 1.2 | 3.7 | 2.6 | 2.6 | 0.6 |
| Total, all species | | 33.0 | 17.1 | 63.4 | 71.2 | 56.7 | 65.3 |
| Dry meadow | | | | | | | |
| *Tullbergia krausbaueri* | | 3.1 | 1.5 | 20.5 | 17.7 | 7.9 | 21.3 |
| *Tetracanthella britannica* | | 6.0 | 23.1 | 63.3 | 20.3 | 54.6 | 25.4 |
| *Folsomia quadrioculata* | | 33.6 | 14.2 | 22.9 | 15.0 | 25.1 | 19.6 |
| Total, all species | | 54.2 | 44.5 | 123.1 | 64.9 | 101.0 | 79.0 |
| Lichen heath | | | | | | | |
| *Xenylla maritima* | 1.5 | 3.9 | 2.6 | 0.7 | 3.6 | 4.1 | 8.1 |
| *Folsomia brevicauda* | 32.1 | 12.6 | 13.2 | 10.3 | 12.0 | 61.9 | 45.0 |
| Total, all species | 35.8 | 21.6 | 19.8 | 16.7 | 19.1 | 74.6 | 61.0 |

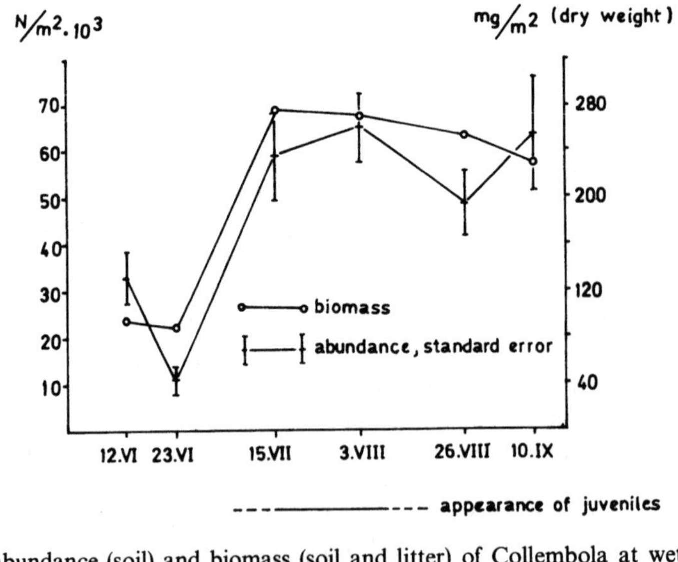


Fig. 1. Total abundance (soil) and biomass (soil and litter) of Collembola at wet meadow in summer 1970

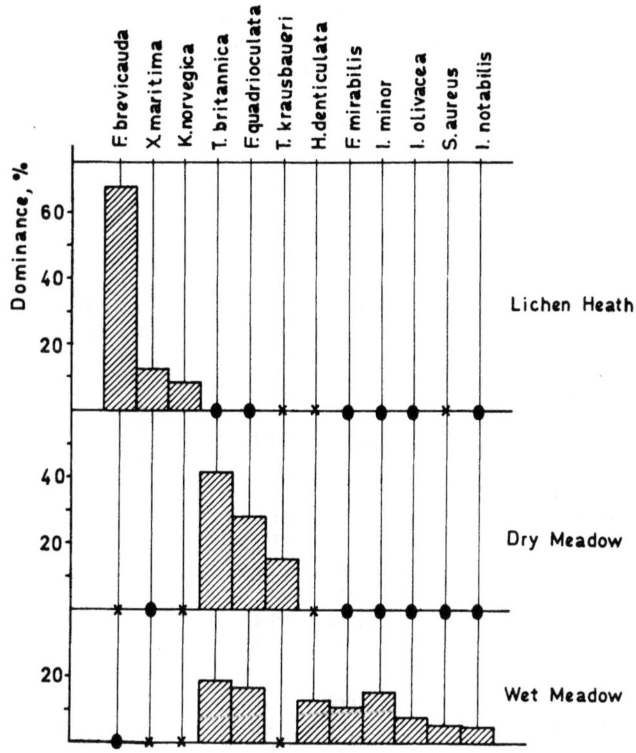


Fig. 2. Species with more than 5% dominance at least on one biotope. Crosses: The species does not occur at the site. Dots: The species occurs with less than 5% dominance

The mean dominance of the most important species is shown in Fig. 2.

For the larger, dominant Collembola species, the biomass has been calculated by applying the body size/weight relationship for a particular species to the size (age) structure of that population (FJELLBERG, in manuscript). The biomass of the smaller and less dominant species has been calculated using a constant individual mean weight chosen to represent the „average size" of each species. The results are summarized in Table 2. All values are dry weights (24 hrs at 105° C). The

Table 2. Mean values of abundance ($N\,m^{-2}$) and biomass ($mg\,m^{-2}$) for the period June–September 1970

| Species | Wet meadow | | Dry meadow | | Lichen heath | |
|---|---|---|---|---|---|---|
| | $N m^{-2}$ | $mg m^{-2}$ | $N m^{-2}$ | $mg m^{-2}$ | $N m^{-2}$ | $mg m^{-2}$ |
| *Hypogastrura purpurescens* | 259 | 33.1 (combined) | 1 | — | — | — |
| *H. scotica* | 278 | | 620 | 2.4 | 1 | — |
| *H. denticulata* | 6614 | | — | — | — | — |
| *Xenylla maritima* | — | — | 34 | 0.1 | 3508 | 5.3 |
| *Willemia anopthalma* | — | — | 1070 | 1.1 | 41 | — |
| *Friesea mirabilis* | 5360 | 9.7 | 2138 | 3.9 | 1 | — |

Table 2 (continued)

| Species | Wet meadow | | Dry meadow | | Lichen heath | |
|---|---|---|---|---|---|---|
| | $N m^{-2}$ | $mg m^{-2}$ | $N m^{-2}$ | $mg m^{-2}$ | $N m^{-2}$ | $mg m^{-2}$ |
| *Anurida forsslundi* | — | — | — | — | 96 | 0.2 |
| *A. pygmaea* | — | — | 50 | — | 259 | 0.3 |
| *Neanura muscorum* | 8 | 0.2 | — | — | — | — |
| *Onychiurus pseudovanderdrifti* | 1 | — | 663 | 6.7 | — | — |
| *Tullbergia krausbaueri* | 42 | — | 11993 | 4.8 | — | — |
| *Karlstejnia norvegica* | — | — | — | — | 1909 | 0.8 |
| *Tetracanthella britannica* | 9543 | 41.5 | 32130 | 120.9 | 46 | 0.1 |
| *T. wahlgreni* | 34 | 0.2 | 1944 | 17.1 | 776 | 6.1 |
| *Anurophorus binoculatus* | — | — | — | — | 931 | 0.3 |
| *Folsomia brevicauda* | 42 | — | — | — | 26732 | 24.0 |
| *F. diplophthalma* | 25 | — | 1714 | 1.7 | — | — |
| *F. quadrioculata* | 8444 | 40.0 | 21717 | 50.0 | 7 | — |
| *F. sensibilis* | — | — | 67 | 0.1 | — | — |
| *Isotomodella pusilla* | 8 | — | 192 | 0.1 | — | — |
| *Isotomiella minor* | 7780 | 21.0 | 1696 | 4.2 | 427 | 1.3 |
| *Isotoma anglicana* | 247 | 3.7 | — | — | — | — |
| *I. neglecta* | 42 | 0.1 | — | — | — | — |
| *I. notabilis* | 2603 | 2.5 | 284 | 0.3 | 1 | — |
| *I. olivacea* | 4028 | 15.4 | 363 | 0.8 | 22 | — |
| *I. pallida* | 293 | 0.3 | — | — | — | — |
| *I. violacea* | — | — | 1 | — | 410 | 2.5 |
| *I. viridis* | 1699 | 42.9 | 115 | 4.5 | 1 | — |
| *Lepidocyrtus cyaneus* | 11 | 0.1 | — | — | 7 | — |
| *L. lignorum* | 235 | 4.7 | 530 | 3.3 | 269 | 5.4 |
| *Sminthurides malmgreni* | 1 | — | — | — | — | — |
| *S. parvulus* | 14 | — | 8 | — | — | — |
| *S. pumilis* | 90 | — | 84 | 0.1 | — | — |
| *S. schoetti* | 632 | 0.4 | — | — | — | — |
| *Sminthurinus aureus* | 2717 | 4.0 | 373 | 0.6 | — | — |
| *Bourletiella clavigera* | 30 | 0.3 | 112 | 1.1 | 1 | — |
| *B. insignis* | 149 | 2.2 | 8 | 0.1 | — | — |
| *B. pistillum* | — | — | — | — | 41 | 0.4 |
| Total | 51229 | 222.3 | 77907 | 223.9 | 35485 | 46.7 |

biomass on the dry meadow fluctuates between 120 and 340 ($\bar{X} = 224$) mg $m^{-2}$. On the wet meadow, it is 90—280 ($\bar{X} = 222$) mg $m^{-2}$, and on the lichen heath 20–90 ($\bar{X} = 47$) mg $m^{-2}$. The fluctuations in biomass follow those of abundance (Fig. 1). For further details see NÆSS et al. (1975).

The Stigstuv values for abundance and biomass fall within the range of the world data summarized by TANAKA (1970).

According to LEWIS and TAYLOR (1967) aggregation can be expressed by the coefficient $b$ in the equation relating variance ($S^2$) and mean ($m$): $S^2 = am^b$. The aggregation of the total number of Collembola in the soil layer (litter and vegetation excluded) has been calculated. The fauna of the lichen heath shows the strongest aggregation, while the fauna of the wet meadow is the most evenly dispersed (Fig. 3).

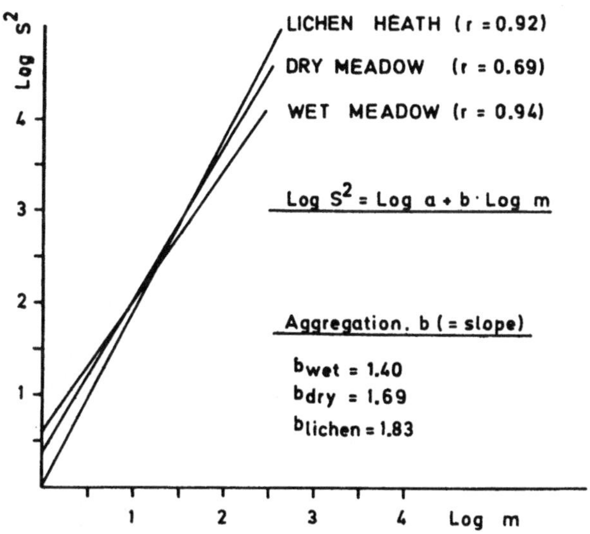


Fig. 3. The correlation between mean ($m$) and variance ($S^2$) of total number of Collembola in soil samples from wet meadow, dry meadow and lichen heath

The relationship between the number of species and the number of specimens of each species in a population can be expressed by Shannon's index (ODUM, 1971): $H' = -\sum \frac{N_i}{N} \log \frac{N_i}{N}$, where $N_i$ is the number of individuals of species $i$, in a sample with $N$ individuals.

The Collembola fauna of the lichen heath has the lowest diversity ($H' = 0.46$ for the period June–September 1970), while the wet meadow shows the highest diversity ($H' = 0.95$). At the dry meadow the Shannon index is calculated to be intermediate ($H' = 0.67$).

It is possibly the high moisture of the wet meadow which makes this biotope the most suitable for Collembola (high diversity, low aggregation).

The life cycle of ten of the most important species has been studied by analysis of the size (age) structure of the population and examination of the genital organs of adults. The following species have a one-year cycle (example: Fig. 4): *Folsomia quadrioculata, Isotoma anglicana, I. viridis, I. notabilis, I. olivacea* and *I. violacea*. These species become reproductive adults before or after their hibernation. A two-year cycle (example: Fig. 5) is found in *Xenylla maritima, Tetracanthella britannica, T. wahlgreni* and *Folsomia brevicauda*. They do not become reproductive adults before they have hibernated twice. Some species (*I. notabilis, I. anglicana, I. violacea*) hatch in June, while the juveniles of most species appear in July/August (*I. viridis, I. olivacea, T. britannica, T. wahlgreni, F. brevicauda, X. maritima*).

JOSSE (1960) examined the life cycles of some Collembola species in the Netherlands and found 2–3 generations during the summer. Apparently the generation time is longer under arctic/alpine conditions.

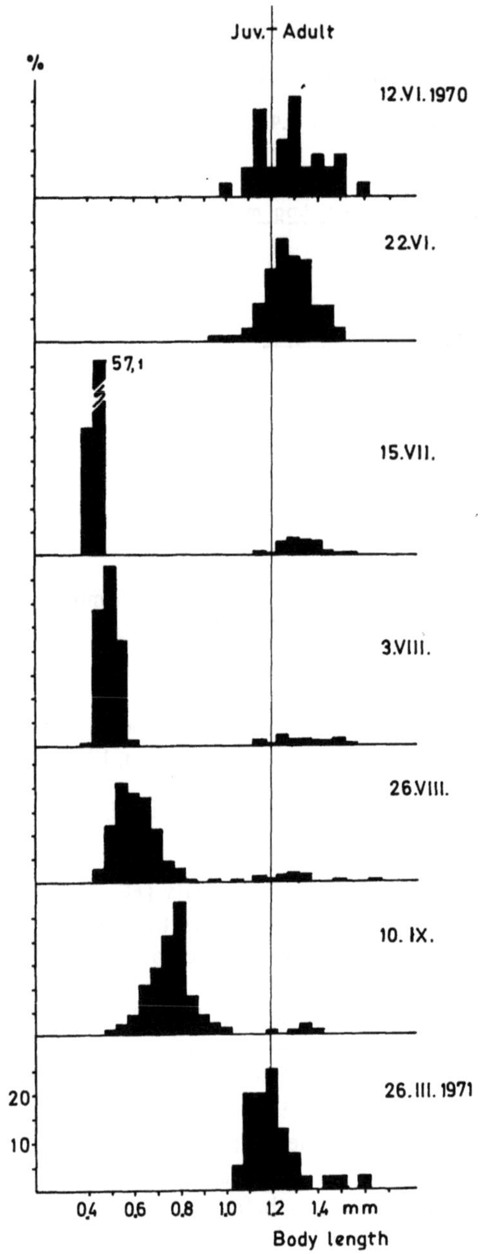


Fig. 4

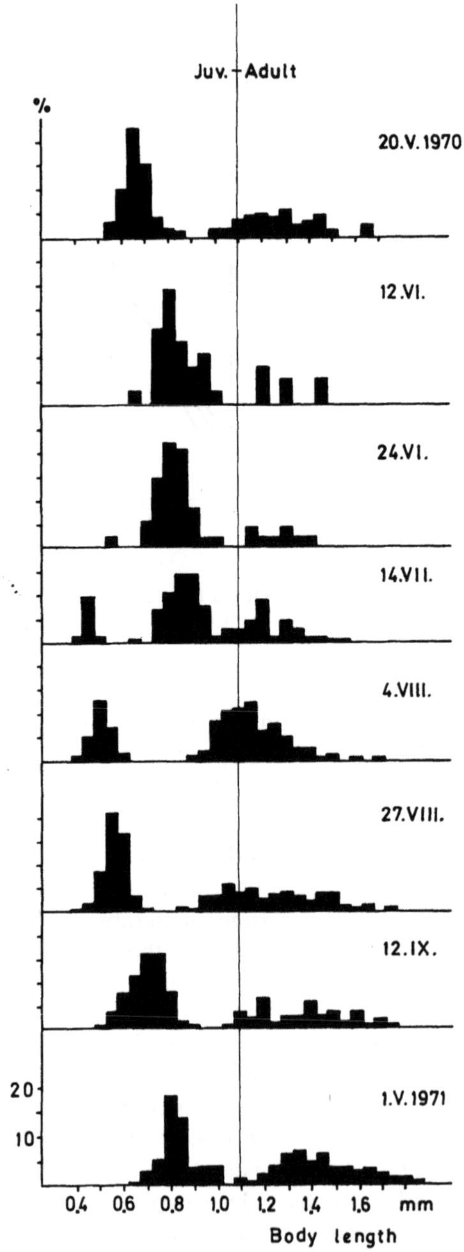


Fig. 5

Fig. 4. Frequency distribution (%) of size classes of *Isotoma olivacea*, a species with a one-year life cycle. The juveniles appear in July and become adults the following spring

Fig. 5. Frequency distribution (%) of size classes of *Tetracanthella wahlgreni*, a species with a two-year life cycle. The juveniles hatch in July and are still juveniles the following spring

## Discussion

Population densities of Collembola are strongly variable from different tundra biotopes within the same area. BENGTSON et al. (1974) found about 20000 ind. $m^{-2}$ in lichen tundra on Spitsbergen, while the abundance in a nearby moss tundra reached about 270000 ind. $m^{-2}$. Within this range also are values from Alaska (EDWARDS, 1967) and Taimyr (ANANJEVA, 1973).

Knowledge of generation time is important for the evaluation of productivity and the biological role of different Collembola species. In temperate regions there are often several generations each year (JOOSSE, 1969, TANAKA, 1970). The production may therefore be rather high despite a small standing crop. The present study has shown that several Collembola species need one or two years before they reproduce under tundra conditions.

## References


ANANJEVA, S. I.: Collembola of Western Taimyr. In: TIKHOMIROV, B. A., MATVEYEVA, N. V. (Eds.): Biogeocenoses of Taimyr tundra and their productivity, **2**, pp. 152–165. Leningrad: Publ. House "NAUKA" 1973.

BENGTSON, S. A., FJELLBERG, A., SOLHØY, T.: Abundance of tundra arthropods in Spitsbergen. Ent. Scand. **5**, 137–145 (1974).

EDWARDS, G. A.: The relationships between weights, volumes and numbers of soil animals. In: GRAFF, O., SATCHELL, J. E. (Eds.): Progress in soil biology, pp. 1–10. Braunschweig: Friedrich Vieweg and Sons 1967.

JOOSSE, E. N. G.: Population structure of some surface dwelling Collembola in a coniferous forest soil. Neth. J. Zool. **19**, 621–634 (1969).

KAURI, H., MOLDUNG, T., SOLHØY, T.: Turnbull and Nicholl's Quick Trap for acquiring standing crop of evertebrates in high mountain grassland communities. Norsk Ent. Tidsskr. **16**, 133–136 (1969).

LEWIS, T., TAYLOR, L. R.: Introduction to experimental ecology. London: Academic Press 1967.

NÆSS, S., STEIGEN, A. L., SOLHØY, T.: Standing crop and calorific content in some invertebrates on Hardangervidda. In: WIELGOLASKI, F. E. (Ed.): Fennoscandian tundra ecosystems, Part 2: Animals and systems analysis, pp. 151–159. Berlin-Heidelberg-New York: Springer 1975.

ODUM, E. P.: Fundamentals of ecology. London: W. B. Saunders 1971.

SOLHØY, T.: Quantitative invertebrate studies in mountain communities at Hardangervidda, South Norway I. Norsk Ent. Tidsskr. **19**, 99–108 (1972).

SONESSON, M., WIELGOLASKI, F. E., KALLIO, P.: Description of Fennoscandian tundra ecosystems. In: WIELGOLASKI, F. E. (Ed.): Fennoscandian tundra ecosystems. Part 1: Plants and microorganisms, pp. 3–28. Berlin-Heidelberg-New York: Springer 1975.

SØRENSEN, T.: A method of establishing groups of equal amplitude in plant society based on similarity of species content. Kgl. Danske Vidensk. Selsk. **5**, 1–34 (1948).

TANAKA, M.: The bio-economics on the populations of *Isotoma trispinata* MacGillivray (Collembola, Isotomidae) and *Onychiurus* (*Protaphorura*) sp. (Collembola, Onychiuridae) in a grassfield. Publ. Amakusa Mar. Biol. Lab. **2**, 51–120 (1970).

# Dynamics of Thysanoptera Populations on Hardangervidda

E. K. KJELLSEN

Two species of Thysanoptera, *Anapothrips secticornis* (Trybom) and *Aptinothrips stylifer* (Trybom), are found at the IBP sites in the Stigstuv area, described by SONESSON et al. (1975). The species are relatively abundant on the dry and wet meadows. On the lichen heath, only occasional specimens of *A. stylifer* are found.

## Material and Methods

The material was collected with a modified quick-trap apparatus and a suction pump. The populations were analyzed on the basis of material sampled in 1969–1970 and supplementary samples were obtained from cut specimens of various plant species. Modified Tullgren funnels were applied for extraction. See KAURI et al. (1969) and SOLHØY (1972) for details.

*A. secticornis*, reared in the laboratory in order to study its life-cycle, was placed singly in glass tubes saturated with humidity and closed with tight lids. Most tubes were placed in a room at a constant temperature of +4° C and some at a constant temperature of +20° C. The ratio light/dark in a 24 hrs period was 16/8 in the summer, 12/12 in the autumn and 0/24 in the winter and spring when the ground was covered with snow. In summer, the tubes were placed at room temperatures for 6 hrs daily, decreasing successively during the late summer and autumn to 4, 2, and 0 hrs. 6 samples representing 0.5 $m^2$ were collected on each sampling date in 1969, and 12 samples (each 0.25 $m^2$) were collected in 1970.

## Results and Discussion

Mean abundance of the two species sampled showed large standard deviations, indicating a high degree of aggregation. There were great differences in abundance during the season, especially on the dry meadow in 1969 (Table 1). On the wet meadow in 1969, and on both areas in 1970, the two species were most abundant in August. The highest density on the dry meadow in 1969 occurred in mid July, but *A. stylifer* reached almost the same level in mid September.

The maximum number at both sites, at least for *A. secticornis*, seemed to be lower in 1970 than in 1969, and preliminary results from the 1971 samplings on

Table 1. Mean number of individuals ($n$) and biomass (μg) per $m^2$ of *Anapothrips secticornis* and *Aptinothrips stylifer* at 3 Stigstuv sites

| | Dry meadow 1969 | | | | | | | | | |
|---|---|---|---|---|---|---|---|---|---|---|
| | July 1 | | July 17 | | Aug. 8 | | Aug. 26 | | Sept. 18 | |
| | *n* | μg | *n* | μg | *n* | μg | *n* | μg | *n* | μg |
| *A. secticornis* | 436 | 1961 | 1050 | 6025 | 22 | 241 | 21 | 290 | 22 | 312 |
| *A. stylifer* | 61 | 323 | 64 | 331 | 37 | 234 | 39 | 278 | 56 | 410 |
| Total | 497 | 2284 | 1114 | 6356 | 59 | 475 | 60 | 568 | 78 | 722 |
| | **Dry meadow 1970** | | | | | | | | | |
| | June 23 | | July 13 | | Aug. 3 | | Aug. 25 | | Sept. 10 | |
| | *n* | μg | *n* | μg | *n* | μg | *n* | μg | *n* | μg |
| *A. secticornis* | 2 | 7 | 21 | 114 | 59 | 372 | 35 | 297 | 26 | 326 |
| *A. stylifer* | 20 | 142 | 34 | 163 | 68 | 369 | 16 | 96 | 12 | 83 |
| Total | 22 | 149 | 55 | 277 | 127 | 741 | 51 | 393 | 38 | 409 |
| | **Wet meadow 1969** | | | | | | | | | |
| | July 2 | | July 18 | | Aug. 9 | | Aug. 27 | | Sept. 19 | |
| | *n* | μg | *n* | μg | *n* | μg | *n* | μg | *n* | μg |
| *A. secticornis* | 14 | 53 | 24 | 103 | 26 | 229 | 77 | 1086 | 6 | 85 |
| *A. stylifer* | 0 | 0 | 12 | 55 | 2 | 10 | 32 | 231 | 7 | 51 |
| Total | 14 | 53 | 36 | 158 | 28 | 239 | 109 | 1317 | 13 | 136 |
| | **Wet meadow 1970** | | | | | | | | | |
| | June 22 | | July 15 | | Aug. 3 | | Aug. 26 | | Sept. 11 | |
| | *n* | μg | *n* | μg | *n* | μg | *n* | μg | *n* | μg |
| *A. secticornis* | 1 | 19 | 16 | 67 | 38 | 236 | 11 | 129 | 7 | 84 |
| *A. stylifer* | 5 | 39 | 14 | 83 | 52 | 234 | 24 | 145 | 38 | 262 |
| Total | 6 | 58 | 30 | 150 | 90 | 470 | 35 | 274 | 45 | 346 |

the dry meadow indicate even lower abundance of *A. secticornis* that year. It is common to find thrips fluctuating greatly in number from one year to another (AHLBERG, 1926), the peaks being caused by favorable environmental conditions. The dry, warm period at the beginning of August 1969 may be one reason for the great decrease in number of *A. secticornis* on the dry meadow in the fall of 1969 and the low values particularly early in 1970. *A. secticornis* was more important than *A. stylifer* on the dry meadow in July 1969 and again in the fall of 1970 (Table 1). The species did not fully recover until the end of the 1970 season, although it is normally the dominant species on dry meadow. *A. secticornis* was also generally the dominant species on wet meadow in 1969, but *A. stylifer* dominated in 1970. The density of *A. stylifer* seemed to be less affected by drought in 1969. This is the only species of thrips found on the dry lichen heath, which also indicates that *A. stylifer* tolerates drier conditions than *A. secticornis*.

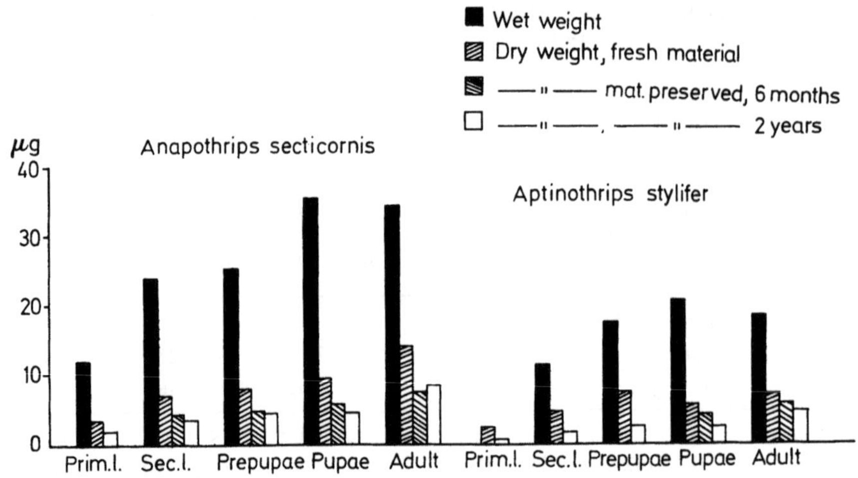


Fig. 1. Mean weights in μg per specimen of two Thysanoptera species. See text for further explanation. Prim. l = Primary larvae; Sec. l = Secondary larvae

Dry weight of fresh material of various stages of the two species and of material preserved in 75% alcohol for 6 months and 2 years respectively, was measured (Fig. 1). There was a significant decrease in dry weight from fresh material to material preserved for 6 months and further to material preserved for 2 years. Total biomass per $m^2$ for each of the species is shown in Table 1. Variations throughout the season show the same general trend as the abundance. Differences are due to changes in the ratio of juveniles to adults during the season.

Both species studied are parthenogenetic in the Stigstuv area.

*A. secticornis* hatches once a year. The species hibernates as adults, and the eggs are hatched after snow melting. Results of the rearing experiments are given in Table 2 and show the considerably longer duration of the stages at low than at higher temperature. Duration of each stage in the field can be estimated based on Table 2 and frequencies given in Fig. 2 for the various stages during the growing season. The following periods are estimated: primary larvae 19–41 days, secondary larvae 25–35 days, prepupae 10–15 days, pupae *ca* 15 days and adults 9–10 months (hibernating).

The samples of *A. stylifer* from quick trap indicate a life cycle similar to that of *A. secticornis*, although adults of *A. stylifer* were found during the whole season. According to PRIESNER (1928), adult females of this species can be found throughout the whole year and larvae from spring to late autumn.

Table 2. Duration (days) of each stage of *Anapothrips secticornis* in rearing experiments at constant temperatures

| | Prim. larvae | Sec. larvae | Prepupae | Pupae | Adult |
|---|---|---|---|---|---|
| + 4° C | >4–9 | 11–30 | 5–12 | 7–14 | 8–10 months |
| +20° C | — | — | 2–3 | <6 | 4–7 months |

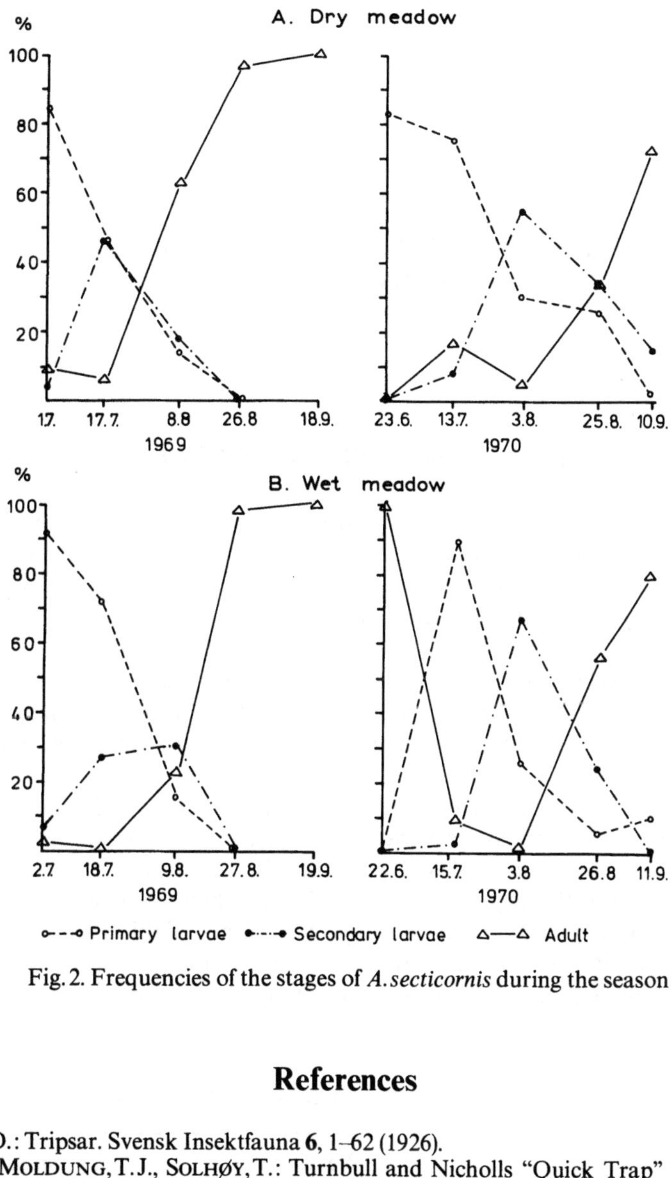


Fig. 2. Frequencies of the stages of *A. secticornis* during the season

# References

AHLBERG, O.: Tripsar. Svensk Insektfauna **6**, 1–62 (1926).

KAURI, H., MOLDUNG, T. J., SOLHØY, T.: Turnbull and Nicholls "Quick Trap" for acquiring standing crop of evertebrates in high mountain grassland communities. Norsk Ent. Tidsskr. **16**, 133–136 (1969).

PRIESNER, H.: Thysanopteren Europas. Wien 1928. Reprint Amsterdam: A. Asher and Co. 1963.

SOLHØY, T.: Quantitative invertebrate studies in mountain communities at Hardangervidda, South Norway. I. Norsk Ent. Tidsskr. **19**, 99–108 (1972).

SONESSON, M., WIELGOLASKI, F. E., KALLIO, P.: Description of Fennoscandian tundra ecosystems. In: WIELGOLASKI, F. E. (Ed.): Fennoscandian tundra ecosystems. Part 1: Plants and microorganisms, pp. 3–28. Berlin-Heidelberg-New York: Springer 1975.

# Dynamics of Aphididae Populations on Hardangervidda

H. TAMBS-LYCHE

## Introduction, Methods and Material

The aphids collected with the quick trap method (KAURI et al., 1969; SOLHØY, 1972) at the Stigstuv IBP alpine tundra dry and wet meadow sites described by SONESSON et al. (1975) comprise 8 species. The population of the wet meadow is sparse, and only material from the dry meadow is considered in the ecological discussion.

In 1969, 6 samples, each covering $^1/_2$ m$^2$, were taken each sampling data, and in the next two years 12 samples, each covering $^1/_4$ m$^2$; total trapped area 3 m$^2$ yr$^{-1}$. As the aphids are very unevenly distributed and often scarce, the numbers are not calculated per m$^2$.

The dominating species (Table 1), *Acyrthosiphon malvae* (Mosley) [= *A. pelargonii* (Kaltenbach)], comprises several subspecies, and the subspecies occurring at Stigstuv has not been identified with certainty. It is, however, observed to live on *Alchemilla alpina* in the dry meadow. Another possible host plant at Stigstuv is *Potentilla crantzii.* Of the remaining species found (Table 1), *Hyperomyzus* is probably *H. boerneri thorsteinni* (Stroyan), a species living on *Euphrasia* spp. *Macrosiphum (Sitobion) avenae* (F.) lives on grasses, *Pterocomma* sp. on *Salix* spp.,

Table 1. The total number of aphids per 3 m$^2$ at Stigstuv, Hardangervidda, total of 5 samplings

| Species (only adults counted) | Wet meadow | | | Dry meadow | | |
|---|---|---|---|---|---|---|
| | 1969 | 1970 | 1971 | 1969 | 1970 | 1971 |
| *Acyrthosiphon malvae* (Mosley) s. lat. | 16 | 14 | 2 | 375 | 127 | 162 |
| *Hyperomyzus boerneri thorsteinni (Stroyan)* | 0 | 3 | 0 | 20 | 45 | 11 |
| *Macrosiphum (Sitobion) avenae* (F.) | 3 | 0 | 2 | 1 | 0 | 2 |
| *Nasonovia* sp. | 0 | 0 | 0 | 0 | 0 | 5 |
| *Amphorophora rubi* (Kaltenbach) | 0 | 0 | 1 | 0 | 0 | 0 |
| *Rhopalosiphum padi* (L.) | 34 | 8 | 7 | 0 | 2 | 19 |
| *Aphis sp.* | 0 | 0 | 3 | 0 | 0 | 0 |
| *Pterocomma sp.* | 0 | 0 | 0 | 0 | 0 | 1 |

and *Amphorophora rubi* (Kaltenbach) on *Rubus* spp., which at Stigstuv would be *Rubus chamaemorus*. All these species are holocyclic.

*Rhopalosiphum padi* (L.) lives on grasses during the summer; the winter host is *Prunus padus*. Since *Prunus padus* is not found at Stigstuv, the aphids must have been wind-transported from some locality where the plant occurs. *P. padus* is found on Hardangervidda at lower altitude and in more sheltered places than at Stigstuv. *Nasonovia* and *Aphis* are not identified as to species and the host plants are unknown.

## Results and Discussion

Figure 1 shows the total number of aphids trapped on 3 $m^2$ in the dry meadow in relation to temperature pentades. The bars on the time-chart for 1970 and 1971 probably illustrate the normal situation, while the high numbers in 1969 are due to special conditions. In June, 1969 temperatures were considerably above normal in southern Norway, giving the aphids a good start from the beginning of the season which resulted in high numbers of fundatrices. Except for a slight fall in temperature around the 1st July, both July and part of August were also warm, enabling the aphids to multiply rapidly. The good start at the beginning of the season 1969 may also be due to the generally warm late August and first half of September 1968, which may have given good conditions for the sexuales and the resulting numbers of winter eggs.

In September 1969, there were very few aphids, which may have been due to the early sharp fall in temperature (the first snow fell on September 20th), but other factors may also have played a role. There were probably few winter eggs giving few fundatrices at the beginning of the 1970 season, even though the

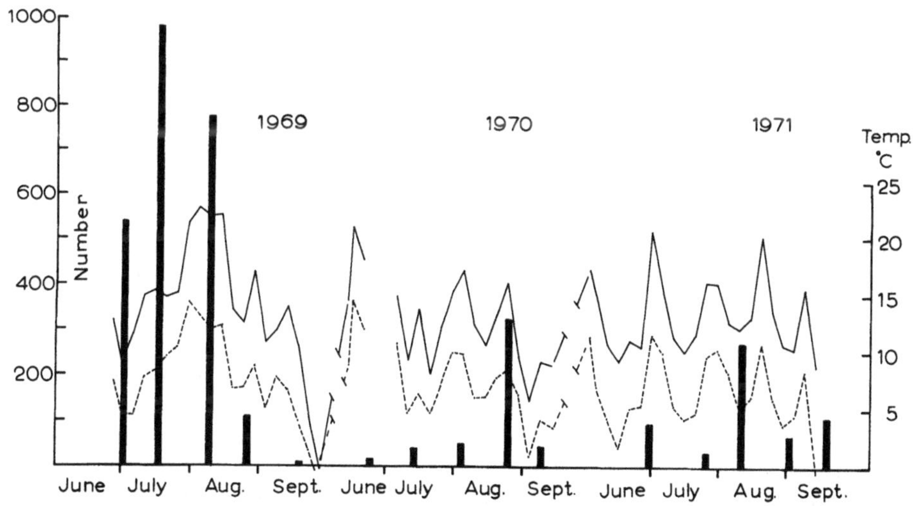


Fig. 1. Total catches of aphids per 3 $m^2$ in dry meadow, Stigstuv, compared with temperature pentades. Unbroken lines: max. temperature, broken lines: mean temperature

Table 2. Total numbers of *Acyrthosiphon malvae* (Mosley) *sensu latiore* on dry meadow

| Date | Adults | | | | | Instar IV | | Instars | | |
|---|---|---|---|---|---|---|---|---|---|---|
| | fund. | apt. | al. | ovip. | ♂♂ | apt. | al. | III | II | I |
| July 1. 1969 | 103 | 0 | 0 | 0 | 0 | 0 | 0 | 7 | 165 | 248 |
| July 17. 1969 | 34 | 76 | 1 | 0 | 0 | 116 | 17 | 292 | 211 | 179 |
| Aug. 8. 1969 | 0 | 55 | 1 | 10 | 39 | 144 | 0 | 258 | 149 | 113 |
| Aug. 26. 1969 | 0 | 34 | 0 | 4 | 18 | 13 | 0 | 9 | 0 | 2 |
| Sept. 18. 1969 | 0 | 1 | 0 | 5 | 0 | 2 | 0 | 1 | 1 | 0 |
| June 23–26. 1970 | 7 | 0 | 0 | 0 | 0 | 3 | 0 | 4 | 0 | 0 |
| July 13. 1970 | 6 | 0 | 0 | 0 | 0 | 0 | 0 | 12 | 19 | 4 |
| Aug. 3– 4. 1970 | 0 | 13 | 0 | 0 | 0 | 7 | 0 | 9 | 15 | 5 |
| Aug. 25–27. 1970 | 0 | 23 | 0 | 1 | 13 | 44 | 0 | 48 | 32 | 16 |
| Sept. 10. 1970 | 0 | 14 | 0 | 16 | 8 | 5 | 0 | 2 | 4 | 2 |
| June 30. 1971 | 15 | 0 | 0 | 0 | 0 | 19 | 0 | 2 | 12 | 4 |
| July 24–26. 1971 | 0 | 3 | 0 | 0 | 0 | 11 | 0 | 5 | 3 | 5 |
| Aug. 10–11. 1971 | 0 | 28 | 0 | 2 | 5 | 22 | 0 | 65 | 84 | 32 |
| Sept. 2– 3. 1971 | 0 | 7 | 0 | 25 | 3 | 18 | 0 | 10 | 4 | 4 |
| Sept. 18. 1971 | 0 | 0 | 0 | 58 | 16 | 14 | 0 | 8 | 2 | 0 |

temperature in June was relatively high. The weather was cold in July 1970 and the aphid population increased very slowly until the end of August. A similar picture was found in 1971, although the population peak was reached earlier, at the beginning of August. September 1971 was relatively warm and the number of sexuales was higher than in previous years. This may also have resulted in a higher population in 1972.

Table 2 shows the dominating species with numbers of the different morphs and larval instars collected on dry meadow throughout the seasons. The most interesting fact is that the sexuales appeared as early as the beginning of August. Around the beginning of July there were only fundatrices; the viviparous apterae appeared around mid-July. In 1970 the viviparous apterae were not found until early August, and the sexuales appeared at the end of the same month.

The occurrence of sexuales early in the season is a feature known also from the arctic. HILLE RIS LAMBERS (1952, 1960) observed oviparae from 6 different species in the end of July and in early August in Greenland. He even recorded fundatrices and oviparae of *Acyrthosiphon brachysiphon* (Hille Ris Lambers) in a sample from 6 July. RICHARDS (1963) observed occurrence of oviparae in arctic Canada of four different species in early August at Baffin Island and 19th July at Ellesmere Island.

In Norway it is possible to find oviparae of some species in August also at lower altitudes, but here the fundatrices appear in spring, so there are several generations during the season. At Stigstuv, as the Table shows, there is generally not time for more than one generation between the fundatrices and the sexuales. The observations of viviparae in September, however, show that sometimes there may be more than three generations totally. HILLE RIS LAMBERS (1952) concluded that there might be 3–4 generations a year in Greenland. The catches from Stigstuv which have made it possible to count the different larval instars and morphs throughout 3 seasons now confirm that under special conditions aphids may complete the whole cycle from egg to egg through three generations.

Production of sexuales has been found to be due to a shortening of the photoperiod combined with a lowering of the temperature acting through a series of maternal switch mechanisms (LEES, 1959, 1960). At Stigstuv, however, the sexuales are found as adults in the beginning of August; and at the time when the photoperiod/temperature will affect the mothers, almost the longest photoperiod and the highest temperatures are found. RICHARDS (1963) called the phenomenon a reversal of the switch mechanism. How this mechanism in aphids really works under these conditions is still to be studied; one possibility is that the aphids have developed genetic varieties with qualities that enable them to survive in places with short seasons.

## References

HILLE RIS LAMBERS, D.: The aphid fauna of Greenland. Medd. Grønland **136**, 1—33 (1952).

HILLE RIS LAMBERS, D.: Additions to the aphid fauna of Greenland. Medd. Grønland **159**, 1–18 (1960).

KAURI, H., MOLDUNG, T.J., SOLHØY, T.: Turnbull and Nicholls "Quick Trap" for acquiring standing crop of evertebrates in high mountain grassland communities. Norsk Ent. Tidsskr. **16**, 133–136 (1969).

LEES, A. D.: The role of photoperiod and temperature in the determination of parthenogenetic and sexual forms in the aphid *Megoura viciae* Buckton. — I. The influence of these factors on apterous virginiparae and their progeny. J. Insect Physiol. **3**, 92–117 (1959).

LEES, A. D.: The role of photoperiod and temperature in the determination of parthenogenetic and sexual forms in the aphid *Megoura viciae* Buckton. — II. The operation of the "interval timer" in young clones. J. Insect Physiol. **4**, 154–175 (1960).

RICHARDS, W. R.: The Aphididae of the Canadian arctic. Canad. Entomol. **95**, 449–464 (1963).

SOLHØY, T.: Quantitative invertebrate studies in mountain communities at Hardangervidda, South Norway. I. Norsk Ent. Tidsskr. **19**, 99–108 (1972).

SONESSON, M., WIELGOLASKI, F. E., KALLIO, P.: Description of Fennoscandian tundra ecosystems. In: WIELGOLASKI, F. E. (Ed.): Fennoscandian tundra ecosystems. Part 1: Plants and microorganisms, pp. 3–28. Berlin-Heidelberg-New York: Springer 1975.

# Occurrence and Role of Different Invertebrate Groups in Alpine Salix herbacea Snowbeds at Hardangervidda

S. HÅGVAR and E. ØSTBYE

On the Hardangervidda mountain plateau, snow covers the ground for more than half of the year (SKARTVEIT et al., 1975). Strong winds during winter blow the hill tops more or less bare of snow which accumulates in crevices and depressions of small slopes where the snow cover may be several meters thick. The snow distribution pattern is rather similar from year to year, and this pattern is reflected in the different plant communities (DAHL, 1975; LYE, 1975). Habitats where snow melts very late, and which have a correspondingly short growth season, are called snowbeds. There is usually a zonation of plant communities from the outer border to the center where snow melts latest or not at all (GJÆREVOLL, 1956).

Most snowbeds at Hardangervidda are of the size order 250–500 $m^2$ and contain an outer zone several meters in breadth, which is totally dominated by *Salix herbacea* L. In the inner parts of this zone, snow may melt more than a month later than in the outermost parts. Only a few species of mosses and lichens are present. Due to the short and dense vegetation, surface temperatures measured with a shaded thermocouple may exceed 40° C on sunny days. Solifluction processes often occur during the melting period. Stones cover roughly 15–20% of the ground, and most are so small that they can be turned by hand.

The present study was made at altitudes between 1200 and 1300 m near Finse (60° 35′ N–7° 31′ E) (SONESSON et al., 1975). In this area, *S. herbacea* snowbeds have a coverage of approximately 23% on north facing and 25% on south facing slopes which is considerably more than the average on the plateau (HESJEDAL, 1975). In the Finse area of Hardangervidda, the vegetation type "oligotrophic dry heath communities" are dominant (SCHMIDT and ØSTBYE, unpubl.). The plant communities near glaciers at Finse are described by ELVEN (1975).

## Methods

15 pit-fall traps, diameter 7 cm, were placed in three rows between stones in the mid/inner *S. herbacea* snowbed (in the inner part of the *Melasoma collaris* zone) 5 m between each trap and row. They were emptied every two weeks throughout the snow-free seasons over three years. Larger invertebrates under stones were

picked by hand. Invertebrates in the vegetation and litter layers were sampled with a suction sampler (quick trap) during the two seasons 1971 and 1972 (KAURI et al., 1969; SOLHØY, 1972). Soil samples were extracted in a high gradient apparatus.

Frozen material was dried and burned in a microbomb calorimeter (PHILLIPSON, 1964), and oxygen consumption measurements were made in constant pressure respirometers (ENGELMANN, 1963).

## Results

**Herbivores.** The most important invertebrate herbivore species in the snow bed habitat is *Melasoma collaris* L. (Col., Chrysomelidae). Both larvae and imagines feed on the leaves of *S. herbacea.* Contrary to many other alpine insects (ØSTBYE et al., 1975), the species has a one-year life cycle. Imagines hibernate among litter and become active as soon as the snow melts. Eggs are laid after a short feeding period, when mainly unopened buds are eaten. These have a considerably higher caloric content ($5405 \pm 279$ cal $g^{-1}$) than developed leaves ($4664 \pm 116$ cal $g^{-1}$). Development from newly laid eggs to hatching of imagines takes about two months. The new generation also has a feeding period before hibernation. A snow-free period of at least two and a half months is required by the species (HÅGVAR, 1975a). The first snow usually falls in late September. Field studies have shown that *M. collaris* utilizes only those parts of *S. herbacea* communitics where snow melts before approximately July 1st. This means that in most snow beds, *M. collaris* only inhabits the outer parts of the *S. herbacea* zone. However, here the host plant may be locally quite defoliated. The attack is always most heavy on the outermost parts, where melt-off occurs first. Up to 400–500 larvae per $m^2$ may develop, but in most populations, the number will be around 100–200 ind. $m^{-2}$, ingesting altogether about 12000–24000 cal $m^{-2}$. The ratio of assimilation to ingestion of the larvae is unusually high: 76, 77, and 65% in the three larval instars, respectively (ratio for the total larval development: 68%). The ratio of production to ingestion is also high: 32%, 31%, and 22%, respectively, in the three larval instars (ratio for the total larval development: 24%) (HÅGVAR, 1975b). The species must be considered as a very effective herbivore. However, the one-year life cycle makes it impossible for the species to colonize the inner (and major) parts of *S. herbacea* communities.

Other herbivore species in this habitat are mainly larvae of Lepidoptera and Hymenoptera (Symphyta), and imagines of *Otiorrhynchus dubius* Strøm. (Col., Curculionidae). Lepidoptera larvae (probably Tortricidae) may locally leave obvious evidence of feeding, but all species are insignificant herbivores compared with *M. collaris.* In the inner zone, where *M. collaris* is absent, invertebrate feeding marks could scarcely be detected. Here the primary production is almost exclusively channeled directly through the decomposition food chain.

**Predators.** Several invertebrate species are known to hide under stones during more or less inactive periods of the day and during egg laying and egg developmental periods (e.g. Araneae), while they occupy habitats between the stones in periods of hunting etc.

Samples taken with a suction sampler during 1971 and 1972 in the inner *S. herbacea* snow-bed zone not inhabited by *M. collaris*, revealed a generally low total density of invertebrates between stones (excluding Collembola and Acari), ranging from 9 to 30 ind. $m^{-2}$ (HÅGVAR et al., 1974). Main predatory groups were Araneae, Opiliones, Carabidae and Staphylinidae. The fauna hiding under stones (excluding Collembola and Acarina) was studied within the inner zones of the snow-bed in August 1969 and in July and August 1970 (HÅGVAR and ØSTBYE, 1972). Spiders dominated under stones in this period, representing 75–83% of the specimens, and 20–42% of the fresh weight, but also many other invertebrate groups occurred under stones, including predatory Opiliones, Carabidae and Staphylinidae. The recorded density of spiders under stones was 22–41 ind. per $m^2$ stones overturned, greatly exceeding the spider density recorded by suction sampler between stones (0.34–4.5 ind. $m^{-2}$). The generally rich fauna under stones in summer may in part be explained by the favorable microclimatic conditions evolved, particularly temperature and humidity. Spiders seemed to prefer stones covering about 400 $cm^2$.

However, high catches (Fig. 1) in the pit-fall traps between stones compared with the generally low invertebrate densities show that there is a rather high activity of specimens moving through the habitat, coming partly from neighbor-

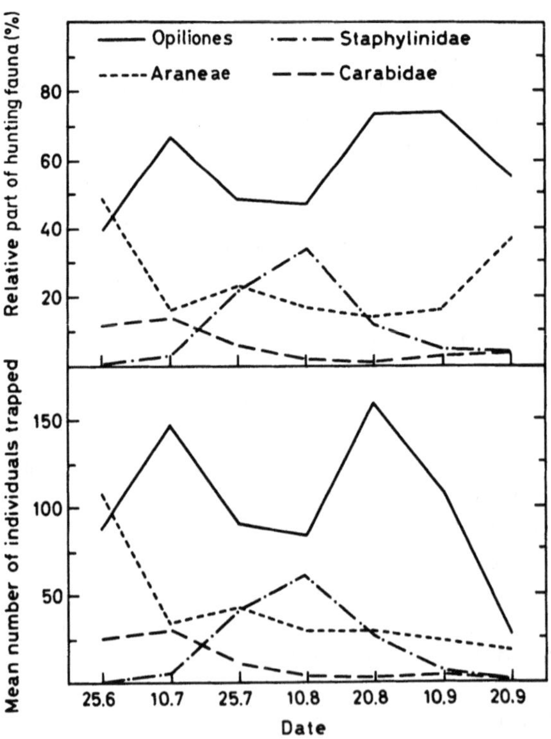


Fig. 1. Mean number and relative abundance of the four dominant predator groups in *S. herbacea* snowbed, as revealed by a pit-fall trapping study (15 traps) during the years 1969–1971

ing habitats even outside the snowbed (e.g. many Opiliones, Araneae, and Carabidae), and partly from under stones.

Of the four main predator groups, Opiliones *(Mitopus morio)* dominates in the pit-falls throughout the entire snow-free season, with the exception of a short period after snow melt (Fig. 1). Araneae is the next dominating group, and seems to be most abundant and active in early spring. The two coleopteran groups, Staphylinidae and Carabidae, occur in much fewer numbers. Staphylinidae have their highest occurrence and thus activity in mid-summer, *viz.* August, while Carabidae are more frequent in spring than later in the season. The latter group is perhaps more dependent on a surplus of melt water for fulfilment of their development. In the Himalayas, they have been found to exhibit their highest activity and development very close to the retiring edge of the melting snow (MANI, 1968). In the area under study, this means the period June to July.

**Decomposers.** The density of Collembola and Acari was measured in the 0–3 cm and 3–6 cm-deep soil layers during 1971 and 1972, the sampling areas being those not used by *M. collaris*, i.e. the late-melting *S. herbacea* snowbeds. Collembola occurred in densities of between 12000 and 57000 ind. $m^{-2}$ in 1971 and between 14000 and 28000 ind. $m^{-2}$ in 1972. The densities of Acari were higher and more stable: 120000–177000 ind. $m^{-2}$ in 1971 and 98000–114000 ind. $m^{-2}$ in 1972. In most cases, more than 93% of the Collembola and Acari occurred in the upper 3 cm layer, including the surface (HÅGVAR et al., 1974).

All Acari are here treated as decomposers, as investigations from Hardangervidda show that predatory forms only account for approximately 5% in terms of numbers and 1% of biomass (SOLHØY, pers. comm.).

The recorded densities of Collembola correspond roughly to or are somewhat lower than results from various other alpine plant communities at Stigstuv about 36 km away, while the density of Acari seems to be high compared with other Hardangervidda alpine habitats investigated (SOLHØY, 1972; SOLHØY et al., 1975). Thus, while the density of larger invertebrates is very low compared with other alpine habitats, Collembola and Acari seem to be well represented in late-melting *S. herbacea* snowbeds.

Among the larger arthropods, dipterous larvae represented the most important decomposer group, more than half of the larger arthropod specimens recorded in the quick-trap samples from July and August 1972 (HÅGVAR et al., 1974).

## Discussion

The actual plant community may by some simplification be regarded as a monoculture of *S. herbacea* which gradually melts free from snow through a period of several weeks. The invertebrate community is limited partly by the low number of host plants, and partly by the short snow-free season in large parts of the habitat. While Thysanoptera and Homoptera are important groups in several other plant communities at Hardangervidda (SOLHØY, 1972; SOLHØY et al., 1975), these herbivore groups are almost missing in the *S. herbacea* snowbed. The main reason probably is the lack of suitable host plants, since the snow-free season is long enough for the development of these groups in the outer parts of the snowbed. The lack of host plants partly explains the low density of medium large, and larger invertebrates.

Monocultures are generally regarded as vulnerable systems, as a herbivore species may reproduce strongly under such favorable food conditions. In fact, *M. collaris* may fully defoliate the *S. herbacea* plants. However, the major part of the community is saved from such detoriation due to the inability of *M. collaris* to develop within the inner parts of the snowbeds.

The relatively high pit-fall catches of larger arthropod predators indicate that *S. herbacea* snowbeds serve as important preying grounds through a great part of the snow-free season at Hardangervidda even for invertebrates migrating from other plant communities. The high surface temperature during sunshine in these habitats is probably an important factor in this connection.

Vertebrate predators, like the meadow pipit [*Anthus pratensis* (L.)], the wheatear [*Oenanthe oenanthe* (L.)] and the snow bunting [*Plectrophenax nivalis* (L.)] nest in or very close to snowbeds, and have also been observed to feed there to a large extent. They may play an important role in the predation of invertebrates.

The inner part of the *S. herbacea* snow bed habitat represents a very simple ecosystem. Here nearly all primary production is channeled directly to the decomposers. This is reflected through the rather high density of Collembola and Acarina, and the dominance of dipterous larvae among the larger invertebrates.

*Acknowledgements.* We wish to thank all those who participated in the field and laboratory work; especially the late J. HAGEN, A. HAGEN, H.-J. SKAR and J. MELÅEN, and M. ESPELAND for improving the English. Financial support was received from the Norwegian IBP.

## References

DAHL, E.: Flora and plant sociology in Fennoscandian tundra areas; In: WIELGOLASKI, F. E. (Ed.): Fennoscandian tundra ecosystems. Part 1: Plants and microorganisms, pp. 62–67. Berlin-Heidelberg-New York: Springer 1975.

ELVEN, R.: Plant communities on recently deglaciated moraines at Finse, southern Norway. In: VIK, R. (Ed.): IBP in Norway. Methods and results. Sections PT-UM Grazing project, Hardangervidda. Botanical investigations, pp. 381–467. Oslo: Norwegian Nat. IBP Comm. 1975.

ENGELMANN, M. D.: A constant pressure respirometer for small arthropods. Ent. News **74**, 181–186 (1963).

GJÆREVOLL, O.: The plant communities of the Scandinavian alpine snowbeds. K. Norske Vidensk. Selsk. Skr. 1956, **1**, 1–405 (1956).

HÅGVAR, S.: Studies on the ecology of *Melasoma collaris* L. (Col., Chrysomelidae) in alpine habitats at Finse, south Norway. Norw. J. Ent. **22**, 31–47 (1975a).

HÅGVAR, S.: Energy budget and growth during the development of *Melasoma collaris* L. (Coleoptera, Chrysomelidae). Oikos **26**, 140–146 (1975b).

HÅGVAR, S., MELÅEN, J., ØSTBYE, E.: Quantitative studies of the invertebrate fauna in an alpine snow-bed community at Finse, south Norway. Norsk Ent. Tidsskr. **21**, 45–51 (1974).

HÅGVAR, S., ØSTBYE, E.: Quantitative and qualitative investigations of the invertebrate fauna under stones (the Hypolithion) in some alpine habitats at Finse, south Norway. Norsk Ent. Tidsskr. **19**, 1–10 (1972).

HESJEDAL, O.: Vegetation mapping at Hardangervidda. In: WIELGOLASKI, F. E. (Ed.): Fennoscandian tundra ecosystems. Part 1: Plants and microorganisms, pp. 74–81. Berlin-Heidelberg-New York: Springer 1975.

KAURI, H., MOLDUNG, T., SOLHØY, T.: Turnbull and Nicholl's 'quick trap' for acquiring standing crop of evertebrates in high mountain grassland communities. Norsk Ent. Tidsskr. **16**, 133–136 (1969).

LYE, K. A.: Survey of the main plant communities on Hardangervidda. In: WIELGOLASKI, F. E. (Ed.): Fennoscandian tundra ecosystems. Part 1: Plants and microorganisms, pp. 68–73. Berlin-Heidelberg-New York: Springer 1975.

MANI, M. S.: Ecology and biogeography of high altitude insects. The Hague; Junk Publ. 1968.

ØSTBYE, E., BERG, A., BLEHR, O., ESPELAND, M., GAARE, E., HAGEN, A., HESJEDAL, O., HÅGVAR, S., KJELVIK, S., LIEN, L., MYSTERUD, I., SANDHAUG, A., SKAR, H.-J., SKARTVEIT, A., SKRE, O., SKOGLAND, T., SOLHØY, T., STENSETH, N. C., WIELGOLASKI, F. E.: Hardangervidda. In: ROSSWALL, T., HEAL, O. W. (Eds.): Structure and function of tundra ecosystems, Ecol. Bull. Vol. **20**, pp. 225–264. Stockholm: Swedish Nat. Sci. Res. Council 1975.

PHILLIPSON, J.: A miniature bomb calorimeter for small biological samples. Oikos **15**, 130–139 (1964).

SKARTVEIT, A., RYDÉN, B. E., KÄRENLAMPI, L.: Climate and hydrology of some Fennoscandian tundra ecosystems. In: WIELGOLASKI, F. E. (Ed.): Fennoscandian tundra ecosystems, Part 1: Plants and microorganisms, pp. 41–53. Berlin-Heidelberg-New York: Springer 1975.

SOLHØY, T.: Quantitative invertebrate studies in mountain communities at Hardangervidda, South Norway. Norsk Ent. Tidsskr. **19**, 99–108 (1972).

SOLHØY, T., KAURI, H., ØSTBYE, E., HAGEN, A., LIEN, L., SKAR, H. J.: Faunal structure of Hardangervidda, Norway. In: WIELGOLASKI, F. E. (Ed.): Fennoscandian tundra ecosystems. Part 2: Animals and systems analysis, pp. 29–45. Berlin-Heidelberg-New York: Springer 1975.

SONESSON, M., WIELGOLASKI, F. E., KALLIO, P.: Description of Fennoscandian tundra ecosystems. In: WIELGOLASKI, F. E. (Ed.): Fennoscandian tundra ecosystems. Part 1: Plants and microorganisms, pp. 3–28. Berlin-Heidelberg-New York: Springer 1975.

# Structure of Small Passerine Communities in Subalpine Birch Forests in Fennoscandia

O. HOGSTAD

Faunistically the breeding avifauna in Fennoscandian subalpine birch forests is well documented, and during the last 20 years many quantitative investigations have been undertaken. Although some of these studies are still in progress, we are lacking long-term investigations revealing the dynamics of the breeding passerine communities. However, despite great annual density fluctuations of most of the species inhabiting the subalpine birch forests, the passerine communities found in the different study areas show great similarities and seem to have qualitative as well as quantitative patterns, characteristic for the subalpine birch forest as a region.

The subalpine birch forest region is a characteristic feature of Fennoscandia and with few exceptions it makes a more of less distinct zone between the coniferous forest and the alpine region (SJÖRS, 1967). The forest, predominated by *Betula pubescens* ssp. *tortuosa*, may be divided in oligotrophic and eutrophic types. The oligotrophic type (here called heath birch forest) is divided in several subtypes where Empetro-Betuletum and Myrtillo-Betuletum probably dominate (cf. MOEN, 1973, pers. comm.). The eutrophic type (here called meadow birch forest) is also divided into subtypes. The dominating types, Betula geraniosum subalpinum (NORDHAGEN, 1943) and Lactucion alpinae, are characterized by a productive field layer rich in species. Often other tree species occur, e.g. *Alnus incana*, *Prunus padus*, and *Sorbus aucuparia*.

## Material

The present paper is based upon 2 Finnish, 3 Swedish and 7 Norwegian studies (Fig. 1). The study areas situated in heath birch forests are (listed from south to north): Vassfaret, Sjodal, Budal, Ammarnäs, Abisko, Inari, and Kevo; those in heath birch intermixed with meadow birch forest: Geilo, Nedal, Stora Sjöfallet; and those in meadow birch forest: Vaksvik, Surnadal, Ammarnäs. The study areas are 10–96 ha in extent.

Except in the Inari study (where estimates are based upon line transects), all density estimates are calculated by means of the territory mapping method of ENEMAR (1959), largely according to the recommendations later outlined by The International Bird Census Committee (IBCC) (SVENSSON, 1970).

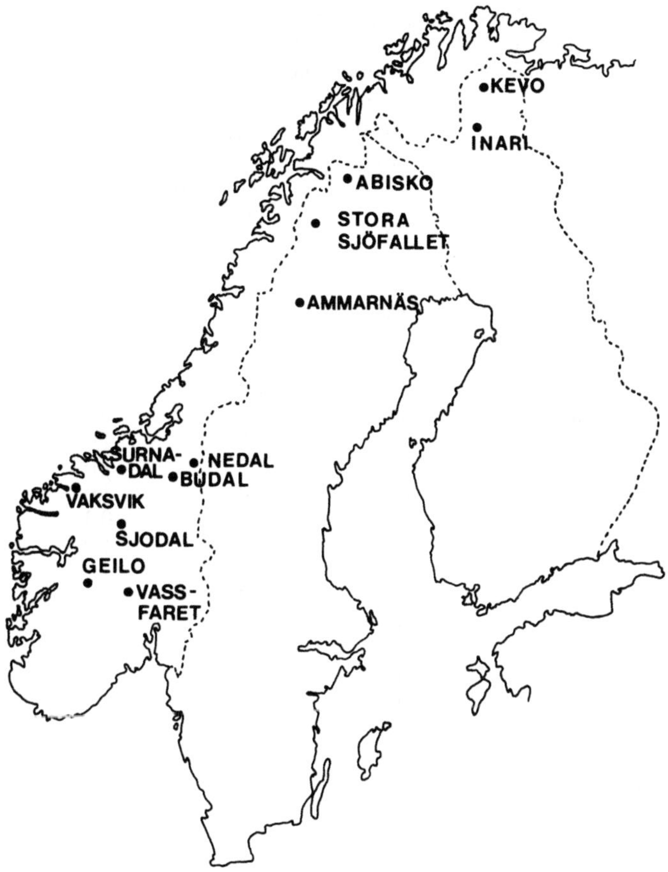


Fig. 1. Locations of the subalpine birch forest study areas. Grouped from south to north: Vassfaret 1970–72 (BREKKE and SELBOE, pers. comm.), Geilo 1964–65 (MYKLEBUST, 1966), Sjodal 1971–73 (BEVANGER, pers. comm.), Vaksvik 1968–70 (YTREBERG, 1972), Budal 1966–74 (HOGSTAD, 1969, unpubl.), Surnadal 1970–71 (RØV, 1972), Nedal 1967–72 (MOKSNES, 1973, pers. comm.), Ammarnäs 1963–64 (ENEMAR, 1964; ENEMAR et al., 1965), Stora Sjöfallet 1965–67 (ERIKSSON et al., 1971), Abisko 1970–74 (ANDERSSON, pers. comm.), Inari 1961 (BAGGE et al., 1963), Kevo 1972 (HAUKIOJA, pers. comm.)

## Results

**Species Composition.** Although the occurrence of some species may be subject to different geographical distribution, the composition of the species in passerine communities in Fennoscandian subalpine birch forests is in its broad features similar (Table 1). The communities are characterized by *Phylloscopus trochilus*, *Fringilla montifringilla*, *Acanthis flammea*, *Emberiza schoeniclus*, *Phoenicurus phoenicurus*, *Turdus iliacus*, *Anthus trivialis*, *Prunella modularis*, and *Luscinia svecica.* In addition, some other species occur fairly regularly in the birch forest, mainly in meadow birch forests, and in heath birch forests intermixed with

Table 1. The constancy of breeding in small passerine species in Fennoscandian subalpine birch forests (1.0 means that the species is present in all study areas). Number of study areas in brackets

| Species | Heath birch forest (7) | Heath and meadow birch forest intermixed (3) | Meadow birch forest (3) | Subalpine forest total (13) |
|---|---|---|---|---|
| *Phylloscopus trochilus* | 1.0 | 1.0 | 1.0 | 1.0 |
| *Fringilla montifringilla* | 1.0 | 1.0 | 1.0 | 1.0 |
| *Acanthis flammea* | 1.0 | 1.0 | 1.0 | 1.0 |
| *Emberiza schoeniclus* | 1.0 | 1.0 | 0.7 | 0.9 |
| *Phoenicurus phoenicurus* | 1.0 | 1.0 | 0.7 | 0.9 |
| *Turdus iliacus* | 0.9 | 1.0 | 1.0 | 0.9 |
| *Anthus trivialis* | 0.7 | 1.0 | 1.0 | 0.8 |
| *Prunella modularis* | 0.7 | 1.0 | 1.0 | 0.8 |
| *Luscinia svecica* | 0.9 | 1.0 | 0.3 | 0.8 |
| *Turdus pilaris* | 0.3 | 1.0 | 1.0 | 0.6 |
| *Turdus philomelos* | 0.4 | 0.7 | 1.0 | 0.6 |
| *Parus montanus* | 0.4 | 0.7 | 1.0 | 0.6 |
| *Ficedula hypoleuca* | 0.3 | 1.0 | 1.0 | 0.6 |
| *Muscicapa striata* | — | 1.0 | 1.0 | 0.5 |
| *Anthus pratensis* | 0.3 | 0.7 | 0.3 | 0.4 |
| *Oenanthe oenanthe* | 0.4 | 0.3 | — | 0.3 |
| *Fringilla coelebs* | — | 0.3 | 1.0 | 0.3 |
| *Motacilla flava* | 0.3 | — | — | 0.2 |
| *Hippolais icterina* | — | — | 0.7 | 0.2 |
| *Sylvia atricapilla* | — | — | 0.7 | 0.2 |
| *Phylloscopus sibilatrix* | — | — | 0.7 | 0.2 |
| *Erithacus rubecula* | — | — | 0.7 | 0.2 |
| *Turdus merula* | — | — | 0.7 | 0.2 |
| *Troglodytes troglodytes* | — | — | 0.7 | 0.2 |
| *Carduelis spinus* | — | — | 0.7 | 0.2 |
| *Motacilla alba* | 0.1 | — | — | 0.1 |
| *Acrocephalus schoenobaenus* | 0.1 | — | — | 0.1 |
| *Sylvia borin* | — | — | 0.3 | 0.1 |
| *Phylloscopus borealis* | 0.1 | — | — | 0.1 |
| *Saxicola rubetra* | — | — | 0.3 | 0.1 |
| *Turdus torquatus* | 0.1 | — | — | 0.1 |
| *Parus major* | — | — | 0.3 | 0.1 |
| *Calcarius lapponicus* | 0.1 | — | — | 0.1 |
| *Pinicola enucleator* | 0.1 | — | — | 0.1 |

meadow birches or other tree species. These latter species seem to be more dependent on special features in the habitat (YTREBERG, 1972). The passerine community in the typical heath birch forest (i.e. without dry, open areas) seems rather stereotyped and is characterized by the nine species listed above. *P. modularis* is not always mentioned among the species found in the Finnish study areas, but has been observed near Kevo by LAINE (1966), see also HAUKIOJA and KOPONEN

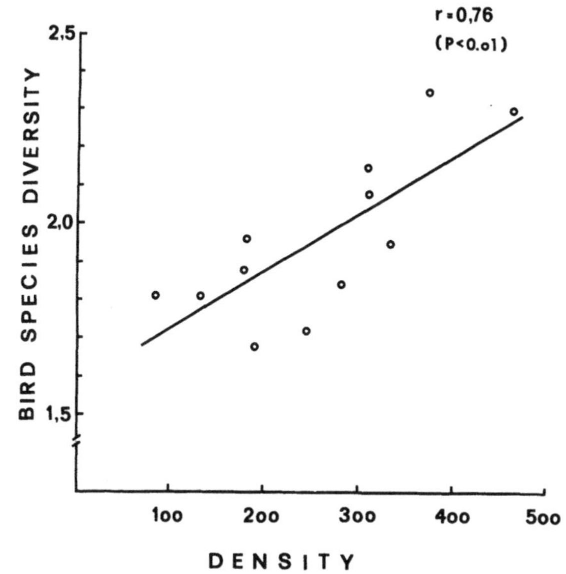


Fig. 2. Relation between mean bird density (terr. km$^{-2}$) and bird species diversity of the passerine communities found in the different study areas

(1975). Species such as *Anthus pratensis*, *Oenanthe oenanthe*, *Motacilla flava*, *Motacilla alba* and *Calcarius lapponicus* also occur in some of the heath birch study areas (Table 1), but these species indicate more open habitas rather than the subalpine forest.

The diversity of bird species (MACARTHUR, 1964) in the passerine communities is calculated at $H' = 1.68$–$1.96$ and $H' = 2.08$–$2.35$ for heath birch and meadow birch forests, respectively. The difference in diversity of birch species between the two types of habitat is statistically significant ($P < 0.01$). The relation between density and diversity of species of the passerine communities is positively correlated (Fig. 2).

**Density.** Although the densities of the passerine communities in the subalpine birch forest in part show great annual variations within each of the study areas, the density estimates are clearly divided in two main groups, i.e. heath birch forests with total passerine densities of 82–300 territories km$^{-2}$, and meadow birch forests with densities of 293–480 territories km$^{-2}$ (Fig. 3). The difference is statistically significant ($P < 0.001$); it is, in fact, probably even more marked, because the relative high density found in Sjodal is undoubtedly caused by elements of tree species other than birch (BEVANGER, pers. comm.). On the other hand, the density value from Inari is probably underestimated because of the method used (line transects). The difference in density values found for heath birch forests and heath birch intermixed with meadow birch is also significant ($P < 0.001$), while the small difference between the latter habitat and the meadow birch forest is not statistically significant ($P > 0.05$).

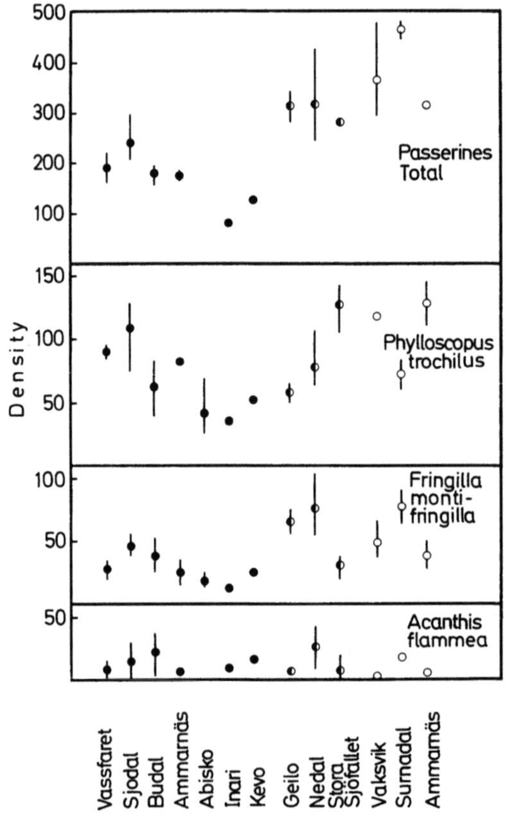


Fig. 3. Extent of variations (minimum and maximum values) and mean densities (terr. km$^{-2}$) of the passerine communities and most abundant species. Filled circles represent heath birch forests; half filled, mixed heath and meadow birch forests; open circles, meadow birch forests

Densities of the two most abundant species, *Ph. trochilus* and *F. montifringilla*, were also significantly lower in the heath birch than in the meadow birch forest ($P < 0.05$). The one bird species of the subalpine birch forest dependent on seed production of *Betula*, *A. flammea*, fluctuates strongly annually, and may outnumber *F. montifringilla* some years (ENEMAR, 1969; HILDEN, 1969). Density of this species, however, is not different in the heath and meadow birch forest habitats.

Although the remaining species of the communities (listed in Table 1) might show great density variations, their average densities are lower than those of *Ph. trochilus* and *F. montifringilla* and their amplitudes of fluctuation are not usually of the same magnitude.

**Community Structure.** The structure of the passerine communities in each of the study areas (excluding Abisko, due to lack of density figures of *A. flammea*), is

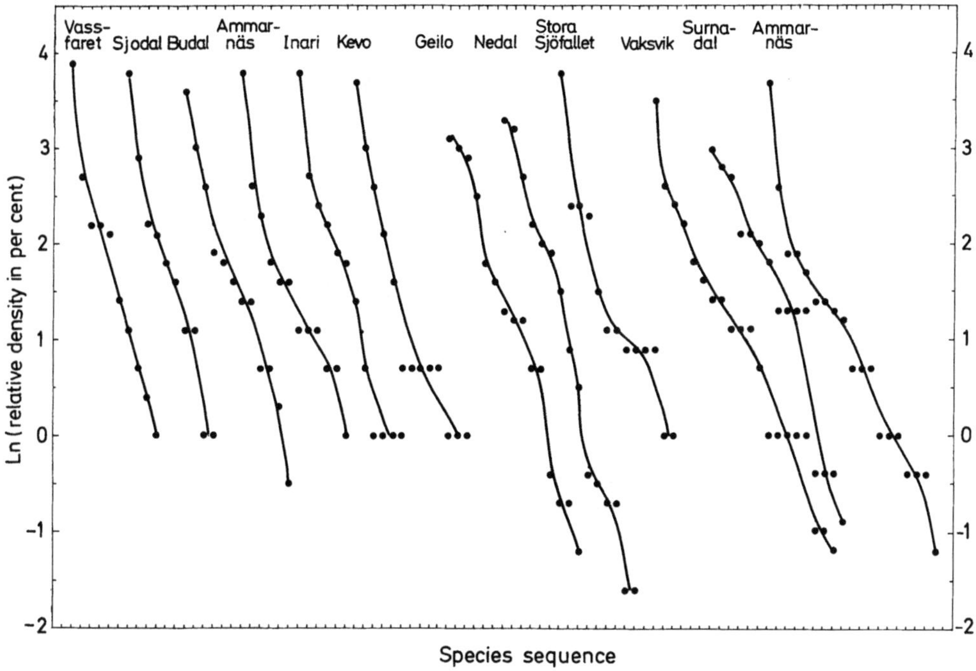


Fig. 4. Species importance curves of the different passerine communities, ranked from the most important to the least important species. In each curve the highest point represents the most numerous species, and the lowest point the least numerous species (WHITTAKER, 1965)

illustrated by species-importance curves (Fig. 4). The curves give a picture of the distribution of the relative position of the species in the community, and further indicate the partitioning of the habitat resources (WHITTAKER, 1965).

The shape of the 6 curves representing heath birch forest communities (Vassfaret–Kevo), is rather similar, and the curves approximate geometric series which indicate communities with relatively few species within rather rigorous environments. The curves for the meadow birch forest communities (Vaksvik–Ammarnäs) are less steep and approximate a more sigmoid form, indicating less severe environments and a higher diversity of bird species.

The difference in the form of these types of curves should suggest a lesser niche differentiation in the heath birch forest, as compared with the meadow birch forest, and that the bird community structure of the former habitat is more influenced by interspecific relations (WHITTAKER, 1965).

The similarity of the passerine communities found in the different study areas within each of the two forest habitats, i.e. heath birch and meadow birch forests, is due to the resemblance in both species composition and the relative density of the species (Figs. 5, 6). The communities are numerically dominated by *Ph. trochilus* and *F. montifringilla*, together making about 50–70% in the heath birch forest, and about 30–55% in meadow birch forest.

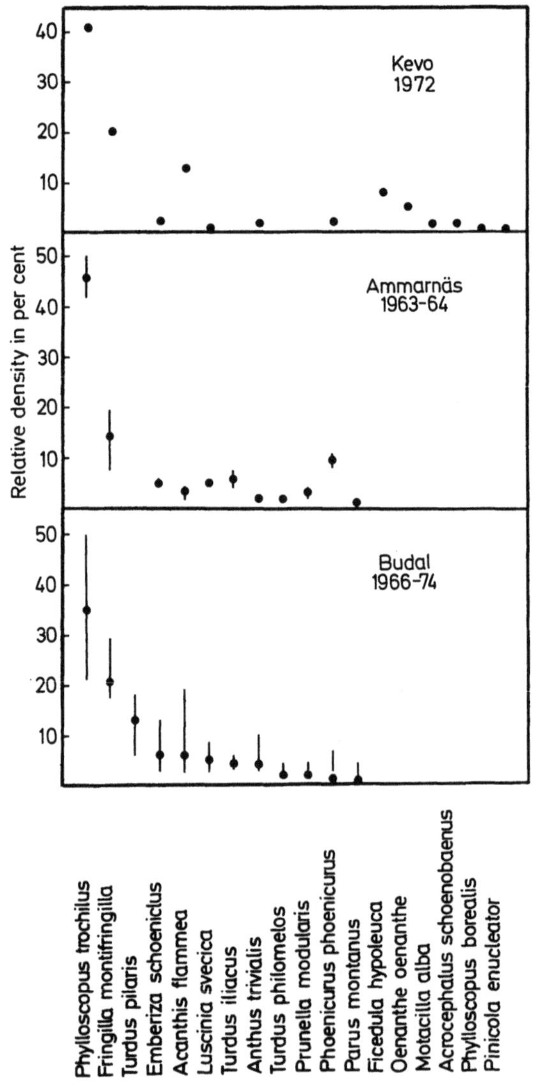


Fig. 5. Extent of variations and mean relative densities of the species in Fennoscandian heath birch-forest passerine communities, represented by Kevo (Finland), Ammarnäs (Sweden), and Budal (Norway)

## Discussion

Despite great fluctuation, especially of the most dominating species, the structure of the passerine community in the subalpine heath birch forests seems to be relatively stable. The community stability is characterized by the species composition which seems remarkably fixed, and by the relative abundance of the various

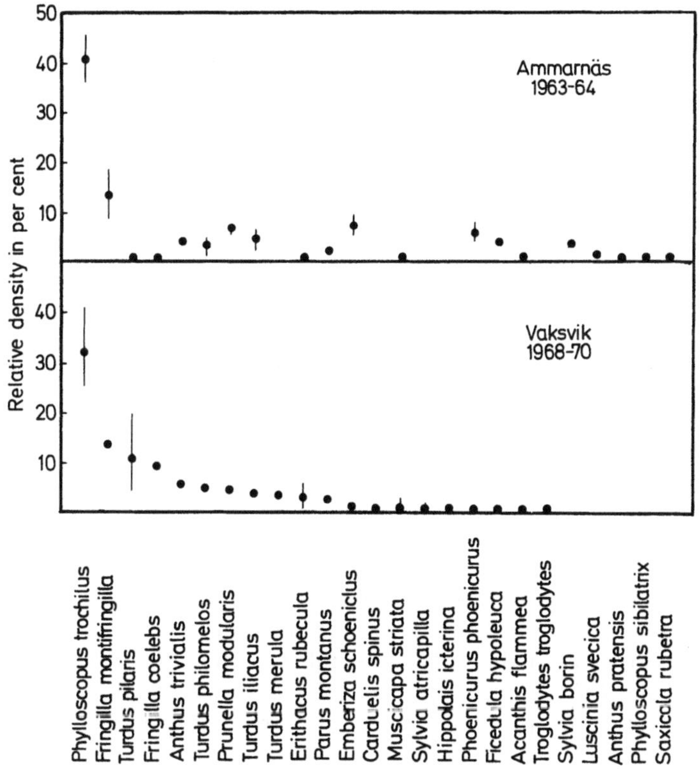


Fig. 6. Extent of variations and mean relative densities of the species in two meadow birch forest passerine communities

species which tend to return quickly from extreme abundance values to an average "normal" equilibrium population size (HOGSTAD unpubl.).

The number of species remain fairly stable in the different heath birch forest study areas, regardless of latitude and altitude. According to MACARTHUR (1972), citing DIAMOND, in press.) species diversity decreases with increasing altitude. The circumstances found in Fennoscandian heath birch forests are probably due to the similarity and rather simple structure of the habitat. As species diversity is similar when the habitat conditions are similar (PATRICK, 1968), it seems that the bird community structure is a result of the structure of the habitat (MACARTHUR and MACARTHUR, 1961).

The meadow birch forests are more complex and are more readily subdivided into several dimensions than the heath birch forests. As there is evidence that bird species are restricted to specific and structurally dissimilar parts of the habitat (TURCEK, 1951; MACARTHUR, 1958; PIELOWSKI, 1961), the more diverse meadow birch forests with taller trees give more opportunities for species to differ in vertical ranges. As bird-species diversity is strongly correlated with foliage-height diversity (MACARTHUR and MACARTHUR, 1961; RECHER, 1969; RØV, 1975), the meadow birch forest might therefore contain more species than the heath birch forest. The number of species within a habitat is undoubtedly positively correlated with the general productivity of the habitat, as theoretically demonstrated by MACARTHUR (1970).

The greater density of some species in the meadow birch forest, in relation to the heath birch forest, indicate a greater production of the latter habitat. The far greater number of species in the meadow birch forest, is probably due to a combination of a greater productivity and a more diverse habitat structure. The relationship between the density of the passerine communities and the bird-species diversity in the different study areas is probably a result of the habitat structure and its resources.

The relatively small, but rather constant number of species in the heath birch forest also indicates some resource-limiting factors. When considering the biology of the bird species inhabiting the heath birch forests, it may be suggested that food is the major limiting factor. As the habitat structure is similar, the spatial dispersion of the food is probably also similar (CODY, 1974).

When considering the species with a constancy of 0.7–1.0, it is suggestive that the heath birch forests have no two species of the same genus (Table 1). In fact, only two families have more than one species commonly occurring in the heath birch forests; Fringillidae and Turdidae. The finches *F. montifringilla* and *A. flammea*, however, belong to different subfamilies, Fringillinae and Carduelinae, respectively, while *T. iliacus*, *L. svecica*, and *Ph. phoenicurus* all belong to Turdidae.

The composition of nongeneric species in the heath birch forest passerine community is scarcely due to chance, and the author suggests that it clearly demonstrates that congeneric species have difficulties in coexisting within the structurally rather simple Fennoscandian heath birch forests. As the passerine community of the Fennoscandian heath birch forests predominantly contains the same species, it is reasonable to suggest that the community is saturated with passerine species which can allow coexistence.

From the heath birch forest in Budal, southern Norway, it is found that the small passerine species divide the habitat by occupying different vertical strata as well as by occupying mutual horizontal areas (HOGSTAD, unpubl.). It is therefore likely that the relatively simple structural heath birch forest has limited food resources, and further that one of the main governing factors determining the structure of the small passerine community is competitive interaction due to shortage of food resources.

*Acknowledgements.* I am greatly indebted to Dr. E. HAUKIOJA, University of Turku, Finland; Dr. N.-Å. ANDERSSON, Abisko Research Station, Sweden; S.-R. MYKLEBUST; O. BREKKE and R. SELBOE, University of Oslo; K. BEVANGER and A. MOKSNES, University of Trondheim, who have kindly placed their unpublished data at my disposal. Thanks are also due to N. RØV, University of Trondheim, for valuable discussions concerning the subalpine avifauna.

## References

BAGGE, P., LEHTOVUORI, M., LINDQVIST, O.: Quantitative Untersuchungen über die Vogelfauna der Kirchspiele Inari und Enontekiö in Nordlappland. (Zusammenfassung), Ornis Fenn. **40**, 21–31 (1963).

CODY, M. L.: Competition and the Structure of Bird Communities. New Jersey: Princeton 1974.

ENEMAR, A.: On the determination of the size and composition of a passerine bird population during the breeding season. Vår Fågelvärld, Suppl. **2**, 1–114 (1959).

ENEMAR, A.: Småfågelfaunans täthet och sammansättning i några skogsbiotoper längs övre Vindelälven år 1963. Fauna Flora **59**, 1–23 (1964).

ENEMAR, A.: Gråsiskan *Carduelis flammea* i Ammarnäs-området, Lycksele lappmark, år 1968. Vår Fågelvärld **28**, 230–235 (1969).

ENEMAR, A., HANSON, S. Å., SJÖSTRAND, B.: The composition of the bird fauna in two consecutive breeding seasons in the forests of the Ammarnäs area, Swedish Lapland. Acta Univ. Lund II **5**, 1–11 (1965).

ERIKSSON, C.-G., NILSSON, L., SVENSSON, B.: The passerine bird fauna of some forest habitats in Stora Sjöfallet National Park, Swedish Lapland. Bird Study **18**, 21–26 (1971).

HAUKIOJA, E., KOPONEN, S.: Faunal structure of investigated areas at Kevo, Finland. In: WIELGOLASKI, F. E. (Ed.): Fennoscandian tundra ecosystems. Part 2: Animals and systems analysis, pp. 19–28. Berlin-Heidelberg-New York: Springer 1975.

HILDÉN, O.: Über Vorkommen und Brutbiologie des Birkenzeisigs (*Carduelis flammea*) in Finnisch-Lappland im Sommer 1968. Ornis Fenn. **46**, 93–112 (1969).

HOGSTAD, O.: Breeding bird populations in two subalpine habitats in the middle of Norway during the years 1966–68. Nytt Mag. Zool. **17**, 81–91 (1969).

LAINE, H.: On some influences of the Atlantic and the Arctic Ocean upon the bird fauna of Utsjoki, Finnish Lapland. Ann. Univ. Turku. A. II **36**, 236–258 (1966).

MACARTHUR, R. H.: Population ecology of some warblers of northeastern coniferous forests. Ecology **39**, 599–619 (1958).

MACARTHUR, R. H.: Environmental factors affecting bird species diversity. Am. Naturalist **98**, 387–397 (1964).

MACARTHUR, R. H.: Patterns of species diversity. Biol. Rev. Cambridge Phil. Soc. **40**, 511–533 (1965).

MACARTHUR, R. H.: Species packing and competitive equilibria for many species. Theor. Pop. Biol. **1**, 1–11 (1970).

MACARTHUR, R. H.: Geographical ecology. New York-Evanston-San Francisco-London: Harper and Row 1972.

MACARTHUR, R. H., MACARTHUR, J. W.: On bird species diversity. Ecology **42**, 594–598 (1961).

MOEN, A.: Erfaringer fra vegetasjonskartleggingen i Trøndelagsområdet, med hovedvekt på myrenhetene. IBP i Norden **11**, 93–109 (1973).

MOKSNES, A.: Quantitative surveys of the breeding bird populations in some subalpine and alpine habitats in the Nedal area in central Norway (1967–71). Norwegian J. Zool. **21**, 113–138 (1973).

MYKLEBUST, S.-R.: Populasjonstettheten av småfugl i hekketiden i det sub-alpine bjørkebeltet ved Geilo, Sør-Norge. Forplantningsbiologi hos løvsanger, *Phylloscopus trochilus* (L.). University of Bergen: Unpubl. thesis 1966.

NORDHAGEN, R.: Sikilsdalen og Norges fjellbeiter. Skr. Bergens Museum **22**, 1–607 (1943).

PATRICK, R.: The structure of diatom communities in similar ecological conditions. Am. Naturalist **102**, 173–184 (1968).

PIELOWSKI, Z.: Untersuchungen über die Struktur der Vogelgesellschaften einiger Waldbiotope. Vogelwelt **82**, 65–83 (1961).

RECHER, H. F.: Bird species diversity and habitat diversity in Australia and North America. Am. Naturalist **103**, 75–80 (1969).

RØV, N.: Stasjonær hekkebestand av spurvefugler (Passeriformes) i tre løvskogshabitater i Surnadal på Nordmøre, 1970–71. University of Oslo: Unpubl. thesis 1972.

RØV, N.: Breeding bird community structure and species diversity along an ecological gradient in deciduous forest in Western Norway. Orn. Scand. **6**, 1–14 (1975).

SJÖRS, H.: Nordisk växtgeografi. Stockholm: Svenska Bokförlaget 1967.

SVENSSON, S. (Ed.): Bird census work and environmental monitoring. Symposium Ammarnäs 1969. Bull. Ecol. Res. Commun. **9** (1970).

Turcek, F.: On the stratification of the avian population of the Querceto-Carpinetum forest communities in southern Slovakia (English summary). Sylvia **13**, 71–86 (1951).

Whittaker, R. H.: Dominance and diversity in land plant communities. Science **147**, 250–260 (1965).

Ytreberg, N.-J.: The stationary passerine populations in the breeding season, 1968–1970, in two mountain forest habitats on the west coast of southern Norway. Norwegian J. Zool. **20**, 61–89 (1972).

# Density Variations of Bird Populations in High Mountain Habitats, Hardangervidda

L. LIEN*, E. ØSTBYE*, A. HAGEN, H.-J. SKAR, and D. SVALASTOG

Five study areas situated in the mid- and low-alpine zones (DAHL, 1975) on the mountain plateau of Hardangervidda were chosen for surveys of density variations of bird populations, carried out from 1967 to 1973, in an attempt to reveal a possible cyclic pattern. Four of the areas are located at Finse (60° 35′ N–7° 31′ E) in the northwestern part of Hardangervidda. The fifth was at Stigstuv (60° 18′ N–7° 40′ E), 35 km SSE of Finse (SONESSON et al., 1975). The areas are situated at an altitude between 1200 and 1470 m. Detailed vegetation maps and climatic data are given in LIEN et al. (1970, 1974), where results of investigations with birds are also presented.

Four of the areas are 100-m-wide transects, varying from 1.35 to 3.20 km in length, stretching from the bottom of a valley to a mountain top. A sedimentation flat of 1 $km^2$ makes up the fifth.

## Methods

Two to three observers surveyed the transects by walking 30–40 m apart along marked trails. The presence of singing males in addition to other signs of territorial activity determined the number of territories. Based on calculations of discovery chance (SEIERSTAD et al., 1965), at least two observations from six surveys within the same locality were regarded as sufficient for plotting a territory (LIEN et al., 1970).

Two to six observers walking 10–30 m apart surveyed the sedimentation flat. Because of the variations in the water level of the meander system, a marked trail was not employed. Nest counts supplemented with visual observations of birds were used to estimate the wader population.

After observation, each species was plotted on maps. At least six surveys were usually carried out in each area during the first two weeks of July. For detailed description of methods see LIEN et al. (1974).

Data concerning the fluctuations in the small mammal fauna (for comparison with birds) are based on trapping studies carried out during the summer period in five characteristic and predominant vegetation types of the Finse area.

---

* Responsible for the manuscript.

Table 1. The number of passerine territories km$^{-2}$ in the four

| | The Blåisen transect 0.315 km$^2$ | | | | | | |
|---|---|---|---|---|---|---|---|
| | 1967 | 1968 | 1969 | 1970 | 1971 | 1972 | 1973 |
| Skylark, *Alauda arvensis* | | | | | | | |
| Shore-lark, *Eremophila alpestris* | | | | | | | |
| Fieldfare, *Turdus pilaris* | | | | 6 | 6 | 3 | 3 |
| Wheatear, *Oenanthe oenanthe* | 6 | 9 | 21 | 18 | 14 | 9 | 23 |
| Bluethroat, *Luscinia suecica* | | | | | | | |
| Meadow-pipit, *Anthus pratensis* | 10 | 15 | 24 | 41 | 19 | 25 | 28 |
| White wagtail, *Motacilla alba* | 3 | | | | | | |
| Lapland-bunting, *Calcarius lapponicus* | | | | | 3 | 1 | |
| Snow-bunting, *Plectrophenax nivalis* | 6 | | 9 | 16 | 15 | 15 | 13 |
| Passeriformes | 25 | 24 | 54 | 81 | 57 | 53 | 67 |

## Results and Discussion

The calculated number of passerine territories in the four transects is shown in Table 1. The Blåisen and Hansbu transects have a northern exposure, while the other two have a southern exposure.

The transect-line survey described, seemed to be more accurate for determination of the number of small passerine territories than for birds with larger territories. Data concerning non-passerines are therefore not included. The pattern of altitude distribution varied greatly between different species and years (LIEN et al., 1974). The number of passerines and non-passerines from the sedimetation flat is given in Table 2.

During the investigation periods none of the study areas was completely free of snow. This is shown by the figures for the percentage of snow cover in the transects and dates for total disappearance of snow from the (Finsefetene) sedimentation flat (Table 3). Comparisons between yearly variations in the bird and small mammal fauna are given in Fig. 1.

The densities of both passerines and waders showed a wide variation from year to year. For example in 1970, the year with the greatest bird density and probably also breeding population, the number of territories was about three times that of 1968, the minimum season (Tables 1 and 2). With the exception of the 1971 and 1972 figures from the Stigstuv transect, the variations in density followed a certain pattern which may indicate a cyclic fluctuation (Fig. 1). It is interesting to note that from a regional point of view the same pattern of variation was shown in a mid-alpine area in nothern Sweden (ALM et al., 1965, 1966; S. SVENSSON, pers. comm.). A small peak for passerines in 1970 was recorded from alpine areas in Trøndelag, central Norway, by MOKSNES (1973a, b).

Figure 1 also suggests that a correlation may exist between the possible bird fluctuations and a typical small mammal fluctuation in the Finse area, although the amplitude of the small mammal fluctuations is greater. The passerine peak mentioned in central Norway also coincided with a small mammal high in that

transects: Blåisen, Hansbu, Stigstuv and Tverå, 1967—1973

| The Hansbu transect 0.135 km$^2$ | | | | | | | The Stigstuv transect 0.320 km$^2$ | | | | The Tverå transect 0.260 km$^2$ | | | |
|---|---|---|---|---|---|---|---|---|---|---|---|---|---|---|
| 1967 | 1968 | 1969 | 1970 | 1971 | 1972 | 1973 | 1969 | 1970 | 1971 | 1972 | 1970 | 1971 | 1972 | 1973 |
| | | | | | | | | | | 3 | | | 6 | 2 |
| | | | | | | | | 3 | 3 | 5 | 8 | 6 | 6 | 11 |
| | | | | | | | | 3 | 3 | | 4 | 4 | 2 | 4 |
| 7 | 15 | 22 | 33 | 28 | 18 | 7 | 16 | 6 | 8 | 6 | 21 | 23 | 18 | 18 |
| | | | | | | | | | 3 | | | | | |
| 28 | 17 | 20 | 53 | 47 | 30 | 32 | 31 | 33 | 20 | 27 | 55 | 32 | 24 | 37 |
| | | | | | | | | | | | | 2 | | |
| | | 5 | 7 | 5 | | 17 | 9 | 18 | 29 | 33 | | 10 | 1 | 9 |
| 7 | | 4 | 7 | 11 | 4 | 4 | | | | | 2 | 3 | 3 | 4 |
| 42 | 32 | 51 | 100 | 91 | 52 | 60 | 56 | 63 | 66 | 74 | 90 | 80 | 60 | 85 |

Table 2. The number of territories at the Finsefetene sedimentation flat in the years 1967 to 1973

| | Number of territories km$^{-2}$ | | | | | | |
|---|---|---|---|---|---|---|---|
| | 1967 | 1968 | 1969 | 1970 | 1971 | 1972 | 1973 |
| Shore-lark, *Eremophila alpestris* | | | | | | 1 | |
| Wheatear, *Oenanthe oenanthe* | 1 | 1 | 2 | | 1 | 1 | 2 |
| Meadow-pipit, *Anthus pratensis* | 2 | 2 | 8 | 14 | 4 | 6 | 4 |
| Reed-bunting, *Emberiza schoeniclus* | | | | | | | 1 |
| Lapland-bunting, *Calcarius lapponicus* | | | 2 | 3 | 5 | 2 | 4 |
| Passeriformes | 3 | 3 | 12 | 17 | 10 | 10 | 11 |
| Ringed plover, *Charadrius hiaticula* | 1 | 1 | | 1 | 1 | 1 | 1 |
| Golden plover, *Pluvialis apricaria* | | | | | | 1 | |
| Redshank, *Tringa totanus* | | | 1 | | 1 | | |
| Common sandpiper, *Tringa hypoleucos* | 2 | 1 | 4 | 4 | 1 | 1 | |
| Purple sandpiper, *Calidris maritima* | 2 | 2 | 1 | | | | |
| Temminck's stint, *Calidris temminckii* | 1 | 2 | 7 | 8 | 7 | 6 | 8 |
| Dunlin, *Calidris alpina* | | | | 1 | 1 | 1 | |
| Red-necked phalarope, *Phalaropus lobatus* | | | 1 | 1 | 1 | | |
| Common gull, *Larus canus* | 2 | 1 | 4 | 3 | 5 | 2 | 2 |
| Non-passeriformes | 8 | 7 | 18 | 18 | 17 | 12 | 11 |
| Total | 11 | 10 | 30 | 35 | 27 | 22 | 22 |

area (Moksnes 1973a, b). The present material covers the duration of one and a half cycles only, nevertheless it may support the hypothesis of synchronous fluctuations in the densities of small rodents and passerines as stated by Hagen (1952) mainly on the basis of scattered observations and assumptions. He further stated that in low density years, passerines always occurred in "visible amounts", while in peak years they never appeared in such densities as the small rodents, which also means a smaller amplitude for bird fluctuations than for small rodents.

Table 3. The percentage of snow cover in the transects – estimated approximatelly 10 July during the years 1967 to 1973 – together with approximate date when snow finally disappeared from Finsefetene

| | 1967 | 1968 | 1969 | 1970 | 1971 | 1972 | 1973 |
|---|---|---|---|---|---|---|---|
| The Blåisen transect | 50 | 38 | 3 | 2 | 11 | 21 | 22 |
| The Hansbu transect | 32 | 23 | 4 | 5 | 12 | 17 | 20 |
| The Stigstuv transect | — | — | 2 | 2 | 7 | 13 | — |
| The Tverå transect | — | — | — | 3 | 16 | 21 | 27 |
| The Finsefetene flat | 15.7 | 3.7 | 1.6 | 10.6 | 10.6 | 25.5 | 9.7 |

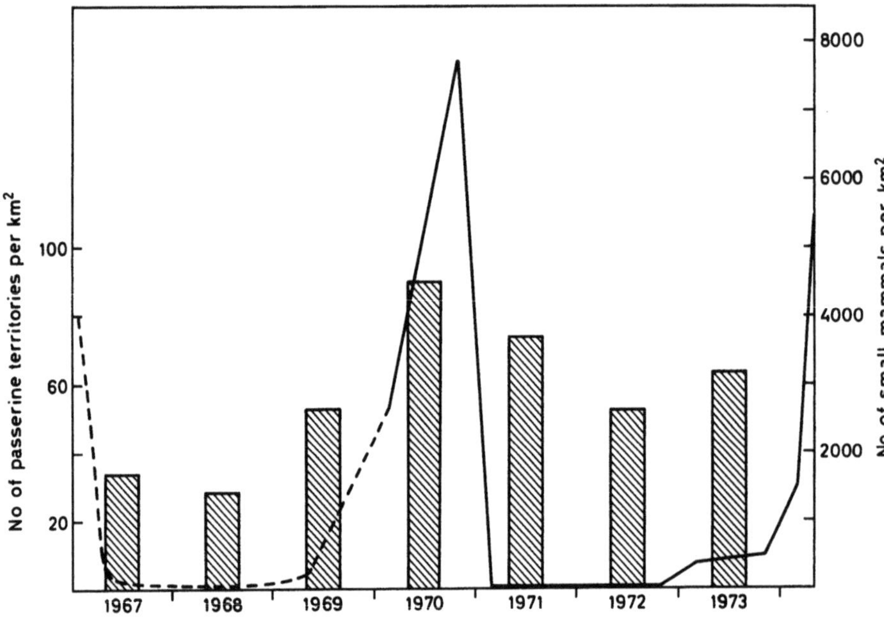


Fig. 1. Variations in the number of small mammals and passerine territories 1967–1973. Data from the Blåisen and Hansbu transects are used to illustrate bird territories. Curve represents small mammals, hatched bars passerine territories

Synchronous fluctuations of small mammals and other groups of both herbivorous and carnivorous animals, e.g. tetraonid birds and birds of prey, are well known in Fennoscandia (e.g. SIIVONEN, 1948; HAGEN 1952, 1960a,b, 1969; LINKOLA and MYLLYMÄKI 1969; MYSTERUD 1970).

The four dominant species: meadow-pipit, wheatear, snow-bunting and lapland-bunting showed generally the same pattern of density variation as the total number of territories for the transects over the different years. The snow bunting is an exception, as it was completely absent in the study plots in 1968, and the lapland bunting was not recorded in the area in the first two years of the survey period. The shore-lark was never found breeding in the northern slope transects. Data from the sedimentation flat suggests that the non-passerines show a similar pattern of density variation as passerines in the other study areas at Finse. The figures for temminck's stint might be termed "number of nests" instead of number of territories, as the present observations support HILDÉN'S (1965) investigation that each pair of birds builds two nests and raises two broods.

Climatic factors may also influence the density of breeding bird populations (LIEN et al., 1974). The percentage of snow cover in the transects (Table 3) is, to some degree, correlated with the densities of breeding birds. The same is illustrated in the correlation between the number of territories and the date of disappearance of snow cover from the sedimentation flat. The long duration of the snow cover in 1967 and 1968 reflected, particularly, climatic influence on the small number of birds in these years. Similarly, 1970, the year with the highest density of passerines in the Finse transects, coincided with the year with the least amount of snow in the breeding season. This does not, however, hold true for the Stigstuv transect. The high amount of snow in 1973 should also have predicted a lower number of territories than was observed in the Finse transects. It is possible that the first areas to become snow-free in that year were those with optimal breeding conditions, and thus could support a high number of territories. However, climatic factors seem to have only a modifying effect on variations in the density of passerine populations. It is also reasonable that the density of a bird population is influenced by the density of the breeding population and their offspring of the previous year. The breeding population of 1973 could be increased by the relatively high density in 1972 compared with the situation in 1968.

This study covers only one and a half amplitudes of the normal 3–4 year cyclic pattern. Of necessity this type of study must be carried out over a longer period if it is to succeed in providing evidence of cyclic variation in the number of bird populations and their possible correlation to other components of a high mountain ecosystem.

*Acknowledgements.* We wish to thank all those who participated in the field work; especially K.M. HAANDE, P.S. HAANDE, O. HOGSTAD, A. KLEMETSEN and A. SKARTVEIT, and M. ESPELAND for improving the English. The editor of the Norwegian Journal of Zoology has kindly given permission to use extracts of papers published in that journal for the present publication. Financial support was received from the Norwegian Research Council for Science and the Humanities and from the Norwegian IBP.

## References

ALM, B., ENEMAR, A., MYHRBERG, H., SVENSSON, S.: The density of birds in two study areas of the alpine region in Southern Lapland in 1964. Acta Univ. Lund II **4**, 1–14 (1965).

ALM, B., MYHRBERG, H., NYHOLM, E., SVENSSON, S.: Densities of birds in alpine heaths. Vår Fågelvärld **25**, 193–207 (1966).

DAHL, E.: Flora and plant sociology in Fennoscandian tundra areas. In: WIELGOLASKI, F. E. (Ed.): Fennoscandian tundra ecosystems, Part 1: Plants and microorganisms, pp. 62–67. Berlin-Heidelberg-New York: Springer 1975.

HAGEN, Y.: Rovfuglene og viltpleien. Oslo: Gyldendal 1952.

HAGEN, Y.: Vekslinger i viltbestanden, pp. 424–443. In: Jakt og fiske i Norge. Vol. 1. Oslo: Norsk Arkivforskning 1960 a.

HAGEN, Y.: Snøugla på Hardangervidda sommeren 1959. Meddr. St. Viltunders., Serie 2, **7**, 1–25 (1960b).

HAGEN, Y.: Norske undersøkelser over avkomproduksjonen hos rovfugler og ugler sett i relasjon til smagnagerbestandens vekslinger. Fauna **22**, 73–126 (1969).

HILDÉN, O.: Zur Brutbiologie des Temminckstrandläufers, *Calidris temminckii* (Leisl.). Ornis Fenn. **42**, 1–5 (1965).

LIEN, L., ØSTBYE, E., HAGEN, A., KLEMETSEN, A., SKAR, H.-J.: Quantitative bird surveys in high mountain habitats, Finse, South Norway, 1967–68. Nytt Mag. Zool. **18**, 245–251 (1970).

LIEN, L., ØSTBYE, E., HOGSTAD, O., HAANDE, K. M., HAANDE, P. S., HAGEN, A., SKAR, H.-J., SKARTVEIT, A., SVALASTOG, D.: Bird surveys in the high mountain habitats of Finse and Stigstuv, Hardangervidda, South Norway, 1967–1972. Norw. J. Zool. **22**, 1–14 (1974).

LINKOLA, P., MYLLYMÄKI, A.: Der Einfluß der Kleinsäugerfluktuationen auf das Brüten einiger kleinsäugerfressender Vögel im südlichen Häme, Mittelfinnland 1952–1966. Ornis Fenn. **46**, 45–78 (1969).

MOKSNES, A.: Quantitative surveys of the breeding bird population in some subalpine and alpine habitats in the Nedal area in central Norway (1967–71). Norw. J. Zool. **21**, 113–138 (1973a).

MOKSNES, A.: Undersøkelser over fuglefaunaen i Nedalsområdet 1972. Foreløpig rapport. Nedalsundersøkelsene. Rapport nr. 2. Zoologisk institutt. Universitetet i Trondheim. 35 pp. (mimeographed). (1973b).

MYSTERUD, I.: Hypothesis concerning characteristics and causes of population movements in Tengmalm's Owl (*Aegolius funereus* (L.)). Nytt Mag. Zool. **18**, 49–74 (1970).

ØSTBYE, E. (Ed.), BERG, A., BLEHR, O., ESPELAND, M., GAARE, E., HAGEN, A., HESJEDAL, O., HÅGVAR, S., KJELVIK, S., LIEN, L., MYSTERUD, I., SANDHAUG, A., SKAR, H.-J., SKARTVEIT, A., SKRE, O., SKOGLAND, T., SOLHØY, T., STENSETH, N. C., WIELGOLASKI, F. E.: Hardangervidda. In: ROSSWALL, T., HEAL, O. W. (Eds.): Structure and function of tundra ecosystems, Ecol. Bull. **20**, pp. 225–264. Stockholm: Swed. Nat. Sci. Res. Council 1975.

SEIERSTAD, S., SEIERSTAD, A., MYSTERUD, I.: Statistical treatment of the inconspicuousness problem in animal population surveys. Nature **206**, 22–23 (1965).

SIIVONEN, L.: Structure of shortcyclic fluctuations in numbers in mammals and birds in the northern parts of the northern hemisphere. Pap. Game Res. **1**, 1–166 (1948).

SONESSON, M., WIELGOLASKI, F. E., KALLIO, P.: Description of Fennoscandian tundra ecosystems. In: WIELGOLASKI, F. E. (Ed.): Fennoscandian tundra ecosystems. Part 1: Plants and microorganisms, pp. 3–28. Berlin-Heidelberg-New York: Springer 1975.

# 1.3. Bioenergetics

## Influence of Various Temperature Approximations on Estimates of Field Respiration in a Mountain Community

T. SOLHØY and A. SKARTVEIT

Many estimates of the annual respiration metabolism in the field of various species or groups of soil and litter invertebrates have been published in recent years. Data are obtained by combining field census data on biomass, site temperature measurements and laboratory estimates of oxygen consumption.

This procedure presents many difficulties and the errors involved seem at present to be incalculable (PHILLIPSON, 1963; WIEGERT, 1965; MACFADYEN, 1967; PETRUSEWICZ and MACFADYEN, 1970).

Temperature has a pronounced effect on the respiratory rates of poikilotherms. Respiration increases nonlinearily up to a certain temperature, following KROGH's curve (KROGH, 1914) more or less closely.

BERTHET (1964) studied 16 different species of adult oribatid mites and derived the following general relationship between weight, temperature and oxygen consumption:

$$y = 18.059 + 0.70\,w - 0.487\,z \qquad (1)$$

where $y$ is $\log_{10}$ of diurnal oxygen consumption in $10^{-3}$ µl; $w$ is $\log_{10}$ of live weight (µg) per individual and $z$ is $10^4\,T^{-1}$ ($T$ is ambient temperature in °K). This formula was confirmed by WEBB (1969) for *Nothrus silvestris* and by LUXTON (pers. comm.) for several other oribatid species.

The nonlinearity of Eq. (1) implies that the mean respiration rate during a period is usually not the same as the respiration rate at the mean temperature of that period. Temperature adaptation would reduce the difference, but NIELSEN (1961) claimed that such adaptation does not seem to occur in most soil animals. Thus, LUDWIG and CABLE (1933) and REMMERT and WÜNDERLING (1970) showed experimentally that alternating temperature in the lower part of the tolerated range accelerates the speed of development in cultures of *Drosophila*.

Obviously the most reliable values of field respiration are calculated from instantaneous values for the temperatures to which the animals are exposed. The present work aims to quantify the errors introduced by using temperature data deviation from this ideal claim.

A study was carried out on data from the lichen heath IBP alpine site at Stigstuv, Hardangervidda. A general site description is given by SONESSON et al. (1975) and some meteorological characteristics are given by SKARTVEIT et al. (1975).

## Materials and Methods

Temperature was measured for 3 years with platinum-resistance thermometers and recorded on magnetic tape every 30 min at the following levels: $T_0$ = *the* temperature at the interface between the litter layer and $A_0$, $T_{10}$ = air temperature 10 cm above ground and $T_{200}$ = air temperature 200 cm above ground (screen height). The thermometers in the air were shielded against radiation.

*Carabodes labyrinthicus* (Mich.) was selected for the study as it is the dominant species by weight of oribatid mites found at the site. On an average, the biomass of adults of this species was about 40% of the total acarine biomass in the vegetation and litter layers. Three years census data have shown that the adults are almost exclusively confined to the litter and lower part of the lichen carpet. $T_0$ is thus thought to be the best estimate of the temperature regime to which the population is exposed.

The mean fresh weight of adults is 18.2 μg with a corresponding dry weight of 6.8 μg. The fresh weight was inserted in Eq.(1) and the resulting oxygen consumption (per individual) was divided by the dry weight to give the diurnal oxygen consumption in μl $O_2$ $mg^{-1}$ dw at any given temperature *T*. These weights are chosen to give a numerical example of the effects of temperature fluctuations on respiration rates calculated from Eq.(1). However, it appears from the Equation that the relative variation of respiration rate with temperature is independent of weight. Thus the results of our calculations apply to other weights as well.

During sunny periods, $T_0$ exceeded 25° C for some part of the day. No data were available on the respiration of oribatid mites at such a high temperature. At high temperatures, however, $Q_{10}$ values generally decline with increasing temperature and this is known to be true for Oribatei (WEBB, 1968). In the present study, therefore, respiration rates were calculated up to $T_0$ 25° C, from which point the 25° maximum was used as a constant rate.

## Results and Discussion

The abbreviation CTR ("calculated true respiration") is hereafter exclusively used for respiration values calculated from half-hourly values of $T_0$. The deviations from CTR of various other computed respiration values are given below. These deviations depend upon temperature spectra and these are not calculated explicitly. Instead, mean values of both temperatures and diurnal ranges of temperature are given in Table 1.

A) Respiration values are calculated from 4 different means of $T_0$: daily means, 5-day means, monthly means and seasonal means. The derivative of the respiration function [Eq.(1)] with respect to temperature increases regularly with increasing temperature. When temperatures applied in Eq.(1) are averages for a period of time, *t*, the calculated respiration sum for, say, one year must decrease whenever *t* increases. Although respiration values for temperatures above 25° C are assigned the 25° C value, this general trend applies to Fig. 1. The calculated

Table 1. Monthly mean values of temperature, $\bar{T}_x$, and diurnal amplitude $\overline{\Delta T_x}$ for $T_0$, $T_{10}$ and $T_{200}$. $N$ is number of days with data in each month

| | | $N$ | $\bar{T}_0$ | $\overline{\Delta T_0}$ | $T_{10}$ | $\overline{\Delta T_{10}}$ | $T_{200}$ | $\overline{\Delta T_{200}}$ |
|---|---|---|---|---|---|---|---|---|
| 1970 | May | 11 | 1.7 | 5.5 | 1.4 | 7.3 | 1.0 | 6.7 |
| | June | 30 | 14.1 | 23.2 | 10.1 | 15.9 | 9.2 | 11.3 |
| | July | 31 | 8.8 | 12.8 | 6.6 | 9.7 | 6.1 | 7.1 |
| | Aug. | 31 | 11.2 | 16.8 | 8.8 | 11.2 | 8.4 | 8.0 |
| | Sept. | 14 | 4.6 | 8.8 | 3.3 | 6.9 | 3.1 | 5.0 |
| 1971 | May | 0 | — | — | — | — | — | — |
| | June | 28 | 8.1 | 13.7 | 5.4 | 10.8 | 4.5 | 7.0 |
| | July | 31 | 10.4 | 12.9 | 8.3 | 10.9 | 7.6 | 7.7 |
| | Aug. | 31 | 9.3 | 12.6 | 7.5 | 11.6 | 7.0 | 7.9 |
| | Sept. | 16 | 5.9 | 11.2 | 4.6 | 11.5 | 4.4 | 8.5 |
| 1972 | May | 0 | — | — | — | — | — | — |
| | June | 24 | 8.5 | 11.3 | 6.1 | 10.4 | 5.4 | 7.2 |
| | July | 31 | 13.3 | 15.1 | 11.0 | 13.1 | 10.1 | 9.3 |
| | Aug. | 31 | 9.1 | 11.9 | 7.2 | 11.0 | 6.7 | 8.0 |
| | Sept. | 16 | 4.6 | 9.8 | 2.7 | 9.0 | 2.4 | 6.8 |

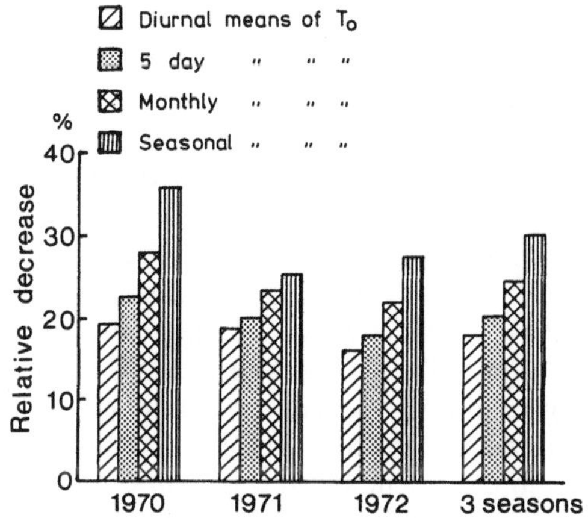


Fig. 1. Percentage decrease relative to CTR of seasonal respiration values calculated from daily, 5-day, monthly and seasonal means of $T_0$ respectively

respiration decreases by about 18% when daily means of $T_0$ are applied in Eq. (1). When 5-day means, monthly means and seasonal means of $T_0$ are applied, the corresponding decreases are about 20%, 24%, and 30% respectively. The relatively large effect of temperature fluctuations during one day as compared with more long-term fluctuations should be noted. The advantage of using daily means of temperature instead of more long-term means is evidently small compared with the effect already introduced by using daily means instead of half-hourly values.

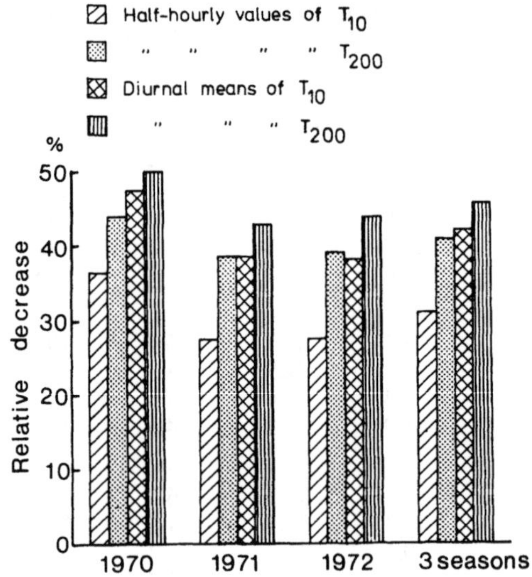


Fig. 2. Percentage decrease relative to CTR of seasonal respiration values calculated from half-hourly values and daily means of $T_{10}$ and $T_{200}$ respectively

B) Respiration values have been calculated both from half-hourly values and from daily means of $T_{10}$ and $T_{200}$. During summer the maximum of both the daily mean and the daily amplitude of temperature are found at ground level. This results in a maximum of calculated respiration at this level. Relative to CTR the calculated respiration is some 31% and 41% lower when half-hourly values of $T_{10}$ and $T_{200}$ respectively are applied in Eq. (1) (Fig. 2). When daily means of $T_{10}$ and $T_{200}$ are applied, the corresponding decreases are about 42% and 46% respectively. The relatively large effect of the difference between the temperature conditions given by $T_0$ and $T_{10}$ should be noted, while half-hourly values and daily means of $T_{200}$ give almost the same estimates of respiration.

It should also be noted that, relative to CTR, half-hourly values of $T_{10}$ lead to slightly greater errors in the calculated respiration than do even seasonal values of $T_0$. Spatial differences in temperature near the ground are therefore, in this case at least, of greater importance than temporal variations.

C) Diurnal values of respiration are obtained by adding respiration values calculated for every 10-min interval according to Eq. (1) and using temperatures calculated from:

$$T = \frac{T_{max} + T_{min}}{2} + \frac{T_{max} - T_{min}}{2} \cdot \sin(2\pi \cdot n/1440)\,, \tag{2}$$

where $n$ is the time of day in minutes, and $T_{max}$ and $T_{min}$ are the observed maximum and minimum of $T_0$ respectively.

Respiration values ($R_{sin}$) calculated by means of Eq. (2) using observed daily extremes of $T_0$ in 1970 exceed CTR by 16% (Table 2). The straightforward

Table 2. The various respiration values µl (mg dw)$^{-1}$ day$^{-1}$ of *C. labyrinthicus* on the lichen heath in 1970, calculated on a monthly (or part of month) basis using (a) Eq. (2) and daily max. and min. of $T_0(R_{sin})$, (b) half-hourly values of $T_0$ (CTR), (c) various means of $T_0$: daily means ($R_1$), 5-day means ($R_5$), monthly means ($R_M$), and the seasonal mean ($R_S$). For monthly mean values of temperature, $\bar{T}_0$, and diurnal amplitude, $\Delta\bar{T}_0$, see Table 1

| Period | Respiration µl (mg dw)$^{-1}$ day$^{-1}$ | | | | | |
|---|---|---|---|---|---|---|
| | $R_{sin}$ | CTR | $R_1$ | $R_5$ | $R_M$ | $R_S$ |
| May 21.—31. | 4 | 3 | 3 | 3 | 3 | 8 |
| June 1.—30. | 23 | 21 | 17 | 16 | 14 | 8 |
| July 1.—31. | 11 | 9 | 7 | 7 | 7 | 8 |
| Aug. 1.—31. | 18 | 14 | 10 | 10 | 9 | 8 |
| Sept. 1.—13. | 5 | 4 | 4 | 4 | 4 | 8 |
| Mean | 14.4 | 12.0 | 9.7 | 9.4 | 8.6 | 8.0 |

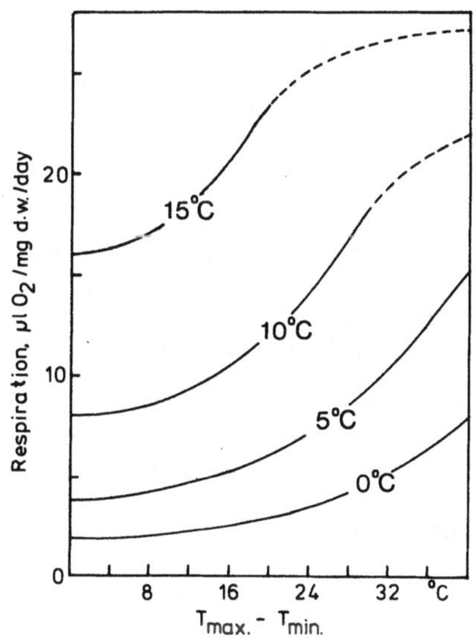


Fig. 3. Relationship between daily values of respiration [Eq. (1)] and daily range ($T_{max}$—$T_{min}$) of an harmonic daily temperature variation for fixed values of mean daily temperature $\frac{T_{max}+T_{min}}{2}$ (see text). The broken lines on the curves indicate maximum temperatures exceeding 25° C

use of only two temperature values per day thus gave more reliable results than any of the approximations shown in Figs. 1 and 2.

Furthermore, results from such calculations (Fig. 3) show clearly that for great values of diurnal range of temperature, the mean diurnal temperature is unreliable for calculating respiration. Thus, for fixed values of mean diurnal temperature between 0° C and 10° C, the calculated respiration doubles as the range of temperature increases from 0° C to upwards of 25° C.

For low values of diurnal range of temperature, on the other hand, the mean diurnal temperatures might work very well.

The study has shown that actual temperature fluctuations are not negligible even under the alpine conditions on Hardangervidda.

## References

BERTHET, P.: L'activité des Oribatides d'une Chenaie. Mem. Inst. Roy. Sci. Nat. Belg. **152**, 1–152 (1964).

KROGH, A.: The quantitative relation between temperature and standard metabolism in animals. Int. Z. Phys.-Chem. Biol. **1**, 491–508 (1914).

LUDWIG, D., CABLE, R. M.: The effect of alternating temperatures on the pupal development of *Drosophila melanogaster* Meigen. Physiol. Zool. **6**, 493–508 (1933).

MACFADYEN. A.: Methods of investigation of productivity of invertebrates in terrestrial ecosystems. In: PETRUSEWICZ, K. (Ed.): Secondary productivity of terrestrial ecosystems II, pp. 383–412. Warszawa-Krakow: Panstwowe Wyd. Naukowe 1967.

NIELSEN, C. O.: Respiratory metabolism of some populations of enchytraeid worms and free living nematodes. Oikos **12**, 17–35 (1961).

PETRUSEWICZ, K., MACFADYEN, A.: Productivity of terrestrial animals. Principles and Methods. Oxford and Edinburgh: Blackwell 1970.

PHILLIPSON, J.: The use of respiratory data in estimating annual respiratory metabolism, with particular reference to *Leiobunum rotundum* (Latr.) (Phalangiida). Oikos **14**, 212–223 (1963).

REMMERT, H., WÜNDERLING, K.: Temperature differences between arctic and alpine meadows and their ecological significance. Oecologia (Berl.) **4**, 208–210 (1970).

SKARTVEIT, A., RYDÉN, B. E., KÄRENLAMPI, L.: Climate and hydrology of some Fennoscandian tundra ecosystems. In: WIELGOLASKI, F. E. (Ed): Fennoscandian tundra ecosystems. Part 1: Plants and microorganisms, pp. 41–53. Berlin-Heidelberg-New York: Springer 1975.

SONESSON, M., WIELGOLASKI, F. E., KALLIO, P.: Description of Fennoscandian tundra ecosystems. In: WIELGOLASKI, F. E. (Ed.): Fennoscandian tundra ecosystems. Part 1: Plants and microorganisms, pp. 3–28. Berlin-Heidelberg-New York: Springer 1975.

WEBB, N.: Comparative studies on population metabolism in soil arthropods. University of Wales, Swansea: Ph. D. Thesis 1968.

WEBB, N.: The respiratory metabolism of *Nothrus silvestris* Nicholet (Acari). Oikos **20**, 294–299 (1969).

WIEGERT, R. G.: Energy dynamics of the grasshopper populations in old field and alfalfa field ecosystems. Oikos **16**, 161–176 (1965).

# Respiration Rates, Winter and Summer Activity in Collembola on Hardangervidda

H. KAURI, T. J. MOLDUNG, and A. FJELLBERG

In order to determine the duration of periods of activity in alpine invertebrate populations at the IBP sites Stigstuv, Hardangervidda, and to investigate respiration rates, regular measurements were taken throughout the year for various species (STEIGEN, 1975). This is a description of consumption for two Collembola species, *Isotoma viridis* Bourlet and *Tetracanthella britannica* Cassagnau collected at the Stigstuv wet meadow site. Collembolas were relatively easy to collect during winter compared to other animals with a lower abundance. It was preferable to work with freshly collected material than laboratory cultures. For a description of the site, see SOLHØY (1972) and SONESSON et al. (1975).

## Material and Methods

Collembolas for investigation were collected at 3-week intervals during the snow-free period from June to October, and later once a month except for November and December. For soil-surface temperature, see Fig. 1. Other climatic

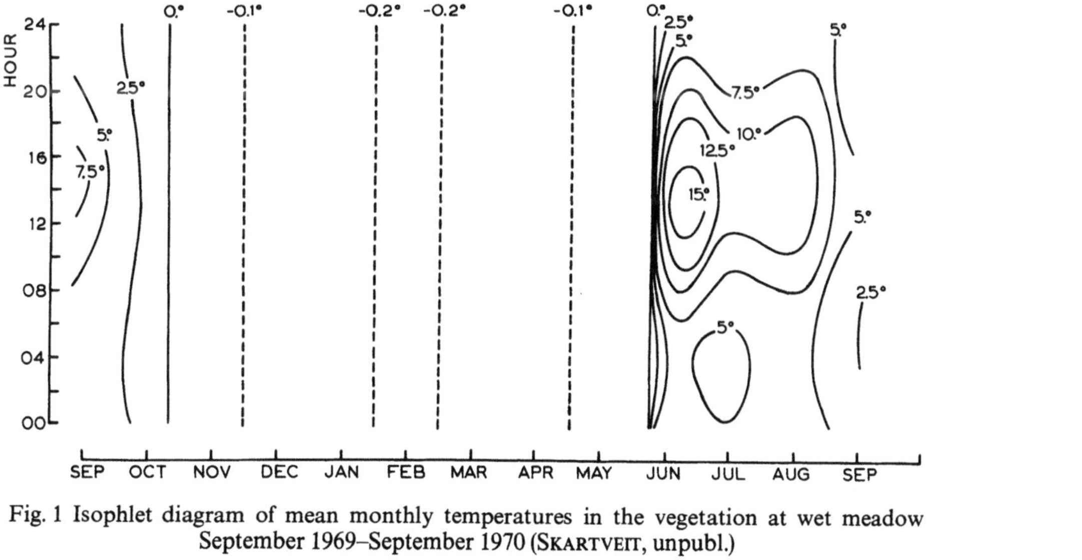


Fig. 1 Isophlet diagram of mean monthly temperatures in the vegetation at wet meadow September 1969–September 1970 (SKARTVEIT, unpubl.)

parameters are given by SKARTVEIT et al. (1975). The material was transported to the laboratory in a refrigerated box kept at 5° C. Temperature for acclimatization was 10° C for a period of 3–10 days, but usually 6–9 days. All animals could feed during this period.

$O_2$ consumption was registered in a SCHOLANDER volumetric respirometer (SCHOLANDER and IVERSEN, 1958) in water bath at 8°, 13°, 18° and 23° C. Measurements were carried out at 8° C. For other temperatures, see STEIGEN (1975). To avoid too large variation in oxygen consumption due to difference in size, animals weighing 103–123 μg dw (24 hrs, 105° C), mean 118.3 μg, (=2.6–2.8 mm body length) for *I. viridis* and 3.7–4.0 μg dw (=0.9 mm body length) for *T. britannica* were used in the respirometers. The photoperiod during acclimatization was 8/16 hrs day/night in the winter months, and 16/8 hrs in summer. In early and late September respectively, 14/10 and 10/14 hrs day/night were used. The winter photoperiod was started in mid October.

## Results

Oxygen concumption for *I. viridis* per ind. and per μg dw $h^{-1}$ is given in Table 1 as the mean for the months January to August. As the material for September and October does not include individuals of the above-mentioned body size, the oxygen consumption for weight 120μg has been calculated by the equation $y = ax^b$ where $y$ is $O_2$ consumption in μl per ind. and per μg dw $h^{-1}$, $a$ is the constant and $b$ the regression coefficient.

Oxygen consumption of *Tetracanthella britannica* is shown in Table 2. Material was available only in January, February, April, May, July, and August, and the number of individuals available was also limited. Nevertheless, the results appear to follow the same pattern as for *I. viridis*. The species is much smaller, and this results in a considerably higher oxygen consumption than for *I. viridis*.

Table 1. *Isotoma viridis*. Dry weight per individual, $O_2$ consumption at 8°C in μl ind.$^{-1}$ $h^{-1}$ and $O_2$ consumption at 8°C in μl (mg dw)$^{-1}$ $h^{-1}$. $O_2$ consumption in October, January, February, April, May, June contra July, August, September $p < 0.001$

| Month | $n$ | dw μg ind.$^{-1}$ $\bar{x} \pm$ S.D. | $O_2$ μl ind.$^{-1}$ $h^{-1}$ $\bar{x} \pm$ S.D. | $O_2$ μl (mg dw)$^{-1}$ $\bar{x} \pm$ S.D. | Calculated $O_2$ μl (mg dw)$^{-1}$ $h^{-1}$ for ind. weighing 120 μg dw |
|---|---|---|---|---|---|
| January | 4 | 121.7±2.51 | 144±2.01 | 1.190±0.008 | — |
| February | 3 | 119.5±3.55 | 146±2.30 | 1.255±0.015 | — |
| March | 4 | 119.8±3.19 | 144±2.00 | 1.210±0.016 | — |
| April | 10 | 120.6±2.54 | 146±2.54 | 1.214±0.017 | — |
| May | 14 | 119.7±2.98 | 148±5.91 | 1.211±0.031 | 1.205 |
| June | 7 | 113.4±6.15 | 142±4.86 | 1.255±0.101 | 1.183 |
| July | 8 | 114.6±6.57 | 153±4.99 | 1.339±0.036 | — |
| August | 10 | 117.1±6.02 | 163±5.73 | 1.386±0.033 | — |
| September | 14 | 120 | 173 | 1.448 | 1.448 |
| October | 25 | 120 | 124 | 1.036 | 1.036 |

Table 2. Body size and oxygen consumption for *Tetracanthella britannica*

| Month | $n$ | dw µg ind.$^{-1}$ $\bar{x}\pm$S.D. | $O_2$ µl ind.$^{-1}$ h$^{-1}$ $\bar{x}\pm$S.D. | $O_2$ µl (mg dw)$^{-1}$ $\bar{x}\pm$S.D. |
|---|---|---|---|---|
| January | 3 | 3.81±1.1 | 127±0.0 | 3.37±1.02 |
| February | 4 | 3.83±0.8 | 127±1.2 | 3.36±0.38 |
| March | — | — | — | — |
| April | 4 | 3.82±0.7 | 130±5.0 | 3.41±1.30 |
| May | 4 | 3.79±1.1 | 129±1.7 | 3.42±1.11 |
| June | — | — | — | — |
| July | 4 | 3.81±1.2 | 154±3.0 | 4.07±0.65 |
| August | 2 | 3.81±0.4 | 146±5.0 | 3.86±1.06 |

## Discussion

According to Fjellberg (unpubl.) the specimens of *I. viridis* with dry weight 103–123 µg, are imagines with their reproductive period July–August, their second hibernation commencing during autumn. The species lives in the litter layer and in the upper soil layer (raw humus). *T. britannica* with a body weight of 3.7–4.0 µg dw have either just become sexually mature or they are immediately before the onset of maturity. Individuals of this size found during winter are therefore in their second winter. The species is edaphic. Only 2% of the population has been found in the litter layer on the wet meadow (Fjellberg, unpubl.).

The $O_2$ consumption for *I. viridis* is at different levels during summer and winter. Both periods are clearly separated, the consumption during July, August, and September is significantly higher than in October–June values (Table 1, Fig. 2). Between January and June the oxygen consumption shows little variation: From 1.19 in January to 1.2 µl (mg dw)$^{-1}$ h$^{-1}$ in May. During July–September, the respirometer value rises from 1.26 µl mg$^{-1}$ h$^{-1}$ in June to 1.45 µl mg$^{-1}$ h$^{-1}$ in September. Higher oxygen consumption is als found for *T. britannica* in July–August than earlier in the year; however, data for the autumn months has not been obtained (Table 2).

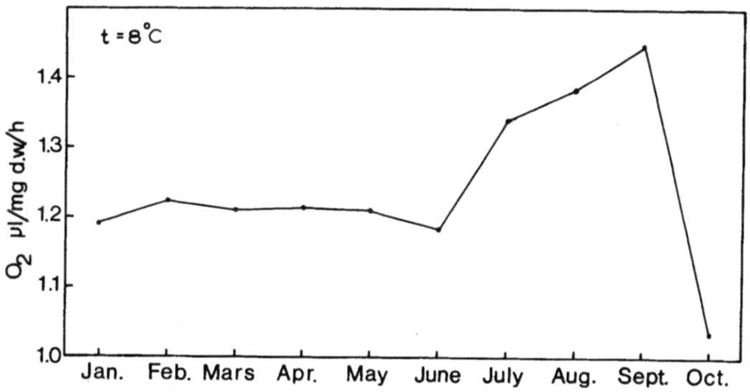


Fig. 2. Oxygen consumption for *Isotoma viridis* at 8° C on wet meadow, Hardangervidda

As the animals studied have been acclimatized at 10° C for several days, the difference in oxygen consumption cannot be attributed to temperature. Therefore, the results must be considered as a type of winter "dormancy" (diapause) (MÜLLER, 1965) or a thermopause (MANSINGH, 1971) of which reduced matabolic capacity visualized by the lower oxygen consumption is an expression. Discovery of the diapause hormone (HASEGAVA, 1957) could suggest some sort of external stimulation if such a hormone influences dormancy. Many observations have also shown that the photoperiod may be such a stimulus (summarized by DANILEVSKII, 1961, 1965). On the other hand DONDALE and LEGENDRE (1971) could not find any connection between photoperiod and "dormancy" for *Pisaura mirabilis* (Clerck). It is also difficult to imagine the photoperiod as a stimulus for edaphic animals such as *I. viridis* and *T. britannica*. However, the importance of photoperiod for the introduction of diapause and its termination in spring has not been determined in the present studies.

For *I. viridis* the dormancy starts in October and is terminated at the end of June or beginning of July—limiting the active period to 3 months. The change from dormancy to summer activity and back again, is fairly abrupt, and takes place over a short period. Compared with other areas with milder, snow-free winters, the period of activity on Hardangervidda is very short, and its start in spring is delayed by several months due to the snow cover (Fig. 1 and SKARTVEIT et al., 1975). Dormancy is terminated towards the end of June instead of in March-April as in lower areas of western Norway with mild atlantic-type winters.

Start and termination of winter dormancy is synchronous for many groups of animals, and is not unique for Collembola. The results regarding the periods of activity and winter dormancy agree closely with the data for *Tipula excisa* at Finse (HOFSVANG, 1972). Weight increases after winter start for second instar larvae of this species in May–June, and ceases at the beginning of October.

The oxygen consumption of adult *Pelophila borealis* (Coleoptera, Carabidae) has been investigated by CONRADI-LARSEN and SØMME (1973) at Finse, on the northern edge of Hardangervidda, 1200 m, above sea level. They have found a different pattern in the seasonal variation of respiration. The respiratory quotient

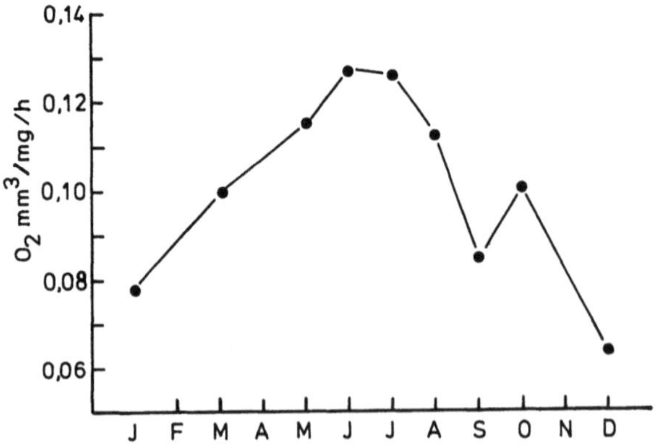


Fig. 3. Oxygen consumption for *Pelophila borealis* at 10° C in $mm^3\ mg^{-1}\ h^{-1}$ calculated with the help of the equations given by CONRADI-LARSEN and SØMME (1973)

decreases from September to December, then increases steadily, reaching a maximum in July and August (Fig. 3). During winter there is thus a continuous change in the physiological activity of this species. The discussed types of winter metabolism suggest different adaptations to the arctic alpine winter conditions.

*Acknowledgements.* We are grateful to T. SOLHØY, who through the difficult winter months managed to maintain sampling on Hardangervidda, also S. G. SIMPSON and B. BAKKE, who have assisted with the respiration measurements.

## References

CONRADI-LARSEN, E. M., SØMME, L.: The overwintering of *Pelophila borealis* Payk. II. Aerobic and anaerobic metabolism. Norsk Ent. Tidsskr. **20**, 325–332 (1973).

DANILEVSKII, A. S.: Photoperiodism and seasonal development of insects (in Russian 1961). Edinburg-London: Oliver and Boyd 1965.

DONDALE, C. D., LEGENDRE: Winter diapause in a Mediterranean population of *Pisaura mirabilis* (Clerck). Bull. Brit. Arach. Soc. **2**, 6–10 (1971).

HASEGAVA, K.: The diapause hormone of the silkworm, *Bombyx mori.* Nature **179**, 1300–1301 (1957).

HOFSVANG, T.: *Tipula excisa* Schum. (Diptera, Tipulidae) life cycle and population dynamics. Norsk Ent. Tidsskr. **19**, 43–48 (1972).

MANSINGH, A.: Physiological classification of dormancies in insects. Can. Ent **103**, 983–1009 (1971).

MÜLLER, H. J.: Probleme der Insektendiapause. Verhandl. Deut. Zool. Ges., Zool. Anz. **29**, Suppl., 191–222 (1965).

SCHOLANDER, P. F., IVERSEN, O.: New design of volumetric respirometer. Scand. J. Clin. Lab. Invest. **10**, 429–431 (1958).

SKARTVEIT, A., RYDÉN, B. E., KÄRENLAMPI, L.: Climate and hydrology of some Fennoscandian tundra ecosystems. In: WIELGOLASKI, F. E. (Ed.): Fennoscandian tundra ecosystems. Part 1: Plants and microorganisms, pp. 41–53. Berlin-Heidelberg-New York: Springer 1975.

SOLHØY, T.: Quantitative invertebrate studies in mountain communities at Hardangervidda, South Norway. I. Norsk Ent. Tidsskr. **19**, 99–108 (1972).

SONESSON, M., WIELGOLASKI, F. E., KALLIO, P.: Description of Fennoscandian tundra ecosystems. In: WIELGOLASKI, F. E. (Ed.): Fennoscandian tundra ecosystems. Part 1: Plants and microorganisms, pp. 3–28. Berlin-Heidelberg-New York: Springer 1975.

STEIGEN, A. L.: Respiratory rates and respiratory energy loss in terrestrial invertebrates from Hardangervidda. In: WIELGOLASKI, F. E. (Ed.): Fennoscandian tundra ecosystems. Part 2: Animals and systems analysis, pp. 122–128. Berlin-Heidelberg-New York: Springer 1975.

# Respiratory Rates and Respiratory Energy Loss in Terrestrial Invertebrates from Hardangervidda

A. L. Steigen

Study of energy flow through an ecosystem requires establishing the amount of energy required to maintain the life processes. This is usually done by estimating oxygen consumption.

Respiratory rates have been measured in some representative species of a few invertebrate groups from Stigstuv on the alpine Hardangervidda plateau in southern Norway. For a description of the area, see Sonesson et al. (1975). A brief survey of some of the results from this Norwegian IBP tundra biome project is presented in the present paper.

## Materials and Methods

Animals for the experiments were collected in the active summer period (mostly August–September) by hand, in pit-fall traps, by washing out in water, and by gradient extraction. Representatives of the groups Opiliones, Araneae, Collembola, Thysanoptera, and Homoptera were used in the studies. The animals were acclimatized for a minimum of 48 hrs at 10° C. Prior to the respiration measurements the animals were given food to avoid any effects of hunger on respiratory rate (cf. Zinkler, 1966; Miyashita, 1969). Respiration measurements were carried out at 8°, 13°, 18°, and 23° C. A modified Scholander volumetric respirometer (Scholander and Iversen, 1958) was used and temperatures were kept within ±0.02° C in the water-bath (factory specifications).

Depending on their size, the animals were kept in the respiratory chambers for 3 to 24 hrs. All weights are given as average individual dry weight (dw). The animals were dried at 105° C for 24 hrs, and cooled in an exsiccator prior to weighing. Respiratory rates are expressed as oxygen consumption in μl $(mg\ dw)^{-1}$ $h^{-1}$. Dry weights vary less than fresh weights (fw), since the water content of fresh specimens may vary considerably. In order to calculate the percentage of calories lost through respiration, it is necessary to know the respiratory energy loss (REL) related to dry weight. Most authors, however, have given respiratory rates related to fresh weight which makes direct comparison difficult. Consequently respiratory rates are also given as oxygen consumption in μl $(mg\ fw)^{-1}$ $h^{-1}$ for some species.

The respiratory $Q_{10}$ was calculated from the formula:

$$Q_{10} = \left(\frac{K_1}{K_2}\right)^{\frac{10}{t_1 - t_2}}$$

Where $K_1$ and $K_2$ are the respiratory rates at the temperatures $t_1$ and $t_2$.

The fact that herbivores tend to have higher respiratory quotients (RQ) than carnivores has not been taken into consideration here when calculating the respiratory energy loss (REL). The interpretation of RQ is complex, and it varies in a poikilotherm not only with feeding habits, but also with biological and ecological conditions. The value used here to calculate the REL is quoted from ROEDER (1953) for a resting insect: RQ=0.82, which gives an oxycaloric coefficient of $4.83 \times 10^{-3}$ cal $\mu l^{-1} O_2$.

## Results and Discussion

The respiratory rates of males and females of *Mitopus morio* have been separated (Table 1) as the males have a significantly higher oxygen consumption at 18° C. Respiratory activity of female animals is lower than the values given by PHILLIPSON (1962) (Table 2). This may be due to the fact that the females, collected in the first week of September, were packed with eggs and twice as heavy as the males, or may be a result of different methods in measuring respiration. In the present study, respiration was measured for a period of three to four hours between 0900 and 1400 hrs. PHILLIPSON (1962), however, kept his animals in the respiratory chambers for 48 hrs. Values quoted in Table 2 are for the first 24 hrs and were obtained during the daytime, when the animals may have started to search for food, with a corresponding increase in respiratory rate. Also, the population of *M. morio* at Hardangervidda belongs to the subspecies *M. morio alpinus* (Herbst) and not to the nominate subspecies.

The females have a minimal respiratory $Q_{10}$ at temperatures of 13–18° C (Table 3). At 8–13° C, the males showed the highest $Q_{10}$ for all the animals tested. The $Q_{10}$ obtained for *Pardosa palustris* (Table 3) seems to be consistent with results for *Pardosa* spp. from other authors (Table 2).

Compared with available data on the respiratory rates of Collembola from temperate regions, the present values are high, and are even higher than respiratory rates of springtails in antarctic regions (TILBROOK and BLOCK, 1972). This indicates that animals collected on Hardangervidda have a respiratory rate presumably adapted to temperature conditions in the elevated area.

The only study on thysanopteran respiration known to the author is the one by PRIESNER (1964) on Egyptian desert species, which does not provide a relevant comparison with the results given in Table 1.

In Table 4 values for respiratory energy loss (REL) and the percentage of total body caloric content lost through respiration are given for a 24-hrs period. Energy metabolism, however, is subject to variations according to diurnal alterations in light, temperature and activity patterns, and seasonal variations in photoperiod. To reveal those variations in respiratory rates, oxygen consumption must be

Table 1. Oxygen consumption at different temperatures (± 95% confidence interval), and average individual weight in mg dw of animals collected at Hardangervidda

| Species | $O_2$-consumption in µl per mg fw ($\bar{x}_{fw}$) or dw ($\bar{x}_{dw}$) per hour | | | | | | | | | | | | Average dw in mg ind.$^{-1}$ |
|---|---|---|---|---|---|---|---|---|---|---|---|---|---|
| | 8° C | | | 13° C | | | 18° C | | | 23° C | | | |
| | $n$ | $\bar{x}_{fw}$ | $\bar{x}_{dw}$ | $n$ | $\bar{x}_{fw}$ | $\bar{x}_{dw}$ | $n$ | $\bar{x}_{fw}$ | $\bar{x}_{dw}$ | $n$ | $\bar{x}_{fw}$ | $\bar{x}_{dw}$ | |
| Opiliones | | | | | | | | | | | | | |
| *Mitopus morio* Fabr. | | | | | | | | | | | | | |
| ♂♂ (primo Sept.) | 8 | 0.195 | 0.71 (±0.06) | 18 | 0.432 | 1.76 (±0.19) | 14 | 0.591 | 2.56 (±0.21) | 7 | 0.717 | 3.71 (±0.52) | 9.10 |
| ♀♀ (primo Sept.) | 9 | 0.201 | 0.88 (±0.18) | 8 | 0.345 | 1.59 (±0.20) | 11 | 0.508 | 2.19 (±0.32) | 3 | 0.696 | 3.46 (±0.41) | 18.86 |
| Araneae | | | | | | | | | | | | | |
| *Pardosa palustris* (L.) | 6 | 0.17 | 0.64 (±0.08) | 7 | 0.34 | 1.33 (±0.24) | 8 | 0.59 | 2.30 (±0.21) | 7 | 0.72 | 2.80 (±0.29) | 4.69 |
| Collembola | | | | | | | | | | | | | |
| *Onychiurus pseudovanderdrifti* Gisin | 23 | | 4.12 (±0.24) | 20 | | 8.37 (±0.48) | 8 | | 11.36 (±0.94) | | | | 0.0039 |
| *Isotoma viridis* Bourlet | 32 | 0.30 | 1.44 (±0.24) | 14 | 0.66 | 3.05 (±0.09) | 4 | 1.17 | 4.59 (±0.30) | 12 | 1.98 | 6.10 (±0.41) | 0.110 |
| *Tetracanthella britannica* Cass. | 8 | | 3.87 (±0.22) | 12 | | 8.50 (±0.41) | 8 | | 11.94 (±0.42) | | | | 0.0038 |
| Thysanoptera | | | | | | | | | | | | | |
| *Anapothrips secticornis* (Trybom) | 7 | | 1.51 (±0.18) | 8 | | 2.43 (±0.21) | 10 | | 3.68 (±0.16) | | | | 0.014 |
| Homoptera | | | | | | | | | | | | | |
| *Arctorthezia cataphracta* (Olafsen) | | | | | | | | | | | | | |
| weight less than 0.5 mg | 11 | 0.22 | 0.60 (±0.22) | 17 | 0.44 | 0.83 (±0.14) | 18 | 0.64 | 1.38 (±0.17) | 6 | 0.70 | 1.67 (±0.57) | 0.289 |
| weight more than 0.5 mg | 18 | 0.14 | 0.29 (±0.04) | 14 | 0.32 | 0.57 (±0.12) | 21 | 0.41 | 0.78 (±0.08) | | | | 1.57 |

Table 2. Respiratory parameters obtained from literature for some species. Most values of respiratory $Q_{10}$ have been calculated by the author

| Species | Accl. temp. °C | Resp. temp. °C | µl $O_2$ (mg fw)$^{-1}$ h$^{-1}$ | $Q_{10}$ (temp. int.) | REL (24 hrs)$^{-1}$ | % Loss | Ref. |
|---|---|---|---|---|---|---|---|
| Opiliones | | | | | | | |
| *Mitopus morio* Fabr. | | 16 | 0.626 | | | | Phillipson, 1962 |
| *Oligolophus tridens* (C.L.K.) | | 16 | 0.943 | | | | Phillipson, 1962 |
| *Leiobunum rotundum* (Latr.) | | 16 | 1.5 (ca.) | | | | Phillipson, 1963 |
| Araneae | | | | | | | |
| *Pardosa lenta* (Hentz) | 10 | 10 | 0.031 | 3.77 (10–20°) | | | Anderson, 1970 |
| *Pardosa lenta* (Hentz) | 10 | 20 | 0.117 | | | | Anderson, 1970 |
| *Phidippus regius* (C.L.K.) | 10 | 10 | 0.028 | 3.54 (10–20°) | | | Anderson, 1970 |
| *Phidippus regius* (C.L.K.) | 10 | 20 | 0.099 | | | | Anderson, 1970 |
| Lycosidae, 3 species | | 15–30 | 0.1–1.5 | | 0.0284–<br>0.3976 | 0.48–6.85 | Breymeyer, 1967 |
| *Paradosa pseudoannulata* | 10–25 | 29 | | | 0.1509 | 2.6 | Ito, 1964 |
| *Pardosa T-insignata* (Bös. and Str.) | | 15–25 | | 2.85 (15–25°)[a]<br>1.59 (15–25°)[b]<br>2.22 average | | | Miyashita, 1969 |
| *Pardosa rabida* (Walk.) | 15 | 15 | 0.72 | | | | Van Hook, 1971 |
| *Pardosa rabida* (Walk.) | 20 | 20 | 0.96 | 2.07 (15–25°) | | | Van Hook, 1971 |
| *Pardosa rabida* (Walk.) | 25 | 25 | 1.49 | | | | Van Hook, 1971 |
| *Tarentula kochi* Keyserling | | 20 | 0.25 | 2.3 (15–25°) | | | Hagstrum, 1970 |
| Collembola | | | | | | | |
| *Onychiurus procampatus* Gisin | | 15 | 1.0–0.32 | 3.5–4.0 (ca.) | | | Healey, 1967 |
| *Cryptopygus antarcticus* Willem | 6 | 2 | 0.082 | 1.06 (2–6°) | | | Tilbrook and Block, 1972 |
| *Cryptopygus antarcticus* Willem | 6 | 6 | 0.084 | | | | Tilbrook and Block, 1972 |
| *Cryptopygus antarcticus* Willem | 6 | 10 | 0.147 | 4.05 (6–10°) | | | Tilbrook and Block, 1972 |
| *Isotoma viridis* Bourlet | | 8 | 0.146 | 3.73 (8–18°) | | | Zinkler, 1966 |
| *Isotoma viridis* Bourlet | | 18 | 0.545 | | | | Zinkler, 1966 |
| *Onychiurus armatus* (Tullb.) | | 8 | 0.241 | 2.61 (8–18°) | | | Zinkler, 1966 |
| *Onychiurus armatus* (Tullb.) | | 18 | 0.630 | | | | Zinkler, 1966 |
| *Isotoma trispinata* MacGillivray | | 16 | 0.85–1.7 | 2.94 (5–25°) | | | Tanaka, 1970 |
| *Onychiurus sp.* | | 16 | 0.62–0.86 | 2.78 (5–25°) | | | Tanaka, 1970 |

[a] animals at rest; [b] animals active.

Table 3. Calculated respiratory $Q_{10}$ of animals collected at Hardangervidda

| Species | 8–13° | 8–18° | 13–18° | 13–23° | 18–23° |
|---|---|---|---|---|---|
| *M. morio* ♂♂ | 6.14 | 3.91 | 2.12 | 2.11 | 2.10 |
| *M. morio* ♀♀ | 3.26 | 2.49 | 1.89 | 2.18 | 2.49 |
| *P. palustris* | 4.32 | 3.59 | 2.99 | 2.11 | 1.48 |
| *O. pseudovanderdrifti* | 4.13 | 2.76 | 1.84 | | |
| *I. viridis* | 4.49 | 3.19 | 2.26 | 2.00 | 1.77 |
| *T. britannica* | 4.33 | 3.09 | 2.20 | | |
| *A. secticornis* | 2.59 | 2.44 | 2.29 | | |
| *A. cataphracta* < 0.5 mg | 1.91 | 2.30 | 2.76 | 2.01 | 1.47 |
| *A. cataphracta* ≥ 0.5 mg | 3.86 | 2.69 | 1.87 | | |

Table 4. Respiratory energy loss (REL) when RQ = 0.82, i. e. oxycaloric coefficient = $4.83 \times 10^{-3}$, and percentage of total body caloric content lost through respiration during a 24-h period

| Species | Av. dw (mg ind$^{-1}$) | Av. cal. cont. per ashfree ind. | REL in cal. mg$^{-1}$ dw | | | | % caloric loss through respiration | | | |
|---|---|---|---|---|---|---|---|---|---|---|
| | | | 8° | 13° | 18° | 23° | 8° | 13° | 18° | 23° |
| *M. morio* ♂♂ | 9.10 | 50.41 | 0.082 | 0.204 | 0.297 | 0.430 | 1.5 | 3.7 | 5.4 | 7.8 |
| *M. morio* ♀♀ | 18.86 | 108.33 | 0.102 | 0.184 | 0.254 | 0.401 | 1.8 | 3.2 | 4.4 | 7.0 |
| *P. palustris* | 4.69 | 28.35 | 0.074 | 0.154 | 0.267 | 0.325 | 1.2 | 2.5 | 4.4 | 5.4 |
| *O. pseudovanderdrifti* | 0.0039 | 0.023 | 0.478 | 0.970 | 1.317 | | 8.1 | 16.4 | 22.3 | |
| *I. viridis* | 0.110 | 0.637 | 0.167 | 0.354 | 0.532 | 0.707 | 2.9 | 6.1 | 9.2 | 12.2 |
| *T. britannica* | 0.0038 | 0.022 | 0.449 | 0.933 | 1.384 | | 7.7 | 16.9 | 23.6 | |
| *A. secticornis* | 0.014 | 0.092 | 0.175 | 0.282 | 0.427 | | 2.7 | 4.3 | 6.5 | |
| *A. cataphracta* < 0.5 mg | 0.289 | 2.18 | 0.070 | 0.096 | 0.160 | 0.194 | 0.9 | 1.3 | 2.1 | 2.6 |
| *A. cataphracta* ≥ 0.5 mg | 1.57 | 12.18 | 0.034 | 0.066 | 0.090 | | 0.4 | 0.9 | 1.2 | |

carefully measured for each species throughout its life-cycle. The values for REL in Table 4 are based on day-time respiratory measurements and must, so far, be considered rough estimates.

In *M. morio*, the REL and percentage of loss are higher in the males than in the females (except at 8° C). The values given for *P. palustris* are generally lower than the corresponding ones for *M. morio*.

The REL calculated by BREYMEYER (1967) for three Lycosid species was based on an RQ of 0.70, but even for a predator this figure seems low. The relatively low value obtained by ITO (1964) (see Table 2) may be due to the fact that his measurements were made in the dark (see ANDERSON, 1970). The REL obtained for the Collembola species show two different patterns, the loss being very high for the smaller species, whereas values calculated for *Isotoma viridis* are intermediate between those and *M. morio*, and close to the values given for *Anapothrips secticornis*. The REL of *A. secticornis* from Hardangervidda is low in relation to its size, and this may be due to its low activity.

The coccid *Arctorthezia cataphracta* also has low activity, but the extremely low values for REL may be caused by the inclusion of the respiratory inert wax plates in the dry weight. Individuals of this species weighing less than 0.5 mg have a maximum value of respiratory $Q_{10}$ at 13–18° C, whereas animals weighing more than 0.5 mg show the usual pattern of decreasing $Q_{10}$ with increasing temperature (Table 3). An interpretation of this phenomenon is not possible at present.

Animals with low respiratory $Q_{10}$ are less affected by diurnal and seasonal changes in temperature. In a high mountain area, where temperature amplitudes during the day are very high, a low respiratory $Q_{10}$, which will dampen the metabolic response to temperature fluctuations, may be of considerable adaptive value. RAO and BULLOCK (1954) stated that temperature adaptation in respiratory $Q_{10}$ occurs, and TRIBE and BOWLER (1968) have presented evidence to suggest that the temperature of acclimatization also affects the $Q_{10}$. The opposite view is held by SCHOLANDER et al. (1953). ANDERSON (1970) has given respiratory rates for spiders acclimatized to different temperatures. The $Q_{10}$ values calculated from his results strongly indicate that in spiders they are dependent on the temperature of acclimatization. Low environmental temperatures are also compensated for by relatively higher respiratory rates for given temperatures. This seems to be the case for the spider, *P. palustris*, and the Collembola from Hardangervidda.

*Acknowledgements.* I wish to express my sincere thanks to Professor H. KAURI, leader of the project, for valuable advice and criticism, to A. FJELLBERG for determination of the Collembola species, and to the assistants G. SIMPSON and B. BAKKE.

## References

ANDERSON, J.: Metabolic rates of spiders. Comp. Biochem. Physiol. **33**, 51–72 (1970).

BREYMEYER, A.: Preliminary data for estimating the biological production of wandering spiders. In: PETRUSEWICZ, K. (Ed.): Secondary Productivity of Terrestrial Ecosystems, vol. II, pp. 821–834. Warszawa-Kraków 1967.

HAGSTRUM, D. H.: Physiology of food utilization by the spider *Tarentula kochi* (Araneae: Lycosidae). Ann. Entomol. Soc. Am. **63**, 1305–1308 (1970).

HEALEY, I. N.: The population metabolism of *Onychiurus procampatus* Gisin (Collembola). In: GRAFF, O., SATCHELL, J. E. (Eds.): Progress in Soil Biology, pp. 127–137. Amsterdam: North Holland Publ. Comp. 1967.
ITO, Y.: Preliminary studies on the respiratory energy loss of a spider, *Lycosa pseudoannulata*. Res. Popul. Ecol. **6**, 13–21 (1964).
MIYASHITA, K.: Effects of locomotory activity, temperature and hunger on the respiratory rate of *Lycosa T-insignita* Boes. et Str. (Araneae: Lycosidae). Appl. Ent. Zool. **4**, 105–113 (1969).
PHILLIPSON, J.: Respirometry and the study of energy turnover in natural systems with particular reference to harvest spiders (Phalangiida). Oikos **13**, 311–322 (1962).
PHILLIPSON, J.: The use of respiratory data in estimating annual respiratory metabolism, with particular reference to *Leiobunum rotundum* (Latr.) (Phalangiida). Oikos **14**, 212–223 (1963).
PRIESNER, H.: A monograph of the Thysanoptera of the Egyptian deserts. Publ. Inst. Desért Egypte **13**, 1–549 (1964).
RAO, K. P., BULLOCK, T. H.: $Q_{10}$ as a function of size and habitat temperature in poikilotherms. Am. Naturalist **88**, 33–44 (1954).
ROEDER, K. D.: Insect Physiology. New York: John Wiley and Sons, Inc. 1953.
SCHOLANDER, P. F., FLAGG, W., WALTERS, V., IRVING, L.: Climatic adaptation in arctic and tropical poikilotherms. Physiol. Zool. **26**, 67–92 (1953).
SCHOLANDER, P. F., IVERSEN, O.: New design of volumetric respirometer. Scandinav. J. Clin. Lab. Investigation **10**, 429–431 (1958).
SONESSON, M., WIELGOLASKI, F. E., KALLIO, P.: Description of Fennoscandian tundra ecosystems. In: WIELGOLASKI, F. E. (Ed.): Fennoscandian tundra ecosystems. Part 1: Plants and microorganisms, pp. 3–28. Berlin-Heidelberg-New York: Springer 1975.
TANAKA, M.: The bio-economics of the populations of *Isotoma (Desoria) trispinata* MacGillivray (Collembola; Isotomidae) and *Onychiurus (Protaphorura) sp.* (Collembola; Onychiuridae) in a grassfield. Publ. Amakusa Mar. Biol. Lab. **2**, 51–120 (1970).
TILBROOK, P. J., BLOCK, W.: Oxygen uptake in an Antarctic collembole *Cryptopygus antarcticus*. Oikos **23**, 313–317 (1972).
TRIBE, M. A., BOWLER, K.: Temperature dependence of "standard metabolic rate" in a poikilotherm. Comp. Biochem. Physiol. **25**, 427–436 (1968).
VAN HOOK, R. I., JR.: Energy and nutrient dynamics of spider and orthopteran populations in a grassland ecosystem. Ecol. Monogr. **41**, 1–26 (1971).
ZINKLER, D.: Vergleichende Untersuchungen zur Atmungsphysiologie von Collembolen (Apterygota) und anderen Bodenkleinarthropoden. Z. Vergl. Physiol. **52**, 99–144 (1966).

# Energetics in a Population of Pardosa palustris (L.) (Araneae, Lycosidae) on Hardangervidda

A. L. STEIGEN

The present study is an investigation of the energetics of the spider *Pardosa palustris* (L.) (*Lycosa tarsalis* Thorell) at Hardangervidda, southern Norway. The study was carried out in the field during summer 1972 at an alpine dry-meadow area at Stigstuv, Hardangervidda, described by SONESSON et al. (1975) and SOLHØY (1972). Some meteorological data are shown in Fig. 1 and given by SKARTVEIT et al. (1975). To construct the population energy budget, abundance and lifecycle have been investigated. In the laboratory, respiration and production of feces and silk were quantified, and feeding experiments were carried out at all stages of the life cycle.

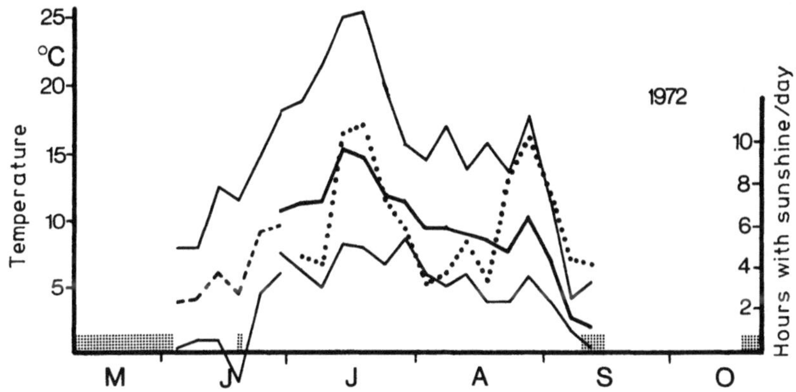


Fig. 1. Fluctuations in temperature at the dry meadow site at Stigstuv in the snow-free period in 1972. Temperatures are pentamax., -min, and -mean at 0 m level. If 0 m values are lacking, 0.1 m air temperature is used. The graph also shows pentadmeans of hours of sunshine/day and periods with snow cover. Solid lines: temp. at 0 m; dashed lines: temp. at 0.1 m; dotted lines: number of hours with sunshine; shaded: period with snow cover

## Material and Methods

Animals were collected qualitatively using pit-fall traps and hand sampling. Animals for respiratory measurements and feeding experiments were chilled down to $+5°$ C and put into separate containers before being transported to the laboratory. Quantitative samples of juvenile spiders were obtained by sifting cut vegetation from ten 0.25 $m^2$ sample areas. The older instars and adults were collected by sampling from an area of 1 × 5 m. Five areas were sampled each time. Samples were taken at noon on sunny days i.e. during periods of high activity in *P. palustris*.

Scholander respirometers (SCHOLANDER and IVERSEN, 1958) were used and measurements carried out in a waterbath, with and accuracy of ±0.2° C (factory specifications). Animals collected for respiratory measurements were acclimatized at 10° C for a minimum of 48 hrs, but usually for 6–9 days. Spiders from feeding experiments were kept at 20° C for 2–3 months before respiratory measurements were taken. All animals tested had consumed food the day prior to the experiments. The photoperiod was light 16 hrs, dark 8 hrs in the laboratory.

All measurements were carried out between 0900 am and 0300 pm. Using animals from feeding experiments, some measurements were taken between 1000 pm and 0300 am to estimate night respiratory rate.

Respiratory rates are given as μl $O_2$ consumed $(mg\ dw)^{-1}\ h^{-1}$ and expressed as an equation. Dry weight of animals from the feeding experiment was estimated from a regression equation obtained by weighing the animals alive and after drying at 105° C for 24 hrs.

When calculating respiratory energy loss (REL) an oxycaloric coefficient of $4.8 \times 10^{-3}$ cal $\mu l^{-1}$ $O_2$ was used.

The respiratory rate/weight relationship equation (HEMMINGSEN, 1960) $y = ax^b$ (y = respiratory rate in μl $O_2$ consumed $animal^{-1}\ h^{-1}$, $x$ = dw, $a$ and $b$ are constants) has been calculated for animals from feeding experiments.

To estimate the amount of feces, excreta and silk produced, animals were put into dram vials where the bottom was covered with calibrated glass discs. The experiment was done at 20° C over 48 hrs. The glass discs were then dried and weighed to find the weight of the ejected material. Feces, excreta and silk were collected from the feeding experiments an caloric content determined.

As it is impossible to separate feces, excreta and silk, they are collectively referred to as ejecta (FU).

Calorific determinations were carried out using a Phillipson microbomb calorimeter.

During feeding experiments, the spiders were kept in separate tubes (PECK and WHITCOMB, 1967) at 20° C and photoperiod light 16 hrs, dark 8 hrs. A stopper of plastic foam was put into one end of the tube, and a stopper of moist cotton into the other (PARRY, 1954). Various sizes of containers were used to increase prey density.

First instar spiderlings were fed by cannibalism (HALLANDER, 1973). The other instars were given Drosophila imagines as prey. To avoid overweight (KESSLER, 1973) animals younger than fifth instar were given five flies per day. Spiders at fifth and later instars were given 10 flies daily after ecdysis until a reduction of consumption rate occurred. Females producing gametes were also provided with 10 flies a day.

Individual energy budgets for a "standard" male and female were calculated from the feeding and ejecta experiments. The sex of spiderlings younger than fifth instar cannot be determined, thus they are not separated according to sex in the budgets.

A week after ecdysis to adult, males and females were put together for mating. This was done to avoid biological effects from lack of copulation (cf. EDGAR, 1971 b; EASON, 1969).

A one-year energy budget for the population was calculated based on population dynamics for 1972.

## Results and Discussion

**Field Investigations.** Instars were determined by measuring carapace width (HAGSTRUM, 1971) and from morphological characters. The first instar is defined as the instar emerging from the egg sac. There were no significant differences in carapace width for the two sexes (cf. EASON, 1969), and both sexes have eight instars.

Table 1. Weight and caloric content of egg sacs and egg-sac silk in *P. palustris* collected at Hardangervidda (SD), $n$ = number of determinations

| | $n$ | Weight mg dw | cal (g ash-free dw)$^{-1}$ | cal. content egg sac silk |
|---|---|---|---|---|
| 1. egg sacs | 22 | 5.25 (1.20) | 6542 (244) | 34.35 |
| 2. egg sacs | 10 | 3.18 (0.50) | 6212 (190) | 19.75 |
| Silk of egg sac | 3 | 0.54 (0.07) | 5501 (137) | 2.97 |

Table 2. Number, weight and caloric content of eggs from egg sacs of *P. palustris* collected at Hardangervidda (SD)

| | Number of eggs | Weight of eggs mg dw | cal egg$^{-1}$ |
|---|---|---|---|
| 1. egg sacs | 50.75 (12.59) | 0.10 | 0.68 |
| 2. egg sacs | 27.78 ( 4.21) | 0.11 | 0.71 |

Table 3. Weight and caloric content of different instars of *P. palustris* collected at Hardangervidda (SD)

| Instar | $n$ | Weight mg dw | $n$ | cal (g ash-free dw)$^{-1}$ | cal animal$^{-1}$ ash-free |
|---|---|---|---|---|---|
| 1. | 50 | 0.13 (0.01) | 5 | 6218 (202) | 0.81 (0.06) |
| 2. | 20 | 0.19 (0.02) | 5 | 6113 (241) | 1.16 (0.12) |
| 3. | 10 | 0.49 (0.05) | 5 | 6015 (235) | 2.94 (0.30) |
| 4. | 10 | 1.09 (0.08) | 5 | 6033 (201) | 6.58 (0.48) |
| 5. ♀♀ | 10 | 3.41 (0.19) | 5 | 5980 (217) | 20.39 (1.14) |
| 6. ♀♀ | 10 | 4.01 (0.31) | 5 | 5977 (251) | 23.96 (1.85) |
| 7. ♀♀ | 10 | 4.63 (0.54) | 5 | 5963 (245) | 27.61 (3.22) |
| 8. ♀♀ | 10 | 5.37 (0.97) | 5 | 5948 (193) | 31.94 (5.77) |
| 5. ♂♂ | 10 | 1.42 (0.14) | 5 | 5971 (202) | 8.48 (0.84) |
| 6. ♂♂ | 10 | 2.77 (0,29) | 5 | 5863 (254) | 16.24 (1.70) |
| 7. ♂♂ | 10 | 3.21 (0.45) | 5 | 5931 (205) | 19.04 (2.67) |
| 8. ♂♂ | 10 | 3.97 (0.43) | 5 | 5761 (249) | 22.87 (2.48) |

In Tables 1–3 weight and caloric content of egg sacs, and different instars and sexes are presented. For adult females the values represent individuals carrying egg sacs.

Where live weight has been transformed to dry weight, the following equation has been used:

$$y = 0.278x + 0.072\,, \qquad r = 0.96,\ p < 0.001$$

$y$ = dry weight, $x$ = live weight.

Size distribution during the 1972 season is shown in Fig. 2, and presence of different instars is shown in Fig. 3. Limits between the year-classes are indicated by roman figures and show hibernation periods during the life-cycle. *P. palustris*

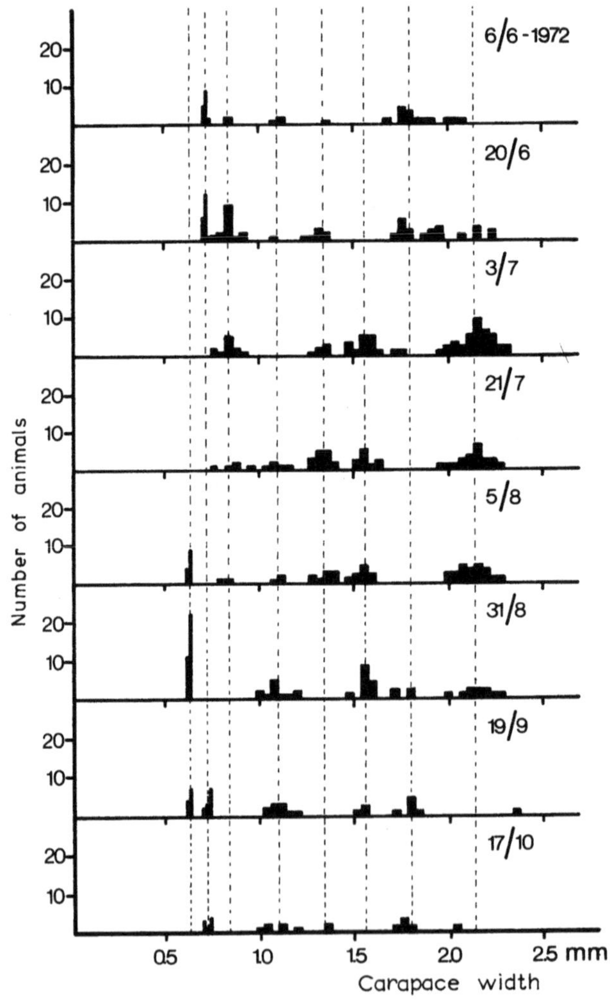


Fig. 2. Graph showing the variations in carapace width of *P. palustris* during summer 1972. Average values for different instars are indicated

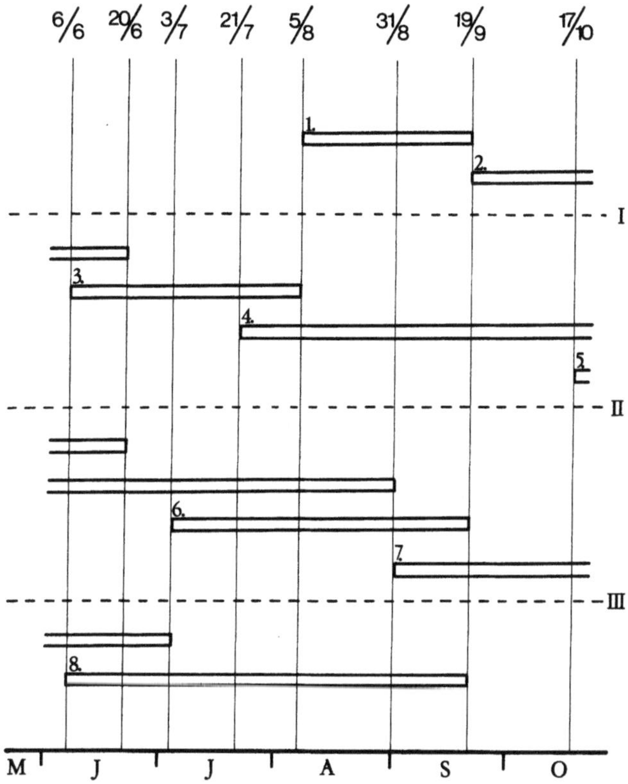


Fig. 3. Qualitative graph of the life cycle of *P. palustris* at Stigstuv. Roman figures indicate the hibernations

overwinters three times before becoming adult, on second, fourth or fith instar, and as subadult. EDGAR (1971a) has shown that *Pardosa lugubris* in central Scotland overwinters twice.

Abundance of *P. palustris* is shown in Tables 4 and 5. The species has a very low abundance in the investigated area, and EDGAR (1971a) notes values ca. 20 times higher for *P. lugubris* in central parts of Scotland.

The survival curve for *P. palustris* on Hardangervidda is a concave, type III curve (DEEVEY, 1947), with greatest mortality in the smaller instars.

*P. palustris* females will construct two egg sacs during summer. The second egg sac is produced after the first week of August. Cocoons collected in September/October show high mortality and contain embryos at all stages of development. Laboratory experiments have shown that at a constant temperature of 10° C, embroynic development ceases and the embryos die. After the production of the second batch, the air temperature exceeds 10° C only for short periods at midday, and in spite of sunning the cocoon, it is unlikely that enough heat is available to complete development (cf. Fig. 1). The presence of varying embryological stages in the second egg sacs is due to a temperature gradient in the cocoon. With this

Table 4. Abundance of different instars of *P. palustris* at Hardangervidda during the season 1972. Estimates of the first four instars are based on sifting $10 \times 0.25\ m^2$ areas of cut vegetation, the others on hand-collecting from $1 \times 5$ m areas

| Date | 1 | | | 2 | | | 3 | | | 4 | | |
|---|---|---|---|---|---|---|---|---|---|---|---|---|
| | N | N/m² | SD | N | N/m² | SD | N | N/m² | SD | N | N/m² | SD |
| June 20 | | 0 | | 11 | 4.40 | 3.50 | 2 | 0.80 | 1.68 | | 0 | |
| July 4 | | 0 | | | 0 | | 4 | 1.60 | 2.80 | | 0 | |
| July 22 | | 0 | | | 0 | | 1 | 0.40 | 1.26 | | 0 | |
| Aug. 5 | 11 | 4.40 | 7.65 | | 0 | | 1 | 0.40 | 1.26 | 1 | 0.40 | 1.26 |
| Aug. 31 | 21 | 8.40 | 10.57 | | 0 | | | 0 | | 1 | 0.40 | 1.26 |
| Sept. 19 | 9 | 3.60 | 4.40 | 3 | 1.20 | 1.93 | | 0 | | 1 | 0.40 | 1.26 |
| Max. abund. | | 8.40 | | | 4.40 | | | 1.60 | | | 0.40 | |

| Date | 5 | | | 6 | | | 7 | | | 8 | | | Total |
|---|---|---|---|---|---|---|---|---|---|---|---|---|---|
| | N | N/m² | SD | N | N/m² | SD | N | N/m² | SD | N | N/m² | SD | |
| June 20 | 3 | 0 | | | 0 | | 11 | 0.44 | 0.18 | 4 | 0.16 | 0.17 | 28 |
| July 4 | 3 | 0.12 | 0.11 | 6 | 0.24 | 0.09 | 1 | 0.04 | 0.09 | 10 | 0.44 | 0.30 | 24 |
| July 22 | 15 | 0.60 | 0.37 | 14 | 0.56 | 0.43 | | 0 | | 7 | 0.28 | 0.23 | 37 |
| Aug. 5 | 8 | 0.32 | 0.30 | 6 | 0.24 | 0.26 | | 0 | | 5 | 0.20 | 0.14 | 32 |
| Aug. 31 | | 0 | | 3 | 0.12 | 0.18 | 3 | 0.12 | 0.11 | 3 | 0.12 | 0.11 | 31 |
| Sept. 19 | | 0 | | | 0 | | 6 | 0.24 | 0.22 | | 0 | | 19 |
| Max. abund. | | 0.60 | | | 0.56 | | | 0.44 | | | 0.40 | | Total 171 |

Table 5. Abundance of females/males of *P. palustris* at 5.–8. instar

| Date / Instar | 5 | 6 | 7 | 8 |
|---|---|---|---|---|
| June 20 | 0 | 0 | 0.24/0.20 | 0.04/0.12 |
| July 4 | 0.08/0.04 | 0.12/0.12 | 0/0.04 | 0.20/0.20 |
| July 22 | 0.24/0.36 | 0.36/0.20 | 0 | 0.12/0.16 |
| Aug. 5 | 0.20/0.12 | 0.08/0.16 | 0 | 0.16/0.04 |
| Aug. 31 | 0 | 0.12/0 | 0/0.12 | 0.12/0 |
| Sept. 19 | 0 | 0 | 0.08/0.16 | 0 |
| Max. abund. ♀/♂ | 0.24/0.36 | 0.36/0.20 | 0.24/0.20 | 0.20/0.20 |

Table 6. Fecundity and calculated natality in *P. palustris* based on 1. egg sacs (SD)

| Year | *n* | Number of eggs | *n* | % of embryos at an earlier stage | Calculated natality female$^{-1}$ | Calculated natality m$^{-2}$yr$^{-1}$ |
|---|---|---|---|---|---|---|
| 1972 | 45 | 48.35 (13.93) | 21 | 1.69 (2.26) | 47.53 (13.69) | 9.51 (2.74) |

background the author assumes that only the first egg sacs will contribute to the recruiting in the population. In Table 6, the fecundity and calculated natality of the population are shown. Egg sac parasitism has not been observed in the population. Kessler and Fokkinga (1973) investigated parasitism in 8 *Pardosa* species, but observed no parasitism in *P. palustris*.

The calculated natality in 1972: 9.51 ind. m$^{-2}$ is in concordance with the maximal observed abundance for first instar spiderlings this year: 8.40 ind. m$^{-2}$.

**Respiratory Measurements.** In Table 7 the respiratory rates and REL for animals collected in the field and from feeding experiments are shown. Animals from feeding experiments reveal a reduction in respiratory rate. This may be due to loss of cold-resistance (Scholander et al., 1953), but may also be an effect of acclimatization to constant temperature in the laboratory population, and a resultingly reduced amplitude in diurnal activity (Anderson, 1970).

Table 7. Compilation of respiratory rates in *P. palustris* from the field and feeding experiments

| | Temp. $t$° C | *n* | Weight mg dw | Respiratory rate µl $O_2$ (mg dw)$^{-1}$ h$^{-1}$ (SD) | REL animal$^{-1}$ day$^{-1}$ in calories |
|---|---|---|---|---|---|
| Animals from the field | 18 | 8 | 4.69 | 2.30 (0.21) | 1.25 |
| Animals from feeding experiments, day rates | 20 | 12 | 4.97 | 1.33 (0.11) | 0.76 |
| Animals from feeding experiments, night rates | 20 | 12 | 4.97 | 1.05 (0.09) | 0.61 |

Respiratory measurements were carried out within a wide weight range in the animals from feeding experiments. A regression equation was calculated for the respiratory rate at 20° C:

$$y = \frac{1.91}{x^{0.245}} \quad r = -0.83, \quad p < 0.001$$

$y = \mu l\ O_2$ consumed (mg dw)$^{-1}$ h$^{-1}$, $x$ = dw in mg.

A reduction in respiratory rate according to reduced activity has been shown for other spiders (cf. ANDERSON, 1970; HAGSTRUM, 1970b).

The equation $y = ax^b$ expresses the relationship between metabolic rate and body weight. In *P. palustris* the equation at 20° C becomes:

$$y = 1.91\, x^{0.75}$$

$y = \mu l\ O_2$ cons. animal$^{-1}$ h$^{-1}$, $x$ = dw in mg. The regression coefficient is equal to HEMMINGSEN'S (1960) universal $b$, and according to BERTALANFFY's metabolic types, *P. palustris* belongs to the intermediate type (BERTALANFFY, (1957).

**Ejecta Production.** The production of ejecta at 20° is expressed by the equation

$$y = \frac{13.40}{1.14^x} x\ 10^{-3} \quad r = -0.76, \ p < 0.02$$

$y$ = mg ejecta produced (mg dw)$^{-1}$ day$^{-1}$, $x$ = dw in mg.

This will give an ejecta production of 0.012 and 0.04 mg (mg dw)$^{-1}$ day$^{-1}$ for animals weighting 1 and 10 mg respectively. Caloric content of the ejecta product is shown in Table 9 and is 23% lower than the caloric content of the ingested prey.

**Individual Energy Budgets.** The energy budget for a "standard" male and female from egg sac emergence to death is shown in Table 8. As first instar spiderlings were fed by cannibalism, individual treatment was excluded, and the results are consequently mean values.

Ecdysis to first instar takes place in the egg sac just before emergence, which is used as reference for the start of the instar.

Caloric equivalents used when calculating the energy budgets are shown in Table 8.

The results of energy budgets of juvenile instars are given in Table 9. During the fifth instar the female of *P. palustris* undergoes a very large increase in body

Table 8. Caloric equivalents used when calculating the energy budgets of *P. palustris*

| | Number of determination | cal (mg ash-free)$^{-1}$ (SD) |
|---|---|---|
| *D. melanogaster* | 8 | 5755 (164) |
| *P. palustris* | 21 | 6096 (199) |
| Exuvia | 10 | 5336 (241) |
| Egg sacs | 10 | 6473 (212) |
| Feces, excreta, silk etc. | 1 | 5003 |
| Food: unchewed, but empty flies | 5 | 5946 (187) |
| chewed flies | 5 | 5156 (109) |

weight. During this instar, which lasts for 44 days, weight increase is close to 300%. Males undergo an extensive growth at the sixth instar. Data for *P. lugubris* indicate that they have a large weight increase during the fourth instar (EDGAR, 1971 b).

EDGAR (1971 b) exposed juvenile *P. lugubris* to a simulated winter period at 4° C. In the present study this has not been done. The juvenile *P. palustris* molted to successive instars without problems, and this indicated that no obligatory diapause occurs in the life cycle of *P. palustris* at Hardangervidda. As the consumption of energy during winter is very low (EDGAR, 1971 b), the energy consumed during hibernation entails only a small fraction of the total energy budget for the population.

The adult instar for females is divided into five periods (Table 9):

A. From molting to adult until the first egg sac is produced.
B. The female carries the first egg sac or first batch spiderlings on her back.
C. From the first batch spiderlings leaving the female until the second egg sac is constructed.
D. The female carries the second egg sac and successively second-batch spiderlings on her back.
E. 10 days after the second batch leaves her back.

The experiment ended after E.

Before constructing the egg sacs, the female shows a great weight increase, but caloric content of the egg sacs is greater than the increase in caloric content of the female. Consequently, the weight of the female is lower after cocoon making than at the beginning of the instar. Some of this weight loss is immediately compensated for by water uptake.

During the period with eggs and spiderlings on her back, the weight of the female decreases, and the utilization of killed prey is reduced. Once the spiderlings have left her back, the female increases her consumption drastically.

All the adult males lived during the 60 days they were fed after molting. In this period only 8% of the consumed food went into production.

**Trophic Indices of the Instars.** Three different trophic indices or "efficiencies" have been calculated for each instar (Table 9).

1. Production/assimilation (P/C), "ecological growth efficiency" (ODUM, 1971).
2. Respiratory energy loss/assimilation (REL/A), i.e. fraction of energy available for metabolism used for respiration.
3. Consumption/food killed (C/FK), "ingestion efficiency" (EDGAR, 1971 b).

P/C decreases with increasing age in *P. palustris*, a general trend in animal metabolism (PHILLIPSON, 1966). The index has a higher value in the "growth instars" (♀♀ fifth instar, ♂♂ sixth instar). REL/A increases with age, but has somewhat lower value in periods of high production. Females carrying egg sacs/spiderlings have a negative production. C/FK is rather low during the first instar and is highest in periods of high production. NAKAMURA (1968) found good correlation between size and capacity of the gut in *Pardosa pseudoannulata* (Bö. and Str.), and it is obvious that capacity of the gut limits consumption in the first and second instars of *P. palustris*. However, it is likely that in the field the size of the prey will be better matched to the size of the spider, and consequently the C/FK-index will be higher.

Table 9. Energy budget of a "standard" male and female *P. palustris*

| Instar | *n* | Duration of instar days (S.D) | Weight at the beginning of the instar mg dw (SD) | Weight increase during the instar mg dw (SD) | Food killed cal. (SD) | Food remains cal. (SD) | Food consumed cal. (SD) |
|---|---|---|---|---|---|---|---|
| 1 | 21 | > 12( 4)[a] | 0.131(0.01) | 0.045(0.01) | 3.30 | 2.74 | 0.56 |
| 2 | 12 | 26( 3) | 0.176(0.01) | 0.265(0.03) | 6.81(0.71) | 3.24(0.31) | 3.56(0.41) |
| 3 | 10 | 25( 3) | 0.441(0.01) | 0.329(0.03) | 7.99(1.23) | 2.81(0.36) | 5.19(0.71) |
| 4 | 6 | 36( 7) | 0.770(0.03) | 0.463(0.11) | 13.51(1.17) | 3.19(0.47) | 10.32(0.93) |
| Total | | 99 | | 1.102 | 31.61 | 11.98 | 19.63 |
| 5♀♀ | 10 | 44( 6) | 1.233(0.09) | 2.444(0.56) | 37.19(2.31) | 4.61(0.90) | 32.58(2.97) |
| 6♀♀ | 10 | 28( 6) | 3.673(0.23) | 0.540(0.31) | 24.32(2.18) | 5.61(0.77) | 18.71(1.54) |
| 7♀♀ | 14 | 86(18) | 4.213(0.62) | 1.289(0.47) | 82.17(9.91) | 16.43(2.03) | 65.74(8.78) |
| Total | | 158 | | 4.273 | 143.68 | 26.65 | 117.03 |
| 5♂♂ | 8 | 35( 9) | 1.233(0.10) | 0.452(0.21) | 15.33(2.28) | 3.64(0.83) | 11.69(1.20) |
| 6♂♂ | 5 | 39(16) | 1.685(0.26) | 1.346(0.44) | 27.15(3.18) | 3.34(0.81) | 23.81(2.03) |
| 7♂♂ | 4 | 64(12) | 3.031(0.29) | 1.517(0.23) | 52.18(6.63) | 11.42(2.07) | 40.77(5.37) |
| Total | | 138 | | 3.315 | 94.66 | 15.40 | 76.27 |
| | Adult | ♀♀ | | | | | |
| A | 10 | 32( 4) | 5.502(0.89) | 5.12 (0.97) | 68.89(8.36) | 11.36(2.61) | 57.53(7.19) |
| B | 10 | 43( 7) | 5.372(0.99) | −0.14 (0.21) | 42.39(5.51) | 11.44(2.18) | 30.95(4.03) |
| C | 10 | 26( 6) | 5.232(1.03) | 3.03 (0.65) | 43.67(6.73) | 6.12(2.16) | 37.57(3.81) |
| D | 9 | 41( 8) | 5.027(1.12) | −0.19 (0.20) | 44.70(5.16) | 12.96(2.39) | 31.74(2.10) |
| E | 9 | 10 | 4.837(0.99) | 1.03 (0.72) | 17.57(2.12) | 4.22(1.03) | 13.35(1.26) |
| Total | | 152 | | 8.85 | 217.22 | 46.10 | 171.14 |
| | Adult | ♂♂ | | | | | |
| | 4 | 60 | 4.590(0.42) | 1.06 (0.39) | 101.86(8.09) | 24.45(2.55) | 77.41(6.02) |

[a] Beginning of instar set to egg sac emergence.

**Annual Population Energy Budget, Hardangervidda.** Many problems arise when extrapolating results from laboratory to field conditions. Petrusewicz and MacFadyen (1970) have discussed some sources of error. The main requirement must be that energy budgets of instars in the laboratory population do not differ too much from development in the field.

Laboratory values for weight and caloric content are in good agreement with field results. Exceptions are males of subadult and adult instars. This may be caused by restricted activity in the cages with corresponding over-weight, or perhaps a consequence of a small number of experimental animals. Weight and caloric content of egg sacs agree with the values from field samples. This indicates that the estimated production from the feeding experiment reflects the field situation.

Respiratory rates calculated from feeding experiments are ca. 40% lower than rates obtained from field respiratory measurements. Diurnal variations (Table 7) will contribute to this discrepancy. Variations in feeding rates of *P. palustris* during the experiments were also found to affect respiratory rate (cf. Miyashita, 1969). Table 10 shows that the constant $b$ in the equation $y = ax^b$ is obtained from feeding experiments at 20° C. Explanations in text

| Cal. content at the beginning of the instar cal. (SD) | Increase in cal. content during the instar cal. (SD) | Exuvia cal. (SD) | Assimilated during the instar cal. (SD) | REL during the instar cal. (SD) | Ejecta cal. (SD) | Production/consumption | REL/Assimilation | Consumption/Food killed |
|---|---|---|---|---|---|---|---|---|
| 0.81(0.06) | 0.26(0.09) | 0.02[b] | 0.44 | 0.18 | 0.12 | 0.46 | 0.41 | 0.17 |
| 1.07(0.06) | 1.61(0.18) | 0.07(0.02) | 3.17(0.27) | 1.56(0.21) | 0.40(0.03) | 0.45 | 0.49 | 0.52 |
| 2.68(0.06) | 1.99(0.18) | 0.19(0.07) | 4.20(0.39) | 2.14(0.46) | 0.99(0.06) | 0.38 | 0.51 | 0.65 |
| 4.67(0.18) | 2.81(0.66) | 0.32(0.12) | 8.21(0.91) | 5.21(0.67) | 2.11(0.18) | 0.27 | 0.63 | 0.76 |
| | 6.67 | 0.60 | 15.99 | 9.09 | 3.62 | 0.34 | 0.57 | 0.62 |
| 7.48(0.55) | 14.81(3.39) | 0.62(0.18) | 26.26(2.43) | 11.13(1.77) | 6.32(0.71) | 0.45 | 0.42 | 0.88 |
| 22.29(1.40) | 3.28(1.88) | 2.38(0.56) | 14.27(1.03) | 10.37(1.51) | 4.44(0.53) | 0.18 | 0.73 | 0.77 |
| 25.57(3.76) | 7.82(2.87) | 2.58(0.42) | 51.27(3.71) | 41.07(5.33) | 14.47(2.71) | 0.12 | 0.80 | 0.80 |
| | 25.91 | 5.58 | 91.80 | 62.74 | 25.23 | 0.22 | 0.68 | 0.81 |
| 7.48(0.59) | 2.74(1.27) | 0.62(0.18) | 8.87(1.33) | 5.81(0.70) | 2.82(0.54) | 0.25 | 0.65 | 0.76 |
| 10.23(1.58) | 8.17(2.67) | 0.89(0.21) | 19.43(1.81) | 10.64(1.16) | 4.38(0.63) | 0.34 | 0.54 | 0.88 |
| 18.40(1.76) | 9.21(1.39) | 1.79(0.39) | 31.01(2.47) | 20.91(2.37) | 9.76(1.31) | 0.23 | 0.67 | 0.78 |
| | 20.12 | 3.30 | 59.31 | 37.36 | 16.96 | 0.26 | 0.63 | 0.81 |
| | | Cal. content of egg sacs | | | | | | |
| 33.39(5.40) | 31.07(5.89) | 36.25(5.29) | 51.60(6.79) | 20.53(2.37) | 5.93(0.71) | 0.54 | 0.39 | 0.86 |
| 32.60(6.01) | − 0.85(1.27) | | 23.33(2.41) | 24.18(2.06) | 7.62(0.94) | −0.02 | 1.03 | 0.73 |
| 31.75(6.25) | 18.39(3.94) | 21.17(3.61) | 32.70(2.16) | 14.31(2.21) | 4.87(0.50) | 0.48 | 0.43 | 0.86 |
| 30.51(6.79) | − 1.15(1.21) | | 24.57(1.99) | 25.72(2.91) | 7.17(0.61) | −0.03 | 1.04 | 0.71 |
| 29.36(6.01) | 6.25(4.37) | | 11.54(0.91) | 5.29(0.73) | 1.81(0.38) | 0.46 | 0.45 | 0.76 |
| | 53.71 | 57.42 | 143.74 | 90.03 | 27.40 | 0.31 | 0.63 | 0.79 |
| 27.86(2.55) | 6.43(2.37) | | 63.64(4.69) | 58.50(5.21) | 13.78(1.55) | 0.08 | 0.92 | 0.76 |

[b] Estimated from collective weighing.

approximately identical for males and females, indicating that the discrepancy is equal in all weight classes of *P. palustris*.

The feeding experiments were carried out at a constant temperature of 20° C. This is an artificial situation to the animals, but investigations performed by REMMERT and WÜNDERLING (1970) have shown that at this temperature the rate of development will not necessarily be affected by constant temperature. Further-

Table 10. The calculated equation $y = ax^b$ for males and females of *P. palustris* estimated from trophic equations and by respiratory measurements at 20° C. $y$ = respiratory rate; $x$ = dw in mg; $a, b$ = constants

| | | |
|---|---|---|
| Calculated from trophic equations: | | |
| Males: | $y = 1.16\ x^{0.71}$ | $r = 0.99 < p = 0.001$ |
| Females: | $y = 1.18\ x^{0.75}$ | $r = 0.99 < p = 0.001$ |
| Respiratory measurements: | | |
| Males/females | $y = 1.91\ x^{0.75}$ | |

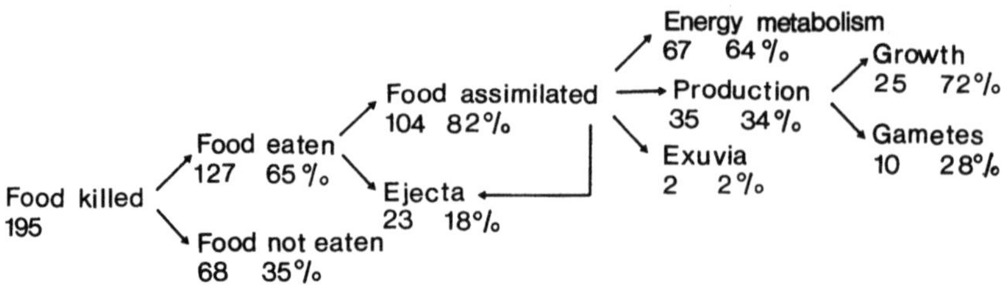


Fig. 4. Annual energy budget of *P. palustris* at a dry meadow site at Stigstuv on Hardangervidda in cal $m^{-2}$ $year^{-1}$. Values are estimated from the population parameters in 1972

more, experiments carried out with fluctuating temperatures may be impossible to reproduce, and comparative analyses may become difficult.

Annual energy budget has been estimated by multiplying the individual energy parameters for each instar with the maximum abundance of that instar.

Cannibalism has not been observed in the population. Diptera and Collembola seem to be the most important prey. In other species, the amount of cannibalism may be rather high (EDGAR, 1969; HALLANDER, 1973). HAGSTRUM (1970a) states that cannibalism is density-dependent in spiders. As the abundance of *P. palustris* on Hardangervidda is very low, there is a good reason to assume that cannibalism is insignificant.

All mortality during feeding experiments occurred in connection with molting, and it is assumed that this is also the case in the field. The contribution of exuvia in the budget is calculated from the abundance of the succeeding instar. In case of mortality, all energy will pass to another trophic level.

According to field observations the average duration of the adult instar of males is set to 60 days. It is assumed that no spiderlings will emerge from the second egg sac, and hence, the period E in the individual energy budget has been omitted when calculating the annual energy budget of adult females.

The annual energy budget for male and female instars in the population in 1972 is shown in Table 11a, b. 2., sub-adult and adult instars render the main contribution to the population.

Total annual energy budget for the population of *P. palustris* at Hardangervidda is shown in Fig. 4. If this budget is compared to the budget of *P. lugubris* in central Scotland (EDGAR, 1971b) the most striking difference is that the detritus component in the budget of *P. palustris* is much greater. This is caused by the low C/FK ratio in the two first instars in *P. palustris*, and the ejecta component which EDGAR does not take into account. Respiration is a little lower and production a little higher in *P. palustris*. Part of the difference may be caused by the lack of a feces estimate in *P. lugubris*, which, in the next step in calculating the energy budget, will act as a surplus in respiration. The fractions of growth and gametes agree in the two budgets.

**Trophic Indices for the Population.** The most commonly used efficiency indices (WIEGERT, 1964), A/C, P/A, and P/C "ecological efficiency" are calculated for the population. In practice, owing to the method of feeding, it is impossible to compare directly the ecological efficiency of a spider with the corresponding index in

Table 11a. Contribution of female instars of *P. palustris* to the total annual energy budget (expressed as cal $m^{-2}$ $yr^{-1}$). Biomass$_1$ (Biom.$_1$) and biomass$_2$ (Biom.$_2$) refers to biomass at beginning and end of instar. Respiration (REL) to surroundings (surr.)

| Instar | From preceding trophic level | | Own trophic level | | | | | To detritus | | | | | To surr. |
|---|---|---|---|---|---|---|---|---|---|---|---|---|---|
| | Food killed | Food cons. | Biom.$_1$ | Biom.$_2$ | Prod. | Ass. | 1.egg sac | Food remains | Ejecta | Exuvia | Egg sacs | Dead animals | REL |
| 1[a] | 13.86 | 2.35 | 3.40 | 4.49 | 1.10 | 1.85 | | 11.51 | 0.51 | 0.05 | | 2.14 | 0.76 |
| 2[a] | 14.98 | 7.83 | 2.36 | 5.90 | 3.54 | 6.98 | | 7.13 | 0.88 | 0.06 | | 3.75 | 3.43 |
| 3[a] | 6.39 | 4.15 | 2.15 | 3.74 | 1.59 | 3.36 | | 2.25 | 0.79 | 0.04 | | 2.80 | 1.71 |
| 4[a] | 2.70 | 2.07 | 0.93 | 1.50 | 0.56 | 1.64 | | 0.64 | 0.42 | 0.06 | | 0.15 | 1.04 |
| 5 | 8.93 | 7.82 | 1.80 | 5.35 | 3.55 | 6.30 | | 1.11 | 1.51 | 0.22 | | 0 | 2.67 |
| 6 | 8.76 | 6.74 | 8.02 | 9.21 | 1.18 | 5.13 | | 2.02 | 1.60 | 0.57 | | 3.07 | 3.73 |
| 7 | 19.72 | 15.78 | 6.14 | 8.01 | 1.88 | 12.30 | | 3.94 | 3.47 | 0.52 | | 1.33 | 9.86 |
| Adult | 39.93 | 31.56 | 6.68 | 5.87 | 9.49 | 26.44 | 6.66 | 8.38 | 5.11 | 0 | 4.82 | 5.87 | 16.95 |
| Total | 115.27 | 78.30 | 31.48 | 44.07 | 22.89 | 64.00 | 6.66 | 36.98 | 14.29 | 1.52 | 4.82 | 19.11 | 40.15 |

[a] As the sex of the males and females at the smaller instars cannot be determined, each of the sexes are considered to contribute with 50% of the budget.

Table 11b. Contribution of male instars of *P. palustris* to the total annual energy budget (expressed as cal $m^{-2}$ $yr^{-1}$). Biomass$_1$ (Biom.$_1$) and biomass$_2$ (Biom.$_2$) refers to biomass at beginning and end of instar. Respiration (REL) to surroundings (surr.)

| Instar | From preceding trophic level | | Own trophic level | | | | To detritus | | | | To surr. |
|---|---|---|---|---|---|---|---|---|---|---|---|
| | Food killed | Food cons. | Biom.$_1$ | Biom.$_2$ | Prod. | Ass. | Food remains | Ejecta | Exuvia | Dead animals | REL |
| 1[a] | 13.86 | 2.35 | 3.40 | 4.49 | 1.10 | 1.85 | 11.51 | 0.51 | 0.05 | 2.14 | 0.76 |
| 2[a] | 14.98 | 7.83 | 2.36 | 5.90 | 3.54 | 6.98 | 7.13 | 0.88 | 0.06 | 3.75 | 3.43 |
| 3[a] | 6.39 | 4.15 | 2.15 | 3.74 | 1.59 | 3.36 | 2.25 | 0.79 | 0.04 | 2.80 | 1.71 |
| 4[a] | 2.70 | 2.07 | 0.93 | 1.50 | 0.56 | 1.64 | 0.64 | 0.42 | 0.06 | 0.15 | 1.04 |
| 5 | 5.52 | 4.21 | 2.69 | 3.68 | 0.99 | 3.19 | 1.31 | 1.02 | 0.12 | 1.63 | 2.09 |
| 6 | 5.43 | 4.76 | 2.05 | 3.68 | 1.63 | 3.88 | 0.67 | 0.88 | 0.18 | 0 | 2.13 |
| 7 | 10.44 | 8.15 | 3.68 | 5.52 | 1.84 | 6.20 | 2.28 | 1.95 | 0.36 | 0 | 4.18 |
| Adult | 20.37 | 15.48 | 5.57 | 6.86 | 1.29 | 12.73 | 4.89 | 2.75 | 0 | 6.86 | 11.70 |
| Total | 79.69 | 49.00 | 22.83 | 35.37 | 12.54 | 39.83 | 30.68 | 9.20 | 0.87 | 17.33 | 27.04 |

[a] As the sex of the males and females at the smaller instars cannot be determined, each of the sexes are considered to contribute with 50% of the budget.

Table 12. Trophic indices for the population of *P. palustris* at Stigstuv, Hardangervidda

| | |
|---|---|
| 1. Assimilation/consumption | 0.82 |
| 2. Production/assimilation | 0.34 |
| 3. Production/consumption, ecological efficiency | 0.28 |
| 4. Production/food killed, trophic efficiency | 0.18 |

another predator. Remains of prey may be considered a feces component in spiders, and it may therefore be reasonable to estimate the ecological efficiency from the caloric content of the prey removed from the prey's trophic level. This procedure however, may cause confusion. Consequently, an index of "trophic efficiency" is defined: production/food killed (P/FK), showing how efficiently calories removed from the prey level are utilized for production at the predator level. The indices are given in Table 12.

VAN HOOK (1971) notes an A/C index of 0.90 in *Pardosa punctulata* (Hentz), which is a little higher than in *P. palustris*. In herbi- and omnivores, VAN HOOK calculated indices of 0.54 and 0.41 respectively. HINTON (1971) obtained an A/C index of 0.41 in the homoptera *Neophilaenus lineatus* (L.). Other data from the literature show that A/C is higher in predators than in herbi- and omnivores.

The P/A index in *P. palustris* is close to values obtained for two species of the grasshopper *Melanoplus* (WIEGERT, 1965), and for *Neophilaenus lineatus* (HINTON, 1971). Calculating P/A for *P. punctulata* reveals an index of 0.30 (VAN HOOK, 1971).

Ecological efficiency of *P. palustris* is close to the value for *P. punctulata* (0.33) presented by VAN HOOK (1971), and calculating the index from the data for *P. lugubris* (EDGAR, 1971b) gives a value of 0.26. KAJAK (1967) noted an P/C index between 0.05 and 0.30 for web-building spiders. For herbi- and omnivores VAN HOOK (1971) obtained values of 0.26 and 0.13 respectively, and HINTON (1971) reports 0.15 in *N. lineatus*.

Trophic efficiency for *P. palustris*, 0.18, is a little lower than the calculated index, 0.21, in *P. lugubris* (EDGAR, 1971b).

*Acknowledgements.* I wish to thank Professor H. KAURI, for valuable advice during the study and criticism of the manuscript. S. J. NÆSS helped me with the caloric determinations. The investigations were carried out at the Museum of Zoology, University of Bergen. Financial support was granted by the Norwegian IBP.

## References

ANDERSON, J. F.: Metabolic rates of spiders (Arach, Aran.) Comp. Biochem. Physiol. **33**, 51–72 (1970).

BERTALANFFY, L.: Quantitative laws in metabolism and growth. Quart. Rev. Biol. **32**, 217—231 (1957).

DEEVEY, E. S.: Life tables for natural populations of animals. Quart. Rev. Biol. **22**, 283–314 (1947).

EASON, R. R.: Life-history and behavior of *Pardosa lapidicina* Emerton (Arach., Aran., Lycosidae). J. Kansas Entomol. Soc. **42**, 339–360 (1969).

EDGAR, W. D.: Prey and predators of the Wolf spider *Lycosa lugubris*. J. Zool. Lond. **159**, 405–411 (1969).

EDGAR, W. D.: The life cycle, abundance and seasonal movement of the wolf spider *Pardosa lugubris* in Central Scotland. J. Anim. Ecol. **40**, 303–322 (1971 a).

EDGAR, W. D.: Aspects of the ecological energetics of the Wolf spider *Pardosa* (Lycosa) *lugubris* (Walckenaer). Oecologia **7**, 136–154 (1971 b).

HAGSTRUM, D. W.: Ecological energetics of the spider *Tarentula kochi* (Araneae: Lycosidae). Ann. Entomol. Soc. Am. **63**, 1297–1304 (1970 a).

HAGSTRUM, D. W.: Physiology of food utilization by the spider *Tarentula kochi* (Araneae: Lycosidae). Ann. Entomol. Soc. Am. **63**, 1305–1308 (1970 b).

HAGSTRUM, D. W.: Carapace width as a tool for evualating the rate of development of spiders in the laboratory and the field. Ann. Entomol. Soc. of Am. **64**, 757–760 (1971).

HALLANDER, H.: Prey, cannibalism and microhabitat selection in the wolf spiders *Pardosa chelata* O. F. Müller and *Pardosa pullata* Clerck. Oikos **21**, 337–340 (1973).

HEMMINGSEN, A. M.: Energy metabolism as related to body size and respiratory surfaces, and its evolution. Copenhagen: Rep. Steno Hosp. 1960.

HINTON, J. M.: Energy flow in a natural population of *Neophilaenus lineatus* (Homoptera). Oikos **22**, 155–171 (1971).

KAJAK, A.: Productivity of some populations of web spiders In: PETRUSEWICZ, K. (Ed.): Secondary productivity of terrestrial ecosystems, pp. 807–820. Warszawa-Kraków, Vol. II, 1967.

KESSLER, A.: A comparative study of the production of eggs in eight *Pardosa* species in the field (Aran., Lyc.) Tijdschr. Entomol. **116**, 23–41 (1973).

KESSLER, A., FOKKINGA, A.: Hymenopterous parasites in egg sacs of spiders of the genus Pardosa (Aran., Lyc.) Tijdschr. Entomol., **116**, 43–61 (1973).

MIYASHITA, K.: Effects of locomotory activity, temperature and hunger on the respiratory rate of *Lycosa T-insignita*. Appl. Entomol. Zool. **4**, 105–113 (1969).

NAKAMURA, K.: The ingestion in Wolf spiders I. Capacity of gut of *Lycosa pseudoannulata*. Res. Popul. Ecol. **X**, 45–53 (1968).

ODUM, E. P.: Fundamentals of ecology, 3. ed. Philadelphia: W. B. Saunders 1971.

PARRY, D. A.: On the drinking of soil capillary water by spiders. J. Exp. Biol. **31**, 218–227 (1954).

PECK, W. B., WHITCOMB, W. H.: An adaptable method for rearing spiders and channibalistic incects. Turtox News **45**, 242–244 (1967).

PETRUSEWICZ, K., MACFADYEN, A.: Productivity of terrestrial animals. IBP Handbook **13**. Oxford: Blackwell Sci. Publ. 1970.

PHILLIPSON, J.: Ecological energetics. London: Edw. Arnold 1966.

REMMERT, H., WÜNDERLING, K.: Temperature differences between arctic and alpine meadows and their ecological significance. Oecologia **4**, 208–210 (1970).

SCHOLANDER, P. F., FLAGG, W., WALTERS, V., IRVING, L.: Climatic adaptation in arctic and tropical poikilotherms. Phys. Zool. **26**, 67–92 (1953).

SCHOLANDER, P. F., IVERSEN, O.: New design of volumetric respirometer. Scand. J. Clin. Lab. Invest. **10**, 429–431 (1958).

SKARTVEIT, A., RYDÉN, B. E., KÄRENLAMPI, L.: Climate and hydrology of some Fennoscandian tundra ecosystems. In: WIELGOLASKI, F. E. (Ed.): Fennoscandian tundra ecosystems. Part 1: Plants and microorganisms, pp. 41–53. Berlin-Heidelberg-New York: Springer 1975.

SOLHØY, T.: Quantitative invertebrate studies in mountain communities at Hardangervidda, South Norway. Norsk Ent. Tidsskr. **19**, 99–108 (1972).

SONESSON, M., WIELGOLASKI, F. E., KALLIO, P.: Description of Fennoscandian tundra ecosystems. In: WIELGOLASKI, F. E. (Ed.): Fennoscandian tundra ecosystems. Part 1: Plants and microorganisms, pp. 3–28. Berlin-Heidelberg-New York: Springer 1975.

VAN HOOK, R. I. JR.: Energy and nutrient dynamics of spider and orthopteran populations in a grassland ecosystem. Ecol. Monogr. **41**, 1–26 (1971).

WIEGERT. R. G.: Population energetics of meadow spittlebugs (*Philaenus spumarius* L.) as affected by migration and habitat. Ecol. Monogr. **34**, 217–241 (1964).

WIEGERT, R. G.: Energy dynamics of the grasshopper populations in old field and alfalfa field ecosystems. Oikos **16**, 161–176 (1965).

# Life Cycle and Energy Budget of Tipula excisa (Schum.) (Diptera, Tipulidae)

T. HOFSVANG

Tipulidae form an important and dominant insect group in tundra ecosystems. The larvae of the soil-dwelling species are of special interest. These larvae are the most active saprovore invertebrates living in tundra soil. They seem to have a function analogous to earth-worms in taiga forest podzol (GHILAROV, 1963). Various IBP investigations in tundra areas indicate that the larvae of crane flies represent a relatively high percentage of the total insect biomass. On the arctic tundra the larvae are heavily preyed upon by birds (CHERNOV, 1967; HOLMES and PITELKA, 1968). Tipulidae also constitute an important part of the prey of the meadow-pipit *(Anthus pratensis)* on Hardangervidda (HÅGVAR and ØSTBYE, 1976). For more details on the life cycle and energy budget of *Tipula excisa*, see HOFSVANG (1972, 1973, 1974).

## Life Cycle and Larval Population

In the Finse area of Hardangervidda eleven species of Tipulidae are recorded. *T. excisa* Schum., *Tipula invenusta* Ried. and *Tipula subnordicornis* Zett. are the three most abundant (HOFSVANG, 1974). The larvae of *T. invenusta* and *T. subnordicornis* seem to live in aquatic bogs, while the larvae of *T. excisa* are predominant in moist soil.

During the years 1969–1971 the life cycle of *T. excisa* was investigated at a site with eutrophic vegetation, high soil moisture, and well-developed tussocks (HOFSVANG, 1972). Two clearly distinct size groups of larvae were found throughout most of the year, except for the period when pupae and adults of the species were recorded. These two coexisting cohorts of larvae indicated that *T. excisa* has a two-yearly life cycle, Fig. 1 shows the mean weights of the larvae during the life cycle and the mean weights of both pupae and adults. The lower part of the Figure shows the duration of the different instars. In *T. excisa*, the larvae could be separated into the different instars by measuring the diameter of the spiracular discs. There were four larval instars, a number usually found in crane flies.

Laboratory experiments were performed to determine more exactly the sudden reduction in weight in the last larval instar prior to pupation and the exact length of the pupal stage. Larvae, after having spent the second winter in the field, were kept in the laboratory in soil from their habitat at two different tempera-

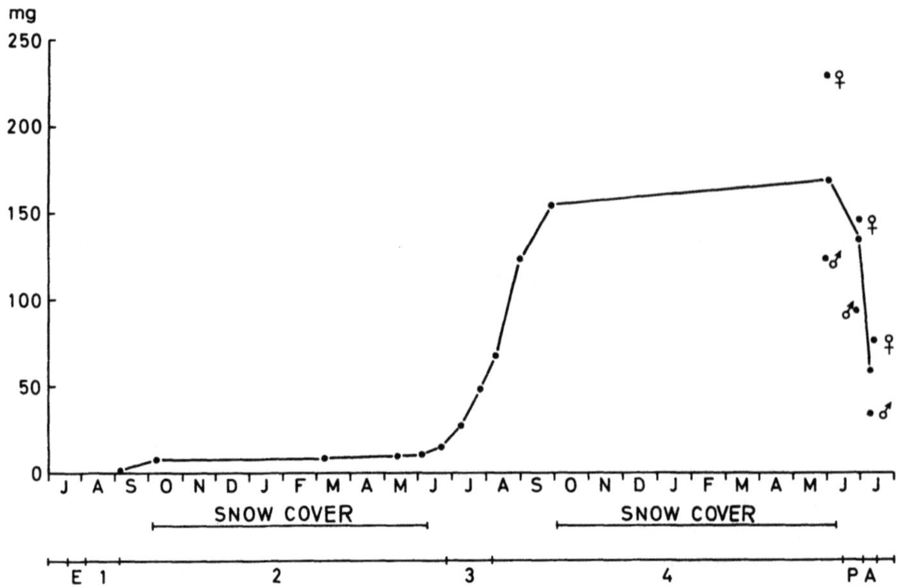


Fig. 1. The mean weights of four larval instars, pupae and adults of *Tipula excisa*. The mean weight of males and females of larvae just before pupation, pupae and adults are also indicated. The lower part of the Figure shows the duration of the different stages and the periods of snow cover

tures. The experiment showed that the greatest part of the weight reduction occurred in the larval stage, the mean weights during the pupal stage being relatively stable. Figure 1 also indicates the mean weights of the sexes on the first day of both the pupal and adult stages. The sexes were easily separated at both these stages. All attempts to separate sexes at other larval stages, however, failed, but it became clear that the great variation in weight of the larval instars was due to sex. Maximum weights of larvae developing into males and females, respectively in the pupal stage, are indicated in Fig. 1.

Tipulidae are usually reported to have a complete life cycle of one year (COULSON, 1962; LAUGHLIN, 1967), compared to the two-yearly life cycle for *T. excisa* in the Finse area. Some other species of Tipulidae from arctic areas have also been reported to have a developmental period lasting more than one year. MACLEAN (1973) reports that *Pedicia hannai antennata* have a larval development requiring at least four years. CHERNOV and SAVCHENKO (1965) cite that *Tipula carnifrons* has a two-yearly life cycle on the arctic plateau of the USSR. MACLEAN (1973) claims that the same species have a developmental period even longer than that found in *P. hannai antennata* in northern Alaska. To extend the life cycle over several years seems to be a general adaptation to low temperatures in insects of alpine and arctic areas.

A modified model of the hot-water process described by MILNE et al. (1958) was used to extract the larvae of *T. excisa* from the soil (HOFSVANG, 1972). Table 1 shows the density estimates of larvae at the tussock site during the greater part of the developmental period. This method of extraction was not used to estimate the

Table 1. Estimated density of larvae of *Tipula excisa* from tussock field, Autumn 1969 to June 1971

| Date | Sampling | No. of samples | No. of larvae | Density $m^{-2} \pm SD$ |
|---|---|---|---|---|
| Autumn 1969 | A | — | — | 69[a] |
| 2–6 June 1970 | B | 40 | 169 | 47± 7 |
| 18–22 June | C | 16 | 72 | 47±11 |
| 30 June–4 July | D | 32 | 141 | 49± 8 |
| 19–23 July | E | 32 | 142 | 49± 8 |
| 8–12 Aug. | F | 32 | 82 | 28± 6 |
| 31 Aug.–3 Sept. | G | 32 | 100 | 35± 7 |
| 25–28 Sept. | H | 32 | 92 | 32± 6 |
| 5–8 June 1971 | I | 32 | 63 | 22± 4 |

[a] Estimate, see text.

density of the larvae before their first winter, as the method is not efficient when the larvae are too small. An approximate estimate was made, assuming the larval mortality during the first winter to be almost equal to the mortality during the second winter. This value is also given in Table 1 (Autumn, 1969).

## Energy Flow

The energy flow or assimilation (A) is given as the sum of the production (P) and the respiration (R) of a population. The energy flow in one generation of *T. excisa* is estimated, and comparisons of the production/respiration ratio with other populations of terrestrial insects are given by Hofsvang (1973).

**Production.** A miniature bomb calorimeter (Phillipson, 1964) was used to determine the caloric equivalents. The results are given in Table 2 showing the different stages of *T. excisa* investigated in this study. There was a great variation in the caloric equivalents during the development. The values decreased during the summer when the one-year-old larvae showed rapid growth. The highest caloric value was found, as expected, in the female pupae.

Table 2. The mean water and ash content and the caloric equivalents in cal (g ash-free dw)$^{-1}$ in the different stages of *T. excisa* ($n$ = number of animals tested)

| | % Water | Ash | Cal $g^{-1} \pm SE$ |
|---|---|---|---|
| Larval instar II | 82 | 3.9 | 5730 ± 61 ($n=5$) |
| Larval instar III | 84 | 7.8 | 5477 ± 54 ($n=5$) |
| Larval instar IV before hibernation | 87 | 4.0 | 5255 ± 84 ($n=9$) |
| Larval instar IV after hibernation | 75 | 4.9 | 5909 ±113 ($n=6$) |
| Pupae, males | 82 | 3.6 | 5412 ± 81 ($n=4$) |
| Pupae, females | 77 | 3.2 | 6165 ± 88 ($n=4$) |
| Adult, males | 59 | 5.2 | 5473 ± 74 ($n=6$) |
| Adult, females | 59 | 7.4 | 5759 ± 54 ($n=6$) |

Table 3. Production and respiration of *T. excisa*

| Developmental stage | Period | Production cal $m^{-2}$ | Respiration cal $m^{-2}$ |
|---|---|---|---|
| Larvae | A–B[a] | 448 | 1599 |
| | B–C | 238 | 187 |
| | C–D | 517 | 206 |
| | D–E | 815 | 398 |
| | E–F | 557 | 559 |
| | F–G | 1175 | 346 |
| | G–H | 687 | 319 |
| | H–I | 328 | 970 |
| Larvae (during weight reduction period before pupation) | 7 days | — | 251 |
| Pupae, males | 10 days | — | 248 |
| Pupae, females | 10 days | — | 225 |
| Adults, males | 14 days | — | 207 |
| Adults, females | 14 days | — | 301 |
| | | 4765 cal $m^{-2}$ $yr^{-1}$ | 5816 cal $m^{-2}$ $yr^{-1}$ |

[a] See Table 1.

The population growth or production of the larvae was calculated using WIEGERT'S formula (1965). Table 3 shows the growth of the larvae between the two periods of larval sampling, converted into cal $m^{-2}$. The study does not include the first larval instar. The total growth from the second instar until the loss of weight begins in the fourth instar just before pupation, is 4765 cal $m^{-2}$. If one assumes that there is no change in the population size from one generation to the next, this figure equals the annual production of *T. excisa*.

**Respiration.** The oxygen consumption of different stages of *T. excisa* was measured at four different temperatures, 5, 10, 15, and 20° C with two different types of respirometers (SCHOLANDER and IVERSEN, 1958; ENGELMANN, 1963). The effect of temperature on oxygen consumption during the different stages is given in Table 4.

The respiration losses of the larval population during the same intervals used in calculation of the production are also presented in Table 3. The respiration of the different developmental stages was corrected for the mean temperature in the

Table 4. Regression equations showing the effect of temperature on oxygen consumption during the different stages of *T. excisa*

| | |
|---|---|
| Larval stage II | $Y = 0.514 + 0.062\,X$ $(r = 0.999)$ |
| Larval stage III | $Y = 0.083 + 0.072\,X$ $(r = 0.977)$ |
| Larval stage IV | $Y = 0.040 + 0.039\,X$ $(r = 0.989)$ |
| Pupae, males | $Y = 0.170 + 0.075\,X$ $(r = 0.998)$ |
| Pupae, females | $Y = 0.050 + 0.035\,X$ $(r = 0.971)$ |
| Adult, males | $Y = 0.215 + 0.101\,X$ $(r = 0.994)$ |
| Adult, females | $Y = -0.095 + 0.093\,X$ $(r = 0.987)$ |

$X$ = temperature in ° C.
$Y$ = oxygen consumption in ml $O_2$ (g dw)$^{-1}$ $h^{-1}$.

habitat between two sampling periods. Microclimatic measurements of the soil temperature at the surface, and at 2 and 5 cm below the surface, were carried out during the investigation, calculated as mean temperature for five days (A. SKARTVEIT, unpublished). In addition, the respiration of the larvae during the period of weight reduction prior to pupation and of males and females of pupae and adults is given in Table 3.

**The P/R Ratio.** The production to respiration (P/R) ratio of the larvae of *T. excisa* during their development from the second larval instar to the pupal stage, is 0.99. MACLEAN (1973) reported from the arctic tundra a total P/R-ratio of 0.59 in larvae of *P. hannai antennata* which required several years for larval development.

When the larvae, pupae, and adults of *T. excisa* are included in the annual respiration (Table 3), the P/R-ratio becomes 0.82. It has been seen from several populations of phytophagous insects studied, that there is a P/R-ratio varying from 50–80% (SMALLEY, 1960; WIEGERT, 1964, 1965; GYLLENBERG, 1969; HINTON, 1971). In spittlebugs and grasshoppers investigated in these studies, the major part of the annual growth occurs during the nymphal stages, in spring and early summer. Growth ceases in the adult stage; they continue, however, to contribute considerably throughout the rest of the summer to the annual respiration of the population. In the spittlebug, *Neophilaenus lineatus*, a P/R-ratio in excess of 100% was found in the nymphal stages in May and early June (HINTON, 1971). QASRAWI (1966), however, found an annual P/R-ratio of 1.09 in the grasshopper *Chorthippus parallelus*. This was because the adults only lived for a relatively short time (quoted from HINTON, 1971).

The larval instars of *T. excisa* represent 95% of the total duration of the life cycle. The non-productive pupal stage lasts only for a few weeks, and during summer when the larvae of *T. excisa* are approximately one year old, about 90% of the larval growth occurs (Fig. 1). In this case the P/R-ratio is more than 100% nearly all the time.

## Conclusion

Soil-dwelling Tipulidae larvae are important decomposers in the tundra ecosystem. The imagines and the larger larvae are important prey for insectivorous birds. Several species of these crane flies have an adaptation which seems to be characteristic for arctic insects, the life cycle being prolonged for several years. The larvae of the different species seem also to be strongly restricted within some of the different types of vegetation on the tundra. These statements indicate that a crane fly, e.g. the dominating *T. excisa*, could possibly serve as a typical model tundra insect.

## References

CHERNOV, Y. I.: The trophic relationships of birds to insects in the tundra zone. (In Russian). Ornitologi'ya **8**, 133–149 (1967).

CHERNOV, Y. I., SAVCHENKO, E. N.: Ecology and preimaginal development phases of the arctic *Tipula (Pterelachisus) carinifrons* Holm. (Diptera, Tipulidae). (In Russian). Zool. Zh. **44**, 777–779 (1965).

COULSON, C. J.: The biology of *Tipula subnodicornis* Zetterstedt, with comparative observations on *Tipula paludosa* Meigen. J. Anim. Ecol. **31**, 1–21 (1962).
ENGELMANN, M. D.: A constant-pressure respirometer for small arthropods. Entomolog. News. **74**, 181–186 (1963).
GHILAROV, M. S.: On the interrelations between soil dwelling invertebrates and soil microorganisms. In: DOEKSEN, J., VAN DER DRIFT, J. (Eds.): Soil organisms, pp. 255–259. Amsterdam: North-Holland Publishing Company 1963.
GYLLENBERG, G.: The energy flow through a *Chorthippus parallelus* (Zett.) (Orthoptera) population on a meadow in Tvaerminne, Finland. Acta Zool. Fennica. **123**, 1–74 (1969).
HÅGVAR, S., ØSTBYE, E.: Food habits of the meadow-pipit, *Anthus pratensis* (L.), in alpine habitats at Hardangervidda, south Norway. Norw. J. Zool. **24**, 25–36 (1976).
HINTON, J. M.: Energy flow in a natural population of *Neophilaenus lineatus* (Homoptera). Oikos **22**, 155–171 (1971).
HOFSVANG, T.: *Tipula excisa* Schum. (Diptera, Tipulidae), life cycle and population dynamics. Norsk Ent. Tidsskr. **19**, 43–48 (1972).
HOFSVANG, T.: Energy flow in *Tipula excisa* Schum. (Diptera, Tipulidae) in a high mountain area, Finse, South Norway. Norw. J. Zool. **21**, 7–16 (1973).
HOFSVANG, T.: Tipulidae (Diptera) from a high mountain area, Finse, South Norway. Norsk Ent. Tidsskr. **21**, 1–4 (1974).
HOLMES, R. T., PITELKA, F. A.: Food overlap among coexisting sandpipers on northern Alaskan tundra. Syst. Zool. **17**, 305–318 (1968).
LAUGHLIN, R.: Biology of *Tipula paludosa*; growth of the larvae in the field. Entomolog. Exp. Appl. **10**, 52–68 (1967).
MACLEAN, S. F. JR.: Life cycle and growth energetics of the Arctic crane fly *Pedicia hannai antennata*. Oikos **24**, 436–443 (1973).
MILNE, A., COGGINS, R. E., LAUGHLIN, R.: The determination of numbers of leatherjackets in sample turves. J. Anim. Ecol. **27**, 125–145 (1958).
PHILLIPSON, J.: A miniature bomb calorimeter for small biological samples. Oikos **15**, 130–139 (1964).
QUASRAWI, H.: A study of the energy flow in a natural population of the grasshopper, *Chorthippus parallelus*, Zett. (Acrididae). Univ. Exeter: Ph. D. Thesis 1966.
SCHOLANDER, P. F., IVERSEN, O.: New design of volumetric respirometer. Scand. J. Clin. Lab. Invest. **10**, 429–431 (1958).
SMALLEY, A. E.: Energy flow of a salt march grasshopper population. Ecology **41**, 672–677 (1960).
WIEGERT, R. G.: Population energetics of meadow spittlebugs (*Philaenus spumaris* L.) as affected by migration and habitat. Ecol. Monogr. **34**, 217–241 (1964).
WIEGERT, R. G.: Energy dynamics of the grasshopper populations in old field and alfalfa field ecosystems. Oikos **16**, 161–176 (1965).

# Standing Crop and Calorific Content in Invertebrates from Hardangervidda

S. J. Næss *, A. L. Steigen *, and T. Solhøy

During investigations of Norwegian IBP at the Hardangervidda alpine plateau, measurements of abundance, standing crop and calorific content of invertebrates were carried out at the dry meadow, wet meadow and lichen heath at Stigstuv (site description in Sonesson et al., 1975). The aim was to reveal any patterns in fluctuations in individuals and community standing crop. It was also of great interest to compare values from Hardangervidda to results obtained in more temperate areas.

## Material and Methods

Most of the invertebrates were sampled applying the "quick trap" method described by Kauri et al. (1969). Larger species were collected by pit-fall traps and by hand.

The material was dried at 105° C for 24 hrs and weighed, thereafter ground and compressed into pellets and dried again before being weighed and burned in a modified Phillipson Microbomb Calorimeter (Phillipson, 1964). The ash remains of the burned sample were weighed, calorific content per g dw and calorific content per g ash-free dw were calculated.

## Results and Discussion

Average calorific content of some invertebrate species from the Stigstuv sites is shown in Table 1. Values are based on samples taken at intervals in the years 1969–1972, and ranges between 2954–7384 cal $(g\ dw)^{-1}$ and 5010–7540 cal (g ash-free $dw)^{-1}$. The difference between cal $(g\ dw)^{-1}$ and cal (g ash-free $dw)^{-1}$ is particularly high in the oribatid genus *Phthiracarus*. The very high ash content (ca. 40%) in this genus is caused by its heavy cuticle (Table 1).

Variations in standing crop of different animal groups per $m^2$ during the season at the sites are presented in Table 2. Values are based on abundance estimates derived from quick-trap samples and are representative for the invertebrate standing crop in vegetation and litter layers only. The maximum standing crops

* Responsible for the manuscript.

Table 1. Average calorific content of some representatives of invertebrate groups from Stigstuv, Hardangervidda

| | cal (g dw)$^{-1}$ | cal (g ash-free dw)$^{-1}$ | Ash % |
|---|---|---|---|
| Acari | | | |
| Oribatei | | | |
| *Platynothrus* sp. | 5341 | 5551 | 3.7 |
| *Phthiracarus* sp. | 2954 | 5010 | 41.0 |
| *Paradamaeus clavipes* | 5232 | 5551 | 5.6 |
| *Liebstadia similis* | 5361 | 5626 | 4.7 |
| *Eupelops* sp. | 5561 | 5951 | 6.4 |
| *Carabodes labyrinthicus* | 5071 | 5361 | 5.4 |
| *Camisia biurus* | 6396 | 6591 | 3.0 |
| Gamasina | | | |
| *Pergamasus* sp. | 5975 | 6265 | 4.4 |
| Trombididae | 6573 | 6913 | 4.9 |
| Araneae | | | |
| *Hilaira frigida* ♀ | 5585 | 5768 | 3.2 |
| *Pardosa palustris* ♀ | 5949 | 6062 | 1.7 |
| *P. palustris* ♂ | 5761 | 5864 | 2.2 |
| Opiliones | | | |
| *Mitopus morio* ♀ | 5532 | 5744 | 3.7 |
| *M. morio* ♂ | 5383 | 5539 | 2.8 |
| Collembola | | | |
| *Isotoma viridis* | 5744 | 5874 | 2.4 |
| *I. olivacea* | 5716 | 6014 | 4.9 |
| Thysanoptera | | | |
| *Anapothrips secticornis* | 6405 | 6524 | 1.8 |
| Homoptera | | | |
| *Arctorthezia cataphracta* | 7384 | 7540 | 1.8 |
| Coleoptera | | | |
| Carabidae | | | |
| *Amara praetermissa* | 5997 | 6099 | 1.7 |
| *Calathus melanocephalus* | 5880 | 6068 | 3.2 |
| *Notiophilus aquaticus* | 6081 | 6385 | 3.2 |
| *Patrobus atrorufus* | 5874 | 5942 | 1.1 |
| *P. septentrionis* | 5858 | 5953 | 1.6 |
| Staphylinidae | | | |
| *Arpedium brachypterum* | 5965 | 6118 | 2.5 |
| *Atheta alpestris* | 5085 | 5726 | 11.2 |
| *A. arctica* | 6314 | 6554 | 3.7 |
| *A. graminicola* | 6270 | 6418 | 2.3 |
| *A. islandica* | 7182 | 7466 | 3.8 |
| *Oxypoda annularis* | 7183 | 7478 | 3.9 |
| *Quedius boopoides* | 6353 | 6520 | 2.6 |
| *Q. fulvicollis* | 6596 | 6815 | 3.2 |
| *Q. boops* | 6220 | 6419 | 3.1 |
| Hydrophilidae | | | |
| *Helophorus glacialis* | 5910 | 5958 | 0.8 |
| Curculionidae | | | |
| *Otiorrhynchus dubius* | 5644 | 5755 | 2.3 |

Table 2. Standing crop through the season in mg dw $m^{-2}$ in vegetation/litter layers on dry and wet meadow sites, Stigstuv

| Dry meadow 1970 | June 23 | July 13 | Aug. 3 | Aug. 25 | Sept. 10 |
|---|---|---|---|---|---|
| Acari | 34.5 | 27.1 | 24.6 | 24.5 | 16.2 |
| Araneae | 2.1 | 6.9 | 3.9 | 6.7 | 3.0 |
| Opiliones | 0.9 | 2.0 | 2.0 | 1.0 | 0.4 |
| Collembola | 13.8 | 15.6 | 3.5 | 11.2 | 19.2 |
| Hemiptera | 5.6 | 5.6 | 6.4 | 10.5 | 9.9 |
| Thysanoptera | 0.3 | 0.1 | 0.3 | 0.2 | 0.1 |
| Mecoptera | | 0.1 | 0.1 | 0.9 | 1.6 |
| Psocoptera | | | | | |
| Coleoptera | 17.6 | 21.7 | 8.7 | 7.8 | 11.9 |
| Lepidoptera | 6.1 | 16.4 | 4.0 | 11.2 | 35.2 |
| Diptera | 19.3 | 36.3 | 0.5 | 23.7 | 15.2 |
| Hymenoptera | 0.5 | 4.0 | 3.6 | 1.0 | 0.2 |
| Sum | 100.6 | 135.9 | 57.7 | 98.8 | 112.9 |

| Dry meadow 1971 | June 30 | July 26 | Aug. 10 | Sept. 2 | Sept. 18 |
|---|---|---|---|---|---|
| Acari | 59.6 | 73.5 | 52.0 | 48.6 | 15.9 |
| Araneae | 3.3 | 2.8 | 7.2 | 1.7 | 7.0 |
| Opiliones | 0.2 | 10.9 | 4.9 | 2.6 | 15.2 |
| Collembola | 15.3 | 30.3 | 34.8 | 35.2 | 30.4 |
| Hemiptera | 12.3 | 8.0 | 13.0 | 9.3 | 24.2 |
| Thysanoptera | 0.2 | 0.2 | 0.2 | 0.2 | 0.1 |
| Mecoptera | | * | | 1.1 | 2.3 |
| Psocoptera | | | * | * | |
| Coleoptera | 36.0 | 18.9 | 10.5 | 7.2 | 4.9 |
| Lepidoptera | 59.9 | 5.2 | 18.3 | 0.8 | 0.5 |
| Diptera | 3.6 | 8.5 | 10.1 | 8.3 | 23.6 |
| Hymenoptera | 2.7 | 0.3 | 32.0 | 7.7 | 0.2 |
| Sum | 193.1 | 158.6 | 182.9 | 122.6 | 124.2 |

| Wet meadow 1971 | July 1 | July 22 | Aug. 10 | Sept. 2 | Sept. 18 |
|---|---|---|---|---|---|
| Acari | 58.6 | 51.6 | 40.2 | 45.5 | 48.0 |
| Araneae | 0.1 | 4.5 | 3.3 | 5.9 | 13.5 |
| Opiliones | 2.7 | 22.1 | 65.8 | 9.5 | 6.3 |
| Collembola | 26.5 | 9.0 | 24.3 | 24.8 | 33.0 |
| Hemiptera | 0.1 | 0.2 | 0.3 | 0.6 | 1.4 |
| Thysanoptera | 0.1 | 0.1 | 0.3 | 0.7 | 0.1 |
| Mecoptera | | | 0.1 | * | |
| Psocoptera | | | * | * | |
| Coleoptera | 15.2 | 7.6 | 19.1 | 7.8 | 4.6 |
| Lepidoptera | * | 1.0 | | 1.4 | * |
| Diptera | 0.1 | 3.8 | 1.8 | 2.6 | 5.2 |
| Hymenoptera | * | 5.5 | 13.6 | 27.2 | * |
| Sum | 103.4 | 105.3 | 168.8 | 125.8 | 112.1 |

* Insignificant.

Table 3. Average estimates of abundance and biomass in vegetation/litter layers on dry meadow, wet meadow and lichen heath sites, Stigstuv

| Dry meadow | 1969 | | 1970 | | 1971 | |
|---|---|---|---|---|---|---|
| | $N\ m^{-2}$ | $dw\ mg\ m^{-2}$ | $N\ m^{-2}$ | $dw\ mg\ m^{-2}$ | $N\ m^{-2}$ | $dw\ mg\ m^{-2}$ |
| Acari | 5766 | 19.2 | 6880 | 25.4 | 6860 | 49.9 |
| Araneae | 46 | 3.7 | 54 | 4.5 | 51 | 4.4 |
| Opiliones | 2 | 2.6 | 1 | 1.3 | 1 | 6.8 |
| Collembola | 6130 | 15.2 | 4810 | 12.7 | 8754 | 29.2 |
| Hemiptera | 446 | 7.8 | 405 | 7.6 | 2221 | 13.4 |
| Thysanoptera | 400 | 1.2 | 60 | 0.2 | 25 | 0.2 |
| Mecoptera | 1 | 0.3 | 2 | 0.5 | 2 | 0.7 |
| Psocoptera | * | — | * | — | * | — |
| Coleoptera | 10 | 12.3 | 9 | 13.6 | 11 | 15.5 |
| Lepidoptera | 12 | 13.0 | 10 | 14.6 | 12 | 16.9 |
| Diptera | 42 | 45.1 | 29 | 19.0 | 73 | 10.8 |
| Hymenoptera | 16 | 2.9 | 11 | 1.9 | 12 | 8.6 |
| Sum | 12871 | 123.1 | 12271 | 101.2 | 18021 | 156.3 |
| **Wet meadow** | | | | | | |
| Acari | 6664 | 31.9 | 9445 | 45.2 | 9359 | 48.8 |
| Araneae | 62 | 11.3 | 52 | 9.5 | 35 | 5.5 |
| Opiliones | 2 | 36.6 | 3 | 54.9 | 4 | 21.3 |
| Collembola | 6068 | 33.0 | 5375 | 29.2 | 4492 | 23.5 |
| Hemiptera | 28 | 1.4 | 19 | 1.0 | 12 | 0.5 |
| Thysanoptera | 44 | 0.2 | 36 | 0.2 | 48 | 0.3 |
| Mecoptera | * | — | * | — | * | — |
| Psocoptera | * | — | * | — | * | — |
| Coleoptera | 18 | 15.9 | 20 | 17.6 | 11 | 10.8 |
| Lepidoptera | * | — | 1 | 0.2 | 2 | 0.5 |
| Diptera | 24 | 3.8 | 18 | 2.9 | 13 | 2.7 |
| Hymenoptera | 3 | 20.4 | 6 | 40.7 | 4 | 9.3 |
| Sum | 12913 | 154.5 | 14975 | 201.3 | 13979 | 123.1 |
| **Lichen heath** | | | | | | |
| Acari | 12907 | 42.9 | 16683 | 55.4 | 18058 | 131.8 |
| Araneae | 46 | 3.7 | 43 | 3.4 | 31 | 2.7 |
| Opiliones | | | 1 | 1.3 | 1 | 3.4 |
| Collembola | 4337 | 10.7 | 10343 | 25.6 | 12750 | 42.1 |
| Hemiptera | 76 | 1.4 | 106 | 1.8 | 150 | 0.9 |
| Thysanoptera | 6 | 0.2 | 2 | — | | |
| Mecoptera | | | * | — | | |
| Psocoptera | 19 | — | 6 | — | | |
| Coleoptera | 10 | 12.3 | 5 | 6.1 | 6 | 8.3 |
| Lepidoptera | 6 | 6.5 | 11 | 11.9 | 9 | 12.4 |
| Diptera | 33 | 35.4 | 131 | 140.6 | 59 | 8.7 |
| Hymenoptera | 3 | 0.6 | 1 | 0.2 | 6 | 4.0 |
| Sum | 17443 | 113.6 | 27332 | 246.4 | 31069 | 214.3 |

* Insignificant.

of Acari and Coleoptera in these layers were found relatively early in the season, while the peak of e.g. Opiliones and Hymenoptera occurred somewhat later. Standing crop of Hemiptera was highest in the fall, at the wet meadow this was observed in 1971 also for Araneae. No clear general seasonal trends can be seen from Table 2 for Coleoptera, Lepidoptera and Diptera, but the low standing crop of these groups in early August 1970 at the dry meadow may have been caused by the low temperature and/or high precipitation in July that year (SKARTVEIT et al., 1975). This may also have been true for Collembola at the site in 1970.

Table 3 shows yearly average biomass and abundance estimates for 1969, 1970, and 1971 on the sites. No weighings were done in 1969, hence values for standing crop are estimated from 1970 data. The density-dominant Acari and Collembola (37–51% of total invertebrate numbers on dry meadow, 59% on wet meadow), do not play the same dominant role when standing crop of cal $m^{-2}$ are considered. The biomass of Opiliones and Hymenoptera were as high as for Collembola and Acari at the wet meadow, although only few specimens were found. Diptera and Lepidoptera showed much higher biomass at the drier sites than at the wet meadow, while Coleoptera species were relatively important at all the sites studied.

Some groups (Acari, Collembola, Enchytraeidae) are considerably more numerous in soil than in vegetation and litter layers. As shown in Table 4, total biomass $m^{-2}$ may be up to ten times as high in the soil (i.e. to a depth of 5 cm) as in vegetation and litter. Enchytraeidae were most numerous and showed the highest standing crop in the soil at the wet meadow and were nearly totally absent at the lichen heath. The annual number and biomass range $m^{-2}$ of Enchytraeidae and some other important invertebrate groups at Hardangervidda are given in SOLHØY et al. (1975).

The fluctuations in invertebrate calorific content at the dry meadow in 1970 may have been caused by the very wet climate in Juli that year (SKARTVEIT et al., 1975). This seems to have favored the ability of Collembola to increase their

Table 4. Average estimates of abundance and biomass in Acari, Collembola and Enchytraeidae from the soil layer on different sites at Stigstuv in 1969–72

| | 1969 | | 1970 | | 1971 | | 1972 | |
|---|---|---|---|---|---|---|---|---|
| | N $m^{-2}$ | dw mg $m^{-2}$ | N $m^{-2}$ | dw mg $m^{-2}$ | N $m^{-2}$ | dw mg $m^{-2}$ | N $m^{-2}$ | dw mg $m^{-2}$ |
| Dry meadow: | | | | | | | | |
| Acari | 95660 | 191 | 105217 | 210 | 112000 | 224 | | |
| Collembola | 62104 | 154 | 74830 | 185 | 56000 | 185 | | |
| Enchytraeidae | | | 8760 | 313 | 20029 | 308 | 22894 | 503 |
| Wet meadow: | | | | | | | | |
| Acari | 25440 | 51 | 24120 | 48 | 35714 | 71 | | |
| Collembola | 33292 | 181 | 47467 | 258 | 39678 | 206 | | |
| Enchytraeidae | | | 29714 | 857 | 38964 | 891 | 37512 | 889 |
| Lichen heath: | | | | | | | | |
| Acari | 42570 | 43 | 71509 | 72 | 107430 | 107 | | |
| Collembola | 23440 | 58 | 26006 | 64 | 40570 | 134 | | |

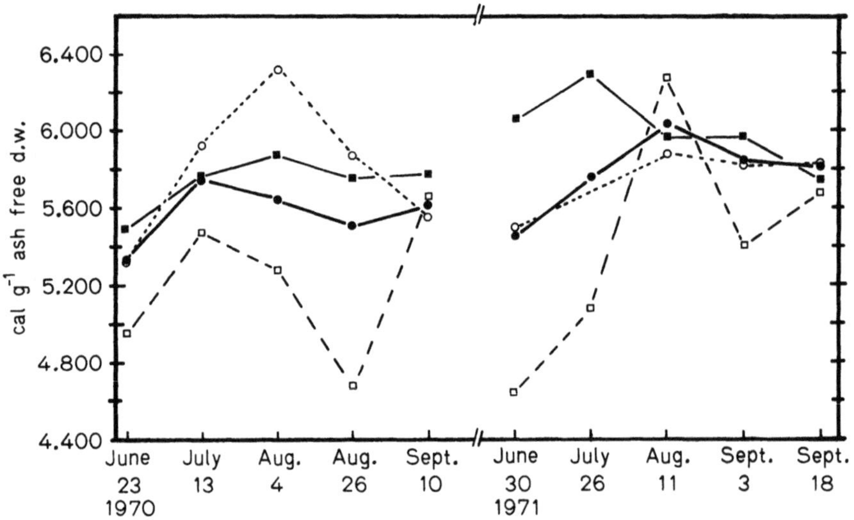


Fig. 1. Variations in ash-free calorific content in invertebrates in litter and vegetation layers at the dry meadow site at Stigstuv in 1970 and 1971. ●——● total invertebrate fauna, ○-----○ collembola, □---□ acari, ■——■ "rest" i.e. remaining invertebrates after the separation of Acari and Collembola

calorific content per g ash-free dw (but notice the reduced biomass of the group per area unit mentioned earlier), and may have had a negative effect on the Acari (Fig. 1). The summer of 1971, however, was very dry, precipitation being only 50–60% of normal. This seems to have reduced the calorific content of Collembola and favored the Acari in comparison with the previous year. A comparison of monthly mean temperatures of the two years (SKARTVEIT et al., 1975) indicates that temperature was not the factor causing the strong fluctuations in 1970. On the wet meadow, which has a stable humid environment, fluctuations were small (Fig. 2). When plants and lichens are considered, seasonal variations seem also to be more obvious at the dry meadow than at the other Norwegian sites (WIELGOLASKI and KJELVIK, 1975). It is interesting to observe that all curves in Figs. 1 and 2 converge to a calorific content between 5550 and 5580 cal (mg ash-free dw)$^{-1}$ approximately at the end of the season. This energy range may represent the optimum utilized by invertebrates in relation to available resources, or it may be the optimal calorific content prior to overwintering, obtained through adaptation. Approximate calorific content of crude protein is 5700 cal (g dw)$^{-1}$ (WEBBER, 1974). This is within the range of calorific content in invertebrates at Hardangervidda in Autumn, and may mean that proteins are of major importance at the end of the season.

Cal m$^{-2}$ for the total invertebrate fauna in vegetation/litter layers also converge in the end of the season in 1970 and 1971 at the dry meadow site, and in 1971 at the wet meadow towards a calorific content of approximately 650 cal m$^{-2}$, although the fluctuations are great earlier in the season (Fig. 3). The fall value may result from optimal utilization of available resources by juvenile invertebrates after death, emigration etc. of adults. Hence, 650 cal m$^{-2}$ may

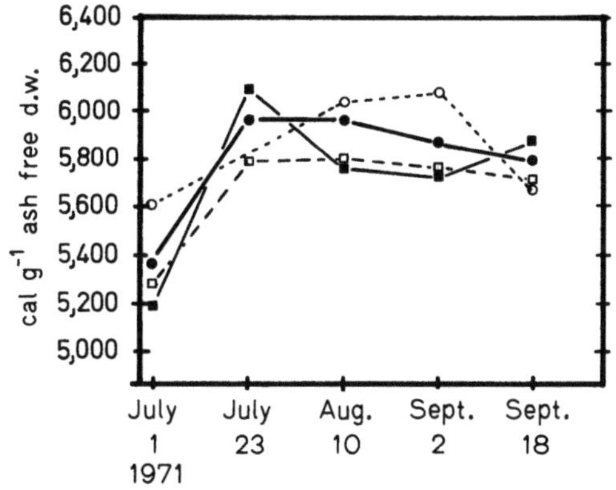


Fig. 2. Variations in ash-free calorific content of invertebrates in litter and vegetation layers at the wet meadow site at Stigstuv in 1971. Symbols, see Fig. 1

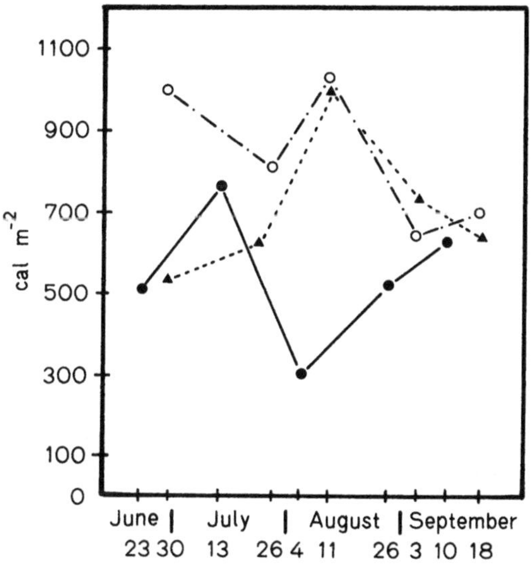


Fig. 3. Variations in cal m$^{-2}$ of the sum of invertebrates in vegetation and litter layers at the dry meadow site in 1970 (●——●) and 1971 (○–·–○) and wet meadow in 1971 (▲---▲)

represent invertebrate carrying capacity of the layers at the end of the productive season and prior to overwintering.

Figure 4 shows the distribution of ash-free calorific content in invertebrates from Hardangervidda also including data given by HOFSVANG (1973), HÅGVAR and ØSTBYE (1974) and STEIGEN (1975), and from temperate regions (EDWARDS et al., 1970; GRIMM, 1973; GYLLENBERG, 1969; HINTON, 1971; PHILLIPSON, 1964; RIEGERT et al., 1974; VAN HOOK, 1971; WIEGERT, 1964, 1965).

The average calorific content of invertebrates from alpine Hardangervidda is ($\pm$S.D.): 6.009($\pm$489.1) cal (g ash-free dw)$^{-1}$, and from temperate areas 5.788 ($\pm$302.6) cal (g ash-free dw)$^{-1}$.

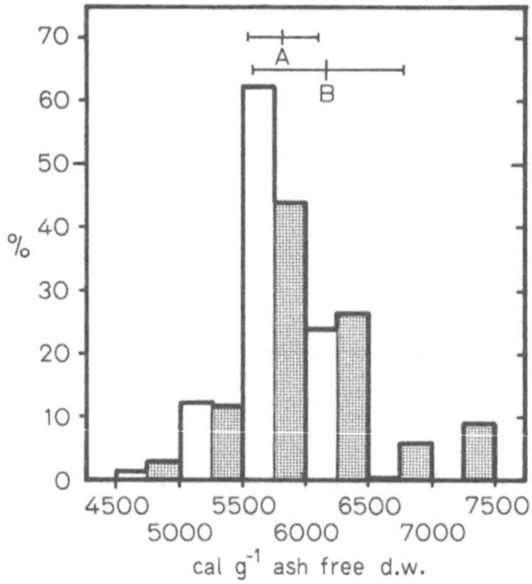


Fig. 4. Distribution of calorific content in invertebrates from Hardangervidda (dotted column) and from temperature regions. A: mean (± SD) for literature data; B: mean (± SD) for Stigstuv data

Values from Hardangervidda are significantly higher ($F = 2.61$, $p < 0.01$) than from temperate regions.

There also seems to be a general tendency to higher calorific values in vascular plants and lichens with latitude and altitude, i.e. with decreasing temperature and increasing radiation (e.g. WIEGOLASKI and KJELVIK, 1975). The high calorific content in invertebrates at Hardangervidda may result from reduced energy metabolism caused by low temperatures. This means that assimilation must be less reduced with temperature than energy metabolism. At present, data of adaptation of assimilation to temperature are lacking for most invertebrate groups.

*Acknowledgements.* The authors wish to express their sincerest thanks to Professor H. KAURI for valuable advice and criticism during the investigations.

## References

EDWARDS, C. A., REICHLE, D. E., CROSSLEY, D. A., JR.: The role of soil inverbetrates in turnover of organic matter and nutrients. In: REICHLE, D. E. (Ed.): Analysis of temperate forest ecosystems. Berlin-Heidelberg-New York: Springer 1970.

GRIMM, R.: Zum Energieumsatz phytophager Insekten im Buchenwald. I. Oecologia **11**, 187–262 (1973).

GYLLENBERG, G.: The energy flow through a *Chorthippus parallelus* (Zett.) (Orthoptera) population on a meadow in Tvärminne, Finland. Acta Zool. Fennica **123**, 1–74 (1969).

HÅGVAR, S., ØSTBYE, E.: Oxygen consumption, caloric values, water and ash content of some dominant terrestrial arthropods from alpine habitats at Finse, South Norway. Norsk Ent. Tidsskr. **21**, 117–126 (1974).

HINTON, J. M.: Energy flow in a natural population of *Neophilaenus lineatus* (Homoptera). Oikos **22**, 155–171 (1971).

HOFSVANG, T.: Energy flow in *Tipula excisa* Schum. (Diptera, Tipulidae) in a high mountain area, Finse, South Norway. Norwegian J. Zool. **21**, 7–16 (1973).

KAURI, H., MOLDUNG, T., SOLHØY, T.: Turnbull and Nicholl's "Quick Trap" for aquiring standing crop of evertebrates in high mountain grassland communities. Norsk Ent. Tidsskr. **16**, 133–136 (1969).

PHILLIPSON, J.: A miniature bomb calorimeter for small biological samples. Oikos **15**, 130–139 (1964).

RIEGERT, P. W., VARLEY, J. L., DUNN, B. C.: Above-ground invertebrates: IV. Populations and energetics of spiders. Technical Report **57**, Matador Project. Saskatoon: Can. Commun. IBP 1974.

SKARTVEIT, A., RYDÉN, B. E., KÄRENLAMPI, L.: Climate and hydrology of some Fennoscandian tundra ecosystems. In: WIELGOLASKI, F. E. (Ed.): Fennoscandian tundra ecosystems. Part 1: Plants and microorganisms, pp. 41–53. Berlin-Heidelberg-New York: Springer 1975.

SOLHØY, T., ØSTBYE, E., KAURI, H., HAGEN, A., LIEN, L., SKAR, H. J.: Faunal structure of Hardangervidda, Norway. In: WIELGOLASKI, F. E. (Ed.): Fennoscandian tundra ecosystems. Part 2: Animals and systems analysis, pp. 29–45. Berlin-Heidelberg-New York: Springer 1975.

SONESSON, M., WIELGOLASKI, F. E., KALLIO, P.: Description of Fennoscandian tundra ecosystems. In: WIELGOLASKI, F. E. (Ed.): Fennoscandian tundra ecosystems. Part 1: Plants and microorganisms, pp. 3–28. Berlin-Heidelberg-New York: Springer 1975.

STEIGEN, A. L.: Energetics in a population of *Pardosa palustris* (L.) (Araneae, Lycosidae) on Hardangervidda. In: WIELGOLASKI, F. E. (Ed.): Fennoscandian tundra ecosystems. Part 2: Animals and systems analysis, pp. 129–144. Berlin-Heidelberg-New York: Springer 1975.

VAN HOOK, R. I., JR.: Energy and nutrient dynamics of spider and orthopteran populations in a grassland ecosystem. Ecol. Monogr. **41**, 1–26 (1971).

WEBBER, P. J.: Tundra primary productivity. In: IVES, J. B., BARRY, R. G. (Eds.): Arctic and alpine environments. London: Methuen 1974.

WIEGERT, R. G.: Population energetics of meadow spittlebugs (*Philaeunus spumarius* L.) as affected by migration and habitat. Ecol. Monogr. **34**, 217–241 (1964).

WIEGERT, R. G.: Energy dynamics of the grasshopper populations in old field and alfalfa field ecosystems. Oikos **16**, 161–176 (1965).

WIELGOLASKI, F. E., KJELVIK, S.: Energy content and use of solar radiation of Fennoscandian tundra plants. In: WIELGOLASKI, F. E. (Ed.): Fennoscandian tundra ecosystems. Part 1: Plants and microorganisms, pp. 201–207. Berlin-Heidelberg-New York: Springer 1975.

# Food Habits and Body Composition of Adult and Juvenile Meadow Pipit [Anthus pratensis (L.)]

H.-J. SKAR, S. HÅGVAR, A. HAGEN, and E. ØSTBYE

Among the main topics of Norwegian IBP tundra biome studies at Hårdangervidda, described by SONESSON et al. (1975), are investigations of energy flow and mineral cycling. Numerous calorimetric investigations on the consumer component of ecosystems have been carried out (PETRUSEWICZ and MACFADYEN, 1970), while less attention has been given to measurement of element content and feeding habits.

From May to October the meadow pipit *(Anthus pratensis)* is one of the most dominant species in the alpine regions of Norway (LIEN et al., 1970, 1974, 1975; HAFTORN, 1971). This paper deals with the body weight of the meadow pipit during this period (the birds mostly being collected in the Finse and Stigstuv area), percentage of water and ash content, caloric equivalents, and the content of some chemical elements (N, Zn, P, Ca, Na, K, Mg, Mn, Fe). The values are related to season, sex and age. Stomach content of adults from June to September has been analyzed, together with the stomach content of nestlings from three clutches. Detailed descriptions of this study have been published (SKAR et al., 1972) or are in press (HAGEN et al., in press; HÅGVAR and ØSTBYE, 1976). In this paper the stomach analyses are described by HÅGVAR, and the caloric and element analyses by SKAR.

## Materials and Methods

The stomach content analyses are based on 88 adults (15–27 each month from June to September) and 16 nestlings collected in July and August. Identification of the invertebrate fragments performed by comparison with preparations of the major invertebrate groups occurring on the Hardangervidda plateau. The minimum number of specimens present within each group was noted, and the length of each specimen estimated. Later, the weight was estimated on the basis of the length/weight-relationships worked out for field invertebrates. Most of the plant diaspores eaten could be identified directly, as they were practically undamaged. The number and weight of stones and bone fragments in each stomach were also noted.

The water content was determined by drying the carcasses in a vacuum oven (70° C). The ash content was determined by ashing at 500° C, or by determining the ash residue after a calorimetric analysis when a sample of the dry homogenized carcass was burned in a bomb calorimeter. The elements Ca, Mg, Mn, and Fe were determined by means of an atomic absorption spectrophotometer, K and Na by emission with the same instrument, P with a grating photometer and N by the semi-micro-KJELDAHL method.

The adult meadow pipits were sorted according to sex and month of capture, the juveniles according to age. The juveniles were divided into five groups, four of these groups covering four days of age each, the fifth covering the 16th to 24th day. The age of the birds in the first three groups was very accurately determined. At 12 to 16 days of age the young birds leave the nest, and these juveniles were grouped according to feather development. The difference between juveniles with well-developed feathers and adults is not quite clear, and these birds were therefore grouped together as adults.

The results in the graphs are given as the mean ± standard deviation (SD).

## Results

Adults. Figure 1 illustrates the relative composition of the stomach content in different months, the upper row being based on number, the lower row on estimated fresh weight of contents. All stomachs contained invertebrates. Plant diaspores were present in only very small amounts in July and August, but to a larger degree in June and September. However, of the total biomass eaten, plant diaspores never exceeded 1%.

Dominant groups as determined by numbers eaten are Coleoptera, Diptera, Hymenoptera, Opiliones, and Araneida. As for fresh weight, Opiliones, Coleoptera, Diptera and the insect larvae dominate.

The food composition changes markedly throughout the season. As far as it is possible to check, these changes correspond well with the changes in the composition of the invertebrate fauna in the field. For example, *Helophorus glacialis* occurs mainly in June and September, Tipulidae swarm heavily in July and September, and Bibionidae in August and September. Other factors may also play a certain role, e.g. the heaviest predation of Opiliones (one species, *Mitopus morio*) in August and September occurs at the time when large, adult specimens are found.

Probably all invertebrates and plant diaspores are swallowed undamaged, and 87% of stomachs from adults contained small stones to help fragmentation. Mean number of stones per stomach containing stones was 13, weighing 28 mg. Small bone fragments were found in three stomachs.

The mean fresh weight for the meadow pipit is 18.1 g, and there ar no significant weight differences between the sexes. The seasonal variation is shown in Fig. 2. On average, 64.1% of fresh weight consists of water (SD = ± 1.4%), whereas ash content is made up of 10.4% of dry weight (SD = ± 1.1%). The caloric values for fresh biomass average 1897 cal $g^{-1}$ (SD = ± 115.6 cal $g^{-1}$), which gives an average of 5285 cal (g dw)$^{-1}$ and 5895 cal (g ash-free dw)$^{-1}$ However, there are significant differences between the sexes in June and July, and also between some

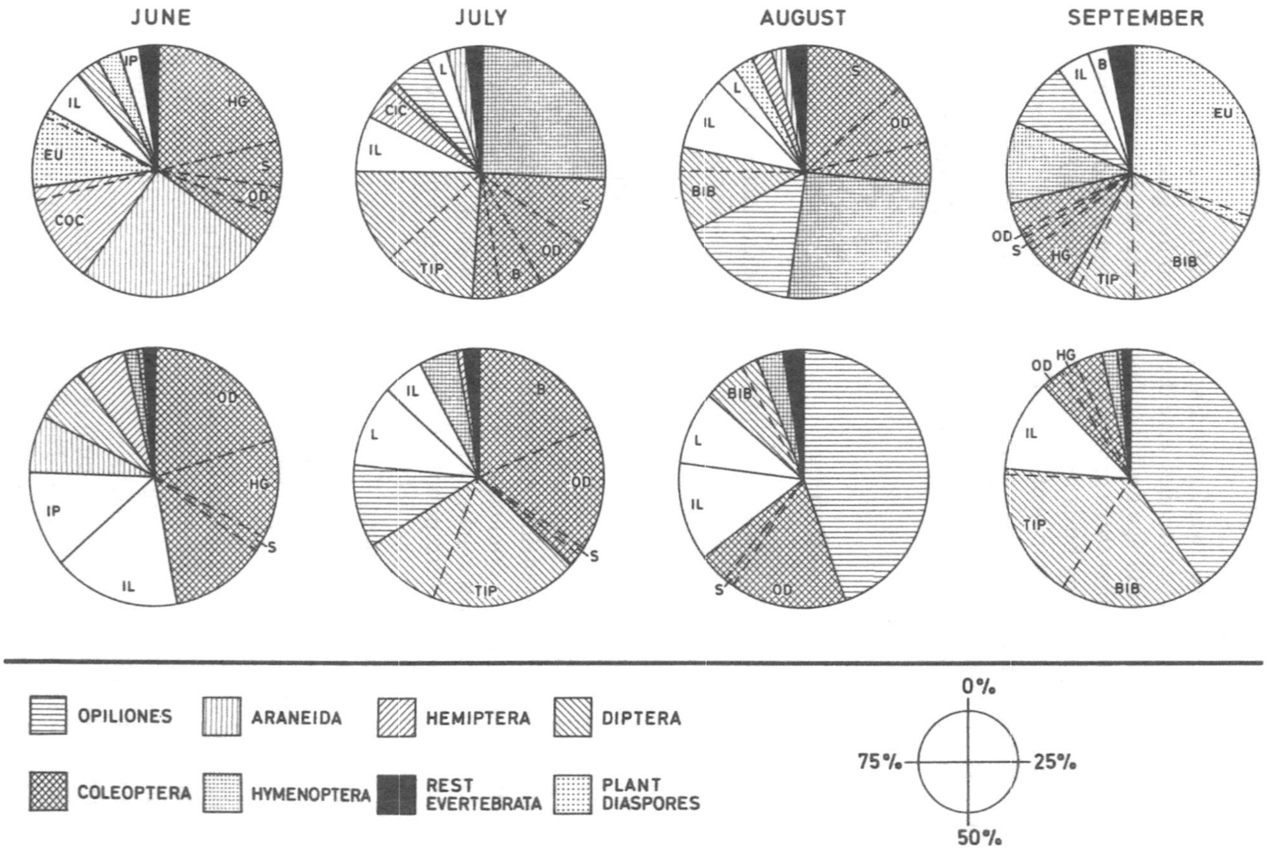


Fig. 1. Composition of stomach content of adult meadow pipits from June to September. Upper row: percentage distribution of the different invertebrate groups and plant diaspores, based on numbers. Lower row: Percentage distribution based on estimated fresh weights. *IP* insect pupae, *IL* insect larvae, *EU Euphrasia* sp., *COC* Coccoidea, *OD Otiorrhynchus dubius* (Curculionidae), *S.* Staphylinidae, *H. G. Helophorus glacialis* (Hydrophiliidae), *B Byrrhus* sp. (Byrrhidae), *L* Lepidoptera, *CIC* Cicadoidea, *TIP* Tipulidae, *BIB* Bibionidae, *B* (on white background) *Boreus* sp. (Mecoptera)

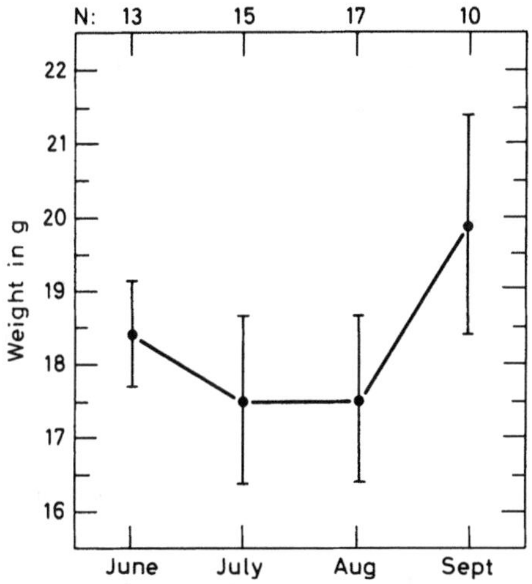


Fig. 2

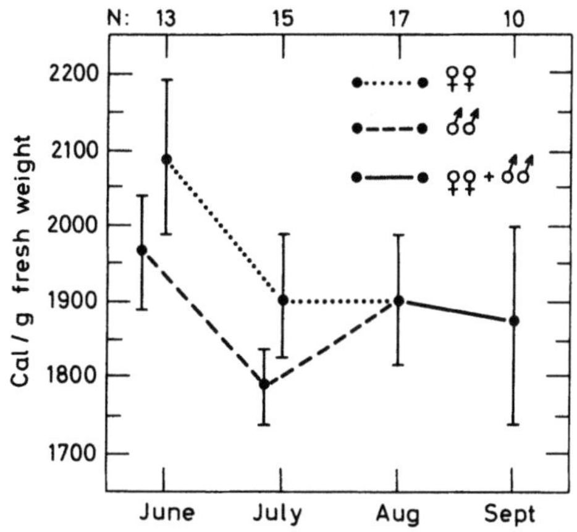


Fig. 3

Fig. 2. Monthly mean fresh weight of adult meadow pipits

Fig. 3. Monthly mean caloric equivalents for fresh biomass in adult meadow pipits

of the monthly mean values (Fig. 3). The average caloric content per individual is 34.5 Kcal.

There seems to be no significant difference in the element content of each sex. The various element contents are shown in Table 1. Variations in P, Ca, N, Zn, and Mg all show approximately the same pattern, with a low point in August (e.g. Fig. 4 A). K, Na, and Mn also show similar variation patterns and have a peak in August (e.g. Fig. 4 B). During the season Fe shows a marked drop.

Juveniles. The food composition of nestlings, 2–10 days old, corresponded approximately to that of adult meadow pipits at that time. A typical feature is, however, that stomachs of nestlings contained more stones than those of adults.

Table 1. The maximal and minimal limits for the mean values of elements analyzed in meadow pipits, in percent of dw

| | Adults | Juveniles |
|---|---|---|
| P | 1.75 – 1.90 | 1.40 – 1.76 |
| Ca | 2.55 – 2.75 | 1.30 – 2.50 |
| N | 10.0 –11.0 | 9.6 –11.2 |
| Zn | 0.0095– 0.0120 | 0.0094– 0.0120 |
| Na | 0.35 – 0.38 | 0.40 – 1.30 |
| K | 0.85 – 0.88 | 0.90 – 1.70 |
| Mg | 0.10 – 0.13 | 0.10 – 0.15 |
| Mn | 0.0010– 0.0014 | 0.0012– 0.0017 |
| Fe | 0.032 – 0.050 | 0.030 – 0.046 |

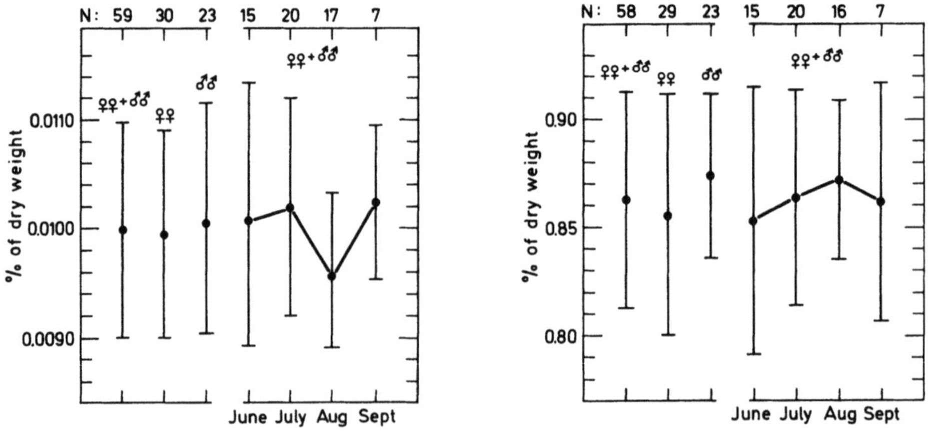


A B

Fig. 4 A and B. Examples of mineral content in relation to dry weight in adult meadow pipits. (A) Values for Zn. (B) Values for K

15 of the 16 stomachs contained stones. The mean number of stones in these 15 stomachs was 19, corresponding to 44 mg. All seven nestlings in one clutch, although only two days old, had stomachs containing stones. The number of stones ranged from 7 to 47, 25 being the mean. Seven of the 16 nestling stomachs contained bone fragments, with a maximum total dry weight of 72 mg, and a maximum number of 12. The largest bone fragment was a 6.5 mm long part of a small rodent jaw with teeth.

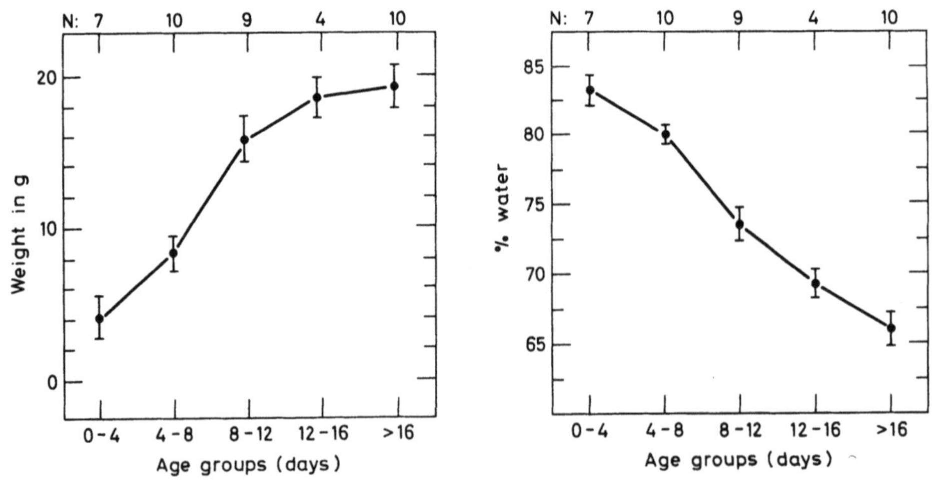


A B

Fig. 5 A and B. Juvenile meadow pipits of different age groups (days). (A) Mean fresh weight. (B) Mean water content in relation to fresh weight

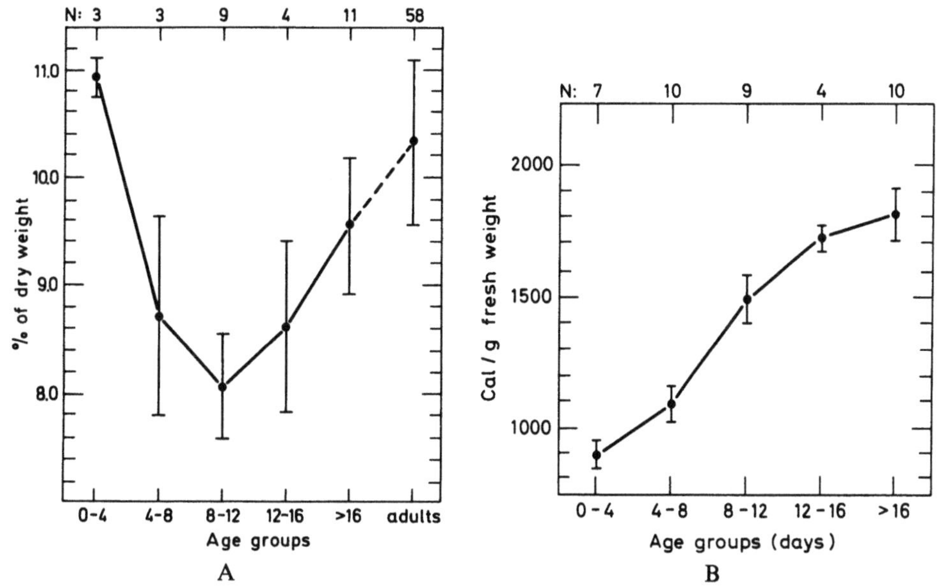


Fig. 6 A and B. Juvenile meadow pipits of different age groups (days). (A) Ash content in relation to dw. (B) Caloric equivalents for fresh biomass

The growth of the meadow pipit is most rapid during the first 12 to 14 days (Fig. 5 A), adult weight being reached at approximately two and a half weeks of age. The relative water content is very high in newly-hatched individuals, but decreases to the adult level in less than three weeks (Fig. 5 B). The ash content shows a drop from hatching to the 10th day, and does not reach adult level in any of the late juvenile groups (Fig. 6 A).

The juveniles are lean during their early development. This, together with higher water content, causes very low caloric values in newly-hatched birds (Fig. 6 B). The fat content relative to dry weight reaches a maximum just before the birds leave the nest. The pronounced increase in caloric values in relation to fresh biomass is, however, caused by the decrease in water content. The juveniles raise their caloric content from approximately 3.5 to 34.0 kcal per individual in two and a half weeks.

Compared with the element content of adults, variations in juveniles are similar or greater (Table 1). A striking feature is that the concentration of all analyzed elements except Fe shows a drop at the age of 4 to 8 days. This drop is clearly reflected in the ash content.

From the age of 4 to 8 days up to over 16 days, Ca and P show an increase (e.g. Fig. 7 A). The variations in N clearly have a pattern similar to Zn (Fig. 7 B), while Fe shows the opposite variation. Na, K, and Mg all show the same pattern (e.g. Fig. 7 C), although Mg shows a small peak in two-week-old juveniles. Mn shows only small variations during development except for the early drop already referred to.

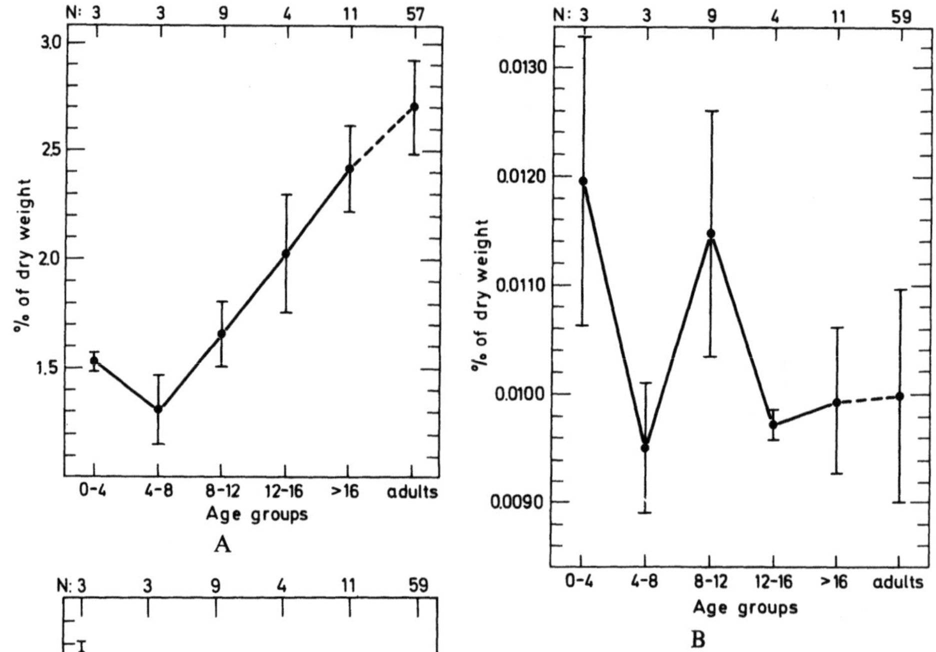


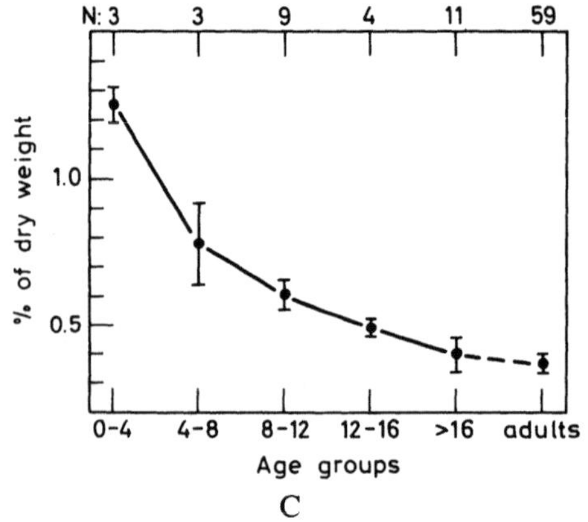


Fig. 7 A–C. Examples of mineral content in relation to dw in juvenile meadow pipits of different age groups (days). (A) Values for Ca. (B) Values for Zn. (C) Values for Na

## Discussion

Adults. Birds feeding on invertebrates in alpine and arctic areas probably cannot concentrate strongly on one or even a few groups. This is partly due to the great changes in weather conditions and to the great variations in available food. Flying insects are only available to any great extent on a limited number of warm, sunny, wind-free days. However, on favorable days massive hatching and swarming may occur, e.g. of Tipulidae, Bibionidae or Chironomidae. The birds will therefore change food habits to correspond with the changes in easily available groups. The ground fauna also changes markedly during the snow-free season. Obviously the meadow pipit is well adapted to the rapidly-changing food conditions characteristic of alpine habitats. Correspondingly, on the Alaskan tundra several species of waders have been found to change their feeding habits radically during summer (HOLMES and PITELKA, 1968).

The stones in the stomach evidently aid fragmentation of the invertebrates and diaspores. Their importance is especially stressed by the fact that a great part of the food consists of Coleoptera which have a very hard exoskeleton. Most identified stones were quartz, which are easy to find in the field because of their light color.

The water content, ash content and caloric equivalents for fresh biomass in meadow pipits are of the same order of magnitude as in some other passerine birds (KALE, 1965; MYRCHA and PINOWSKI, 1970; YARBROUGH and JOHNSTON, 1965; PATTIE, unpublished).

One of the most obvious reasons for weight loss and reduction in the caloric content in adult meadow pipits during June and July seems to be the energy demands of breeding activities. The more extreme caloric reduction in the males can be explained by the territorial behavior before pairing. A corresponding pattern was also found between the sexes in the willow grouse (WEST and MENG, 1968). The weight gain observed in September has no corresponding incrase in caloric content, and is not only caused by fat deposition. It can partly be explained by an increased stomach content. It is possible however, that the fat and protein fractions have a parallel increase, with a smaller increase in water content.

Compared with the caloric content of dry tissue (SKAR et al., 1972), all the elements except Fe show more or less the opposite variation. In relation to fat-free tissue the concentration of these elements will be fairly constant throughout the season. The drop in Fe will, however, increase in relation to fat-free material.

The content of Ca, P, N (protein = N · 6.25 (KLEIBER, 1961)), Mg, Mn and Fe in the meadow pipit is of the same order of magnitude as in the snow bunting (PATTIE, unpublished).

Juveniles. Evidently the nestlings are fed with stones and bone fragments immediately after hatching. The large number of stones stresses their function in fragmenting invertebrates in the stomach. The bone fragments may serve the same purpose. In addition it is possible that the nestlings are able to extract some of the phosphorus and calcium which they sorely need at this stage of their development. MACLEAN (1974) has shown that arctic sandpipers (*Calidris* spp.) used bone fragments of brown lemmings (*Lemmus trimucronatus*) as a source of calcium for egg-laying and body growth.

Meadow pipits leave the nest 12 to 16 days after hatching. In contrast to species dependent on well-developed flying ability when leaving the nest, the meadow pipit reaches a weight peak after leaving the nest, the meadow pipit reaches a weight peak after leaving the nest. This supports RICKLEFS' (1968) hypothesis that birds depending on good flying capabilities reach their peak weight before fledging, while others reach their peak weight after leaving the nest. The marked drop in water content during the growth period is also found in other birds (RICKLEFS, 1967; MYRCHA and PINOWSKI, 1969) and even in small rodents (MYRCHA and WALKOWA, 1968).

The ash content is of a similar amount to that in other passerine birds (RICKLEFS, 1967; MYRCHA and PINOWSKI, 1969), but the variations during the growth period show a different pattern. The adult ash content level was not reached in the first four weeks of life; this is possible caused by the late ossification (ANOROVA, 1965; BAUMAN, 1968).

The caloric content of fresh biomass in newly-hatched passerine birds seems low, i.e. approximately 600 cal $g^{-1}$ (RICKLEFS, 1967; MYRCHA and PINOWSKI, 1969). In meadow pipits approximately two days old 800 cal $g^{-1}$, were found, and there was a tendency to increase with age. By interpolating to the day of hatching, the value for meadow pipits is comparable to that of other passerine birds. During the first 10 to 12 days the caloric value of meadow pipits increases rapidly and then slows down. This same tendency is seen in the red-winged blackbird (RICKLEFS, 1967) and tree sparrow (MYRCHA and PINOWSKI, 1969). In the barn swallow, however, these values show a more regular increase (RICKLEFS, 1967). The caloric content of fresh biomass in meadow pipits does not reach adult values before the juveniles have left the nest. The difference is mainly due to a higher water content in juveniles.

All the analyzed elements in meadow pipits show almost the same mean values as shown for juvenile snow buntings (PATTIE, unpushished). Since fat contains small amounts of minerals, the marked drop in all the elements except for Fe during the first day of life can be partly explained by an increase in the fat content (SKAR et al., 1972). Seen in relation to fat-free dry material, the elements which show an increase during the further growth period (Ca and P) will have this increase accentuated.

Ca and P are the main constituents of bone tissue. The increase in these elements must therefore be seen in relation to ossification. N and Zn show the same variations as in adults, and these elements are possibly linked together in the protein component. In juvenile meadow pipits Na and K show a decrease parallel to the water content. This can be explained by the fact that the predominant osmotically active solutes in vertebrates are Na, K and Cl (BENTLEY, 1971). Seen in relation to fat-free material, Mg and Mn maintain a fairly constant level throughout development.

*Acknowledgements.* We wish to thank all those who have participated in the field and analytical work, especially the late Mrs. J. HAGEN and Mr. D. SVALASTOG, and Mrs. M. ESPELAND for improving the English. Extracts of papers published in the Norwegian Journal of Zoology are included in this paper by kind permission of the Editor. Financial support was received from the Norwegian IBP.

## References

ANOROVA, N. S.: The age changes in birds. Ornithologia **7**, 398–415 (1965).

BAUMAN, V. K.: Calcium and phosphorus, metabolism and regulation in birds. Riga: Zinatne 1968.

BENTLEY, P. J.: Endocrines and osmoregulation. A comparative account of the regulation of water and salt in vertebrates. Zoophysiology and Ecology **1**. Berlin-Heidelberg-New York: Springer 1971.

HAFTORN, S.: Norges fulger. Oslo-Bergen-Tromsø: Universitetsforlaget 1971.

HAGEN, J., HAGEN, A., ØSTBYE, E., SKAR, H.-J.: Some elements in the body of the meadow pipit [*Anthus pratensis* (L.)]. Norw. J. Zool. (in press; 1976).

HÅGVAR, S., ØSTBYE, E.: Food habits of the meadow-pipit, *Anthus pratensis* (L.), in alpine habitats at Hardangervidda, South Norway. Norw. J. Zool. **24**, 25–36 (1976).

HOLMES, R. T., PITELKA, F. A.: Food overlap amongst co-existing sandpipers on northern Alaskan tundra. Syst. Zool. **17**, 305–318 (1968).

KALE, H. W. II: Ecology and bioenergetics of the long-billed Marsh Wren, *Telmatodytes palustris griseus* (Brewster) in Georgia salt marshes. Publs. Nuttall Orn. Club **5**, 1–141 (1965).
KLEIBER, M.: The fire of life. New York-London: John Wiley and Sons, Inc. 1961.
LIEN, L., ØSTBYE, E., HAGEN, A., KLEMETSEN, A., SKAR, H.-J.: Quantitative bird surveys in high mountain habitats, Finse, South Norway, 1967–68. Nytt Mag. Zool. **18**, 245–251 (1970).
LIEN, L., ØSTBYE, E., HAGEN, A., SKAR, H.-J., SVALASTOG, D.: Density variations of bird populations in high mountain habitats, Hardangervidda. In: WIELGOLASKI, F. E. (Ed.): Fennoscandian tundra ecosystems, Part 2: Animals and systems analysis, pp. 105–110. Berlin-Heidelberg-New York: Springer 1975.
LIEN, L., ØSTBYE, E., HOGSTAD, O., HAANDE, K. M., HAANDE, P. S., HAGEN, A., SKAR, H.-J., SKARTVEIT, A., SVALASTOG, D.: Bird surveys in the high mountain habitats of Finse and Stigstuv, Hardangervidda, south Norway, 1967–72. Norw. J. Zool. **22**, 1–14 (1974).
MACLEAN, S. F. JR.: Lemming bones as a source of calcium for arctic sandpipers (*Calidris* spp.). Ibis **116**, 552–557 (1974).
MYRCHA, A., PINOWSKI, J.: Variations in the body composition and the caloric value of nestling tree sparrows (*Passer montanus* L.). Bull. Acad. Pol. Sci. Cl. II. Ser. Sci. Biol. **17**, 475–480 (1969).
MYRCHA, A., PINOWSKI, J.: Weights, body composition, and caloric value of the body during postjuvenal moulting European tree sparrows (*Passer montanus*). Condor **72**, 175–181 (1970).
MYRCHA, A., WALKOWA, W.: Changes in the caloric value of the body during the postnatal development of white mice. Acta theriol. **13**, 391–400 (1968).
PATTIE, D. L.: Canadian high arctic bird bioenergetics and population levels on Devon Island. Unpublished.
PETRUSEWICZ, K., MACFADYEN, A.: Productivity of terrestrial animals, principles and methods. IBP handbook No. **13**. Oxford and Edinburgh: Blackwell Sci. Publ. 1970.
RICKLEFS, R. E.: Relative growth, body constituents, and energy content of nestling barn swallows and red-winged blackbirds. Auk **84**, 560–570 (1967).
RICKLEFS, R. E.: Weight recession in nestling birds. Auk **85**, 30–35 (1968).
SKAR, H.-J., HAGEN, A., ØSTBYE, E.: Caloric values, weight, ash and water content of the body of the meadow pipit [*Anthus pratensis* (L.)]. Norw. J. Zool. **20**, 51–59 (1972).
SONESSON, M., WIELGOLASKI, F. E., KALLIO, P.: Description of Fennoscandian tundra ecosystems. In: WIELGOLASKI, F. E. (Ed.): Fennoscandian tundra ecosystems. Part 1: Plants and microorganisms, pp. 3–28. Berlin-Heidelberg-New York: Springer 1975.
WEST, G. C., MENG, M. S.: Seasonal changes in body weight and fat and the relation of fatty acid to diet in the willow ptarmigan. Wilson Bull. **80**, 426–441 (1968).
YARBROUGH, C. G., JOHNSTON, D. W.: Lipid deposition in wintering and premigratory myrtle warblers. Wilson Bull. **77**, 175–191 (1965).

# Energy Budget of a Population of the Root Vole [Microtus oeconomus (Pall.)] in a High Mountain Habitat, Hardangervidda

A. HAGEN*, E. ØSTBYE, and H.-J. SKAR

In this paper the energy budget of a root vole population (*Microtus oeconomus*) from a high mountain habitat at Finse, Hardangervidda, Norway (60°36′ N, 7° 31′ E) is calculated for the period 1969–1972 including a population high period (August 1969 to January 1971).

The main trapping area is situated around an abandoned summer dairy in the transition zone between low- and mid-alpine regions (DAHL, 1975). The pedology of the higher part of the site is characterized by glacifluvial deposits, sand, gravel and blocks overlaid with a thin layer of organic soil, and the lower part by thick organic deposits upon glacifluvial material (VEUM, unpubl.). Two brooks run through the area which is dominated by oligotrophic plant communities in the drier parts, and by mesotrophic/eutrophic communities where water is more plentiful. Some parts of the area are grazed and fertilized by cattle and sheep.

Metabolic data are taken from literature, all other data are from Norwegian IBP investigations at Finse. A detailed description of the work is given in HAGEN (1973).

## Material and Methods

The trapping grid used in the snow-free season consisted of 10 × 10 trap-lines 10 m apart. A circle radius $2\frac{1}{2}$ m around the intersections was defined as a trapping station. Two live traps were placed at each station.

In winter, 25 traps were placed in five abandoned farm houses once a month for from one to three days.

In summer, trapping periods lasted from four to six days. During the rodent-low of 1971–1972, the period was reduced to three to four days. The trapping started in August 1969 with up to four trap controls a day. Later the controls were reduced to twice daily. The trapping was continued until September 1972. Additional data were obtained by snap-trapping on plots in adjacent areas. The size of the trapping area is found by the HAGEN et al. (1973) method.

* Responsible for the manuscript.

Population size is estimated according to the calendar of catches method (PETRUSEWICZ and ANDRZEJEWSKI, 1962). The formula used is

$$N = R + \frac{1}{n} \sum_{i=1}^{n} E_i$$

where $R$ is the resident population in the trapping period, $E_i$ is the number of ephemerals registered on day $i$, $n$ is the duration of the trapping period in days.

Residents are animals caught at least twice and on separate days. Ephemerals are animals caught once, or at most twice, and then on the same day.

Energy equivalents ($\bar{K}_b$) are determined on animals snap-trapped at other localities during the period of investigation. The animals were burned in an adiabatic bomb-calorimeter.

The mean body weights ($\bar{B}$) are calculated from the first weight registered for each individual during the actual trapping period.

The turnover-rate ($\Theta_N$) is calculated from the time of residency, which is defined as the time between the first and the last catch of an individual plus $2 \times \frac{1}{2}$ month.

The average daily metabolic rate (ADMR) is taken from GRODZIŃSKI (1971) from his data on *M. oeconomus macfarlani.*

His formula for $\text{ADMR} = 23.86 \cdot W^{-0.50}$ ml $O_2$ $g^{-1}$ $h^{-1}$ is identical to $\text{ADMR} = 2.75 \cdot W^{0.50}$ kcal animal$^{-1}$ day$^{-1}$, where $W$ is the body fresh weight in $g$.

The daily energy budget (DEB) [mean metabolic rate ($M$)] was calculated from ADMR by adjusting for reproduction and ambient temperature. The percentage of reproducing females in the population was presumed identical to that recorded for animals snap-trapped in a field 1 km away. The energy cost of pregnancy and lactation was taken from data on *M. arvalis* (TROJAN and WOJCIECHOWSKA, 1967; MIGULA, 1969). Temperature in the microhabitat was measured, and corrections made for out of nest activity based on data for *M. agrestis* (HANSSON and GRODZIŃSKI, 1970).

Production ($P$) is calculated in two ways. (1) Values for shorter periods as a product of number of eliminated animals ($N_e$) (i.e. animals disappeared from the population), the mean caloric equivalents, and the mean body weight ($P = N_e \bar{B} \bar{K}_b$), and (2) for the period of population high from the formula $P = \bar{K}_b(\bar{N} \bar{B} \Theta_N)$ (GRODZIŃSKI et al., 1966). The digestability of forage percentage is taken as 70% from GĘBCZYŃSKA (1970).

Energy flow or assimilation in a small rodent population is expressed by the equation

$$A_T = \sum_T \bar{K}_b(\bar{N} \bar{B} \Theta_N) + M(\bar{N} \bar{B})\, T$$

(GRODZIŃSKI et al., 1966), where $\bar{N}$ is the mean number of animals in the population, $\bar{B}$ mean biomass of an animal (mass), $\Theta_N$ turnover-rate of individuals (time$^{-1}$), $M$ mean metabolic rate (energy mass$^{-1}$ time$^{-1}$), $\bar{K}_b$ mean energy equivalents per animal (energy), $T$ the length of the period (time).

## Results and Discussion

The size of the trapping area was by the HAGEN et al. (1973) method found to be 1.22 ha. Another method (STENSETH et al., 1974) gives $1.01 \pm 0.20$ ha.

In 1969, the root vole population increased to 61 individuals per ha in September. About one third of the animals survived the winter, and in August 1970, the population had a maximum of 69 individuals. The population then decreased and collapsed in the following winter. From February 1971 to September 1972, there were very few animals in the area, perhaps from one to four. Obviously the calendar method gives an underestimate.

The density, calculated from the number of animals and the area to which they belong, is the least reliable parameter in the energy budget.

The total mean of the caloric equivalents was $1.38 \pm 0.01$ kcal $g^{-1}$ fresh weight.

The mean bodyweight ($\bar{B}$) for the period August 1969 to January 1971 was 37.5 g. The turnover-rate ($\Theta_N$) for the same period was 4.78 $yr^{-1}$. The time of residency was 2.51 months. The daily energy budget (DEB), based on data from literature is shown in Table 1.

Production, respiration, assimilation, and consumption for the root vole population and for the rodent community in a south-facing dry-meadow community at Finse is given in Table 2. The values given are for a peak year, a low year, and for an "idealized rodent cycle" consisting of one peak year and three low years.

For the growing seasons 1969–1971 the mean net primary production on a dry meadow measured at another site at Hardangervidda was $2.4 \cdot 10^7$ kcal $ha^{-1}$ $yr^{-1}$,

Table 1. Daily energy budget (DEB) for an average root vole from August 1969 to January 1971. Calculated from $ADMR = 2.75 \cdot W^{0.50}$ kcal $animal^{-1}$ $day^{-1}$ (GRODZIŃSKI, 1971) by adding the cost of reproduction and thermoregulation $(r+t)$ in percent. $DEB = a \cdot \bar{B}^{0.50}$ kcal $animal^{-1}$ $day^{-1}$, where $\bar{B}$ is the mean biomass of an animal in g

| Period | Aug. 69–Sept. 69 | Oct. 69–May 70 | June 70–Aug. 70 | Sept. 70 | Oct. 70–Jan. 71 |
|---|---|---|---|---|---|
| $(r+t)$ | 22 | 0 | 19 | 5 | 0 |
| $a$ | 3.36 | 2.75 | 3.27 | 2.89 | 2.75 |
| $\bar{B}$ | 32.4 | 33.8 | 45.2 | 37.7 | 33.2 |
| DEB | 19.1 | 16.0 | 22.0 | 17.7 | 15.8 |

Table 2. Production ($P$), respiration ($R$), assimilation ($A$), and consumption ($C$) in a small rodent community (lemmings excluded) in a south-facing dry-meadow plant community at Finse in a peak year, a low year, and in an idealized cycle (one peak year + three low years). Kcal $ha^{-1}$ $yr^{-1}$

| | $P$ | $R$ | $A$ | $C$ |
|---|---|---|---|---|
| Peak year | 8000 | 236000 | 244000 | 349000 |
| Low year | 500 | 15000 | 15500 | 22000 |
| Cycle | 9500[a] | 281000[a] | 290500[a] | 415000[a] |

[a] kcal $ha^{-1}$ $cycle^{-1}$.

of which $8.6 \cdot 10^6$ kcal ha$^{-1}$ yr$^{-1}$ consisted of green parts of vascular plants (WIELGOLASKI and KJELVIK, 1975). This means that root voles during a peak year consumed 1.2% of the primary production, or—if we accept the green parts of vascular plants as the value for available food—3.4%.

The peak year 1969–1970 was a medium year. In 1974 the populations were higher, especially those of lemming (*Lemmus lemmus*) (HAGEN, LIEN, ØSTBYE and STENSETH, unpubl.). The consumption of a rodent community in these oligotrophic meadows may amount to 3 to 4% of primary production or up to 10% of the food available in a peak year, whereas in a low year it is next to nothing.

This work is a first attempt to calculate an energy budget for small rodents in this alpine area, and is an approximation. To make a reliable estimate, better census methods are needed, and metabolic investigations must be carried out on animals from the area in question.

*Acknowledgements.* The authors wish to thank all who assisted in the field work or in the laboratory, and to Mrs. M. ESPELAND for improving the English.

## References

DAHL, E.: Flora and plant sociology in Fennoscandian tundra areas. In: WIELGOLASKI, F. E. (Ed.): Fennoscandian tundra ecosystems. Part 1: Plants and microorganisms, pp. 62–67. Berlin-Heidelberg-New York: Springer 1975.

GĘBCZYŃSKA, Z.: Bioenergetics of a root vole population. Acta theriol. **15**, 33–66 (1970).

GRODZIŃSKI, W.: Energy flow through populations of small mammals in the Alaskan taiga forest. Acta theriol. **16**, 231–275 (1971).

GRODZIŃSKI, W., GÓRECKI, A., JANAS, K., MIGULA, P.: Effect of rodents on the primary productivity of alpine meadows in Bieszczady Mountains. Acta theriol. **11**, 419–431 (1966).

HAGEN, A.: Energibudsjett for en fjellrottepopulasjon [*Microtus oeconomus* (Pall.)] i den lav/mellomalpine sone på Finse, Hardangervidda. University of Oslo: Unpubl. thesis 1973.

HAGEN, A., ØSTBYE, E., SKAR, H.-J.: A method for calculating the size of the trapping-area in catch-mark-release censusing of small rodents. Norw. J. Zool. **21**, 59–61 (1973).

HANSSON, L., GRODZIŃSKI, W.: Bioenergetic parameters of the field vole *Microtus agrestis* L. Oikos **21**, 76–82 (1970).

MIGULA, P.: Bioenergetics of pregnancy and lactation in European common vole. Acta theriol. **14**, 167–179 (1969).

PETRUSEWICZ, K., ANDRZEJEWSKI, R.: Natural history of a free-living population of house mice (*Mus musculus* Linnaeus), with particular reference to groupings within the population. Ekol. Pol. A **10**, 85–122 (1962).

STENSETH, N. C., HAGEN, A., ØSTBYE, E., SKAR, H.-J.: A method for calculating the size of the trapping area in capture-recapture studies in small rodents. Norw. J. Zool. **22**, 253–271 (1974).

TROJAN, P., WOJCIECHOWSKA, B.: Resting metabolism rate during pregnancy and lactation in the European common vole—*Microtus arvalis* (Pall.) Ekol. Pol. A **15**, 811–817 (1967).

WIELGOLASKI, F. E., KJELVIK, S.: Energy content and use of solar radiation of Fennoscandian tundra plants. In: WIELGOLASKI, F. E. (Ed.): Fennoscandian tundra ecosystems. Part 1: Plants and microorganisms, pp. 201–207. Berlin-Heidelberg-New York: Springer 1975.

# 1.4. Herbivory Aspects

## On the Ecocatastrophe of Birch Forests Caused by Oporinia autumnata (Bkh.) and the Problem of Reforestation

P. KALLIO and J. LEHTONEN

Subarctic ecosystems are more labile than those in temperate zones. There is, for instance, considerable variation in population densities of animals in different groups (see KALELA, 1966), which may be attributed to fluctuation in primary production (TAST and KALELA, 1971). This influence is normally most pronounced at climatic vegetation border lines. Forest limits of both coniferous and *Betula* trees are indicative of the climatic environmental factor, due to their sensitivity to changes in the environment.

In Fennoscandia, the climatic and human-influenced changes in coniferous tree limits have been studied by several investigators (literature cited e.g. in HUSTICH, 1966; KALLIO et al., 1971). The problems of taxonomy and morphology of birches have been discussed e.g. by VAARAMA and VALANNE (1973) and KALLIO and MÄKINEN (1975), and the problem of birch forest limits in Lapland has been studied by HÄMET-AHTI (1963). However, owing to the small economic importance of these forests only a limited number of studies has been carried out.

The birch forests in northernmost Finland belong to the Fennoscandian subarctic birch forest zone, which is here about 60–180 km wide between the continuous coniferous forest limit and the tundra zone or the coast of the Arctic Ocean. Reindeer management is the chief means of livelihood in many places, particularly in the more continental parts, and human influence has affected these birch forests mainly through reindeer grazing.

During the last few centuries these birch forests have been subjected to great changes caused by the geometrid larvae of the moth *Oporinia autumnata.* Damage has changed birch forest limits and seems to reflect environmental alterations. TENOW (1972) gives a good survey of the history of *Oporinia* damage in Fennoscandia and KALLIO and LEHTONEN (1973) have discussed the problem with reference to Finnish Lapland. The present paper gives a brief survey of the *Oporinia* damage, especially in the 1960s, and examines its effects in northern Finland. Most of this work has been done within the framework of the Finnish IBP-Tundra Biome project in the Kevo area (KALLIO and KÄRENLAMPI, 1971; SONESSON et al., 1975).

## Methods

The damage in the 1960s took place in 1965–1966, but the present studies were mostly carried out later. The damaged areas were first mapped from the air to obtain a broad view and later as thoroughly as possible by excursions. The mapping was done in order to follow the later development of the damaged birch forest ecosystem. The intensity of the damage and the degree of recovery, as indicated by saplings, was estimated with the use of squares of $50 \times 50$ m or straight lines of 100 trees. In these squares and lines, the degree of damage was estimated using the following scale:

0: no damage (i.e. untouched)
1: 1–10% of the tree without leaves
2: 11–80% of the tree without leaves
3: 81–99% of the tree without leaves
4: total damage (no leaves).

The degree of recovery was estimated according to the following scale:

0: no saplings
1: 1–5 saplings
2: 6–20 saplings
3: more than 20 saplings.

The areas between these investigated plots were studied by visual estimation.

Main characteristics of the ground vegetation were also studied in every plot. Nine plots of $20 \times 20$ m were selected as experimental areas and surrounded by fences to exclude the effect of reindeer grazing (Fig. 1). Preliminary attempts have also been made to simulate the *Oporinia* effect by defoliation.

## Results

Figure 1 shows the distribution of the latest *Oporinia* damage in northern Finland. In Utsjoki (northernmost administrative district in Finland, the area studied most thoroughly), damaged forests cover about 1350 km$^2$, which include half the birch forests of the whole area. About 50% of the damaged forests in the area have a damage index of 3.51–4.00; i.e. more than 90% of the green has been destroyed.

Almost half of the damaged area in Utsjoki has a recovery index between 0.00–0.50; i.e. this area will not recover vegetatively (Fig. 2). The rest will recover to some extent. Reforestation by seedlings does not work. The number of seedlings per ha. was on average 100–300 and these have frequently been cut by herbivores, both voles and reindeers. Therefore 6–7 years after the damage, only very scant reforestation by seedlings has taken place. This means that about 650 km$^2$ in Utsjoki and more than 1000 km$^2$ in the whole of northern Finland will change into a treeless tundra. The recent treeless areas, partly of secondary origin, are shown in Fig. 1.

Areas that have been preserved in damage centers are mostly river valleys. Almost all deeper valleys in northernmost Finland are preserved and the contin-

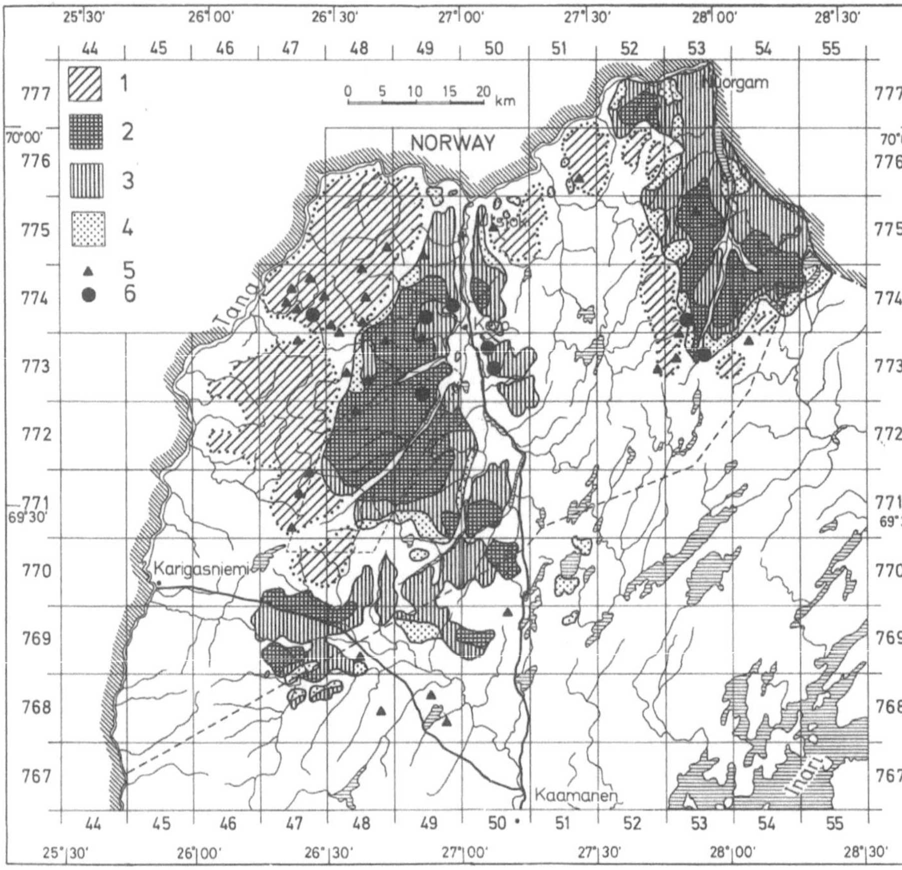


Fig. 1. The *Oporinia* damage areas in northern Finland. Diagonal lines (1) show treeless areas. The recent damage areas are drawn by thick continuous lines and the degree of damage is indicated as follows: (2) 3.51–4.00, (3) 2.51–3.50 and (4) 1.51–2.50. Triangles indicate older birch damage (5), and circles the sites of fences and experimental plots (6)

uous birch damage areas are broken by narrow green belts of some meters to some kilometers in width.

The standing crop of birches in the research area is ca. 8 t $ha^{-1}$ (dry weight, roots excluded), based on the average estimations made at the Kevo IBP birch site (KÄRENLAMPI, 1972; KJELVIK and KÄRENLAMPI, 1975). In the areas of monocormic birches, particularly in the area of the Kevo Nature Park, the crop is ca. 80–90% of the figure given above. On the other hand, still higher values of about 10.5 t $ha^{-1}$ have been estimated in some areas. The standig crop contains about 8% leaves and, in addition, an amount of young twigs, which are suitable for reindeer grazing and for rodents. The consumption of birch leaves by reindeer is ca. 2 kg dw $ha^{-1}$ (HAUKIOJA and KOPONEN, 1975) or 20–25 kg dw $animal^{-1}$ during a summer. It is evident that a relatively large amount of reindeer fodder has been lost in the areas damaged by *Oporinia*.

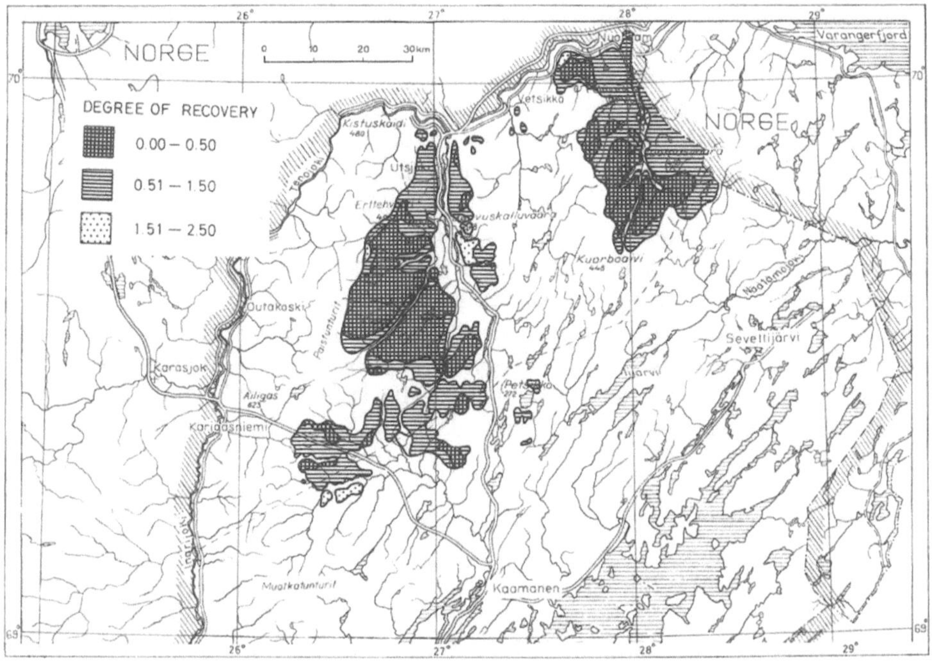


Fig. 2. The degrees of recovery of birches in Utsjoki after *Oporinia* damage in 1965–1966

The new birch biomass which has developed from saplings in the damaged areas is small, particularly when compared to the biomass before the *Oporinia* attack. In the areas where the mean degree of recovery was 1.51–2.50, the new biomass in 1973, i.e. 6–7 years after the damage, was about 300–750 kg dw $ha^{-1}$. Areas with a higher new biomass are very rare. In areas with a mean recovery index of 0.51–1.50, the biomass of new saplings is 20–300 kg dw $ha^{-1}$. If the mean recovery value is 0.00–0.50, the new biomass is less than 20 kg $ha^{-1}$; in most cases it is very small and frequently even absolutely 0.

It is typical for damaged mesic forests that the ground vegetation becomes rich during the first 4–5 years after the damage. In particular, the *Deschampsia flexuosa, Trientalis europaea, Solidago virgaurea* and *Cornus suecica* vegetations are enriched (KÄRENLAMPI and KAUHANEN, 1972). This takes place, however, in mesic forests only. In dry forests, dominated by lichens and *Empetrum*, there are no great changes in the vegetation.

Even in mesic forests the changes are not permanent. The mean cover values of *Deschampsia flexuosa* and *Trientalis europaea* in 20 analyzed plots were 14.2% and 7.0% in 1970. In 1973 these values were respectively 9.5% and 2.0%. In 1970 the *Deschampsia* vegetation was mostly fertile, while only sterile leaves were to be found in 1973. *Trientalis* sprouts were also larger in 1970. The predamage situation is thus restored in the ground vegetation some years after the damage.

The damage in the 1960s represented the most dramatic change known in the landscape of the subarctic ecosystem of Finnish Lapland. The damage caused by

*Oporinia* is not, however, rare. According to TENOW (1972) damage has occurred in Fennoscandia every ten years during the last century. Previous *Oporinia* damage known in northern Finland is shown in Fig. 1, mostly from the years 1927 and 1955. In the western part of this area remains of ancient birch forests are found in an area which is nowadays almost treeless tundra.

*Oporinia*-simulation experiments have been carried out during three summers (1971–1973) by defoliating the experimental trees but leaving the buds untouched. The experiments showed that birches were able to form new leaves after the defoliation any time between bud break and the last week of July. Similarly, trees that were defoliated during two successive summers also developed new leaves and were not killed. The summers of these experiments were quite warm, as shown by the thermal sums (above basic temperature 5° C) at the Kevo Meteorological Station in Utsjoki: 824.4 (1970), 577.7 (1971), 905.5 (1972) and 719.7 (1973), compared to the ten-year average degree $day^{-1}$ values given below.

## Discussion

There is a typical feature in the damage distribution that deeper valleys have been totally preserved. As the imagos of *Oporinia* occurring in enormous masses everywhere in the early autumn of 1965 must have laid eggs also in valleys, one must suppose that the eggs were killed. The topoclimatical observations from the winter 1965–1966 show that there was a very strong thermal inversion between valleys and ridges. In the Utsjoki river valley the temperature was below −30° C for 52 nights, which was the case only 5 times on top of a mountain near by, which is 180 m higher. These two places represent respectively a totally undamaged, and a badly damaged area (KALLIO and LEHTONEN, 1973). It seems likely that very cold winters have a restricting effect on the population development of *Oporinia* in general, and topoclimatic factors can be discerned when studying the geographical distribution of the damages.

The *Oporinia* damage in the 1960s happened during a period when temperatures of the growth seasons were low. The lowest values of thermal sums in Finnish Lapland within this century have been measured during the period 1961–70. The average thermal sums (above basic temperature 5° C) in Inari, about 120 km south of the Kevo Station are as follows:

| | | | | |
|---|---|---|---|---|
| 1911–20 | 690 | 1941–50 | 729 | |
| 1921–30 | 703 | 1951–60 | 716 | |
| 1931–40 | 768 | 1961–70 | 664 | (MIKOLA, 1971) |

At the Kevo Meteorological Station, founded in 1962, the average thermal sum for the years 1962–1970 was 589. The summers 1965 and 1966 have values of 444.4 and 589.3, respectively (KÄRENLAMPI, 1972).

These observations are of interest in the study of birch recovery. The *Oporinia* simulation experiments showed that birches are normally able to develop new leaves after defoliation. This is possibly due to the energy stored in the stem and roots during the previous year. The amount must be dependent on the environmental conditions during that season. Thus, if the previous growth season has been warm, as in the experiments, the birches recover well. If the previous season has been cold and short, the energy storages are poor and the recovery is difficult or impossible. Temperatures can affect the primary production and it seems that low temperatures may reduce birch forest limits by using *Oporinia* as a tool.

The birch in damaged areas is *Betula pubescens* var. *tortuosa*, but there is a very clear introgression activity: genes from *Betula nana* are continuously interfering with the *B. pubescens* population (VAARAMA and VALANNE, 1973). Morphologically there are two types of birches, which have not been analyzed genetically. These types are a bushformed, many-stemmed birch with an ability for vegetative distribution and preservation, and a monocormous, one-stemmed birch with weaker capacity for sapling formation. These two types of birch are differently affected by *Oporinia* damage (KALLIO and MÄKINEN, 1976). Thus *Oporinia* exercizes a selective power on the birches in the area. The extensive earlier damage in northern Finland took place in areas of monocormous birches; in the 1960s the damage occurred in areas of polycormous trees.

Birch leaves and small twigs are important fodder for reindeer (SKUNCKE, 1963; HAUKIOJA and KOPONEN, 1975). This means that reindeer management has suffered severely as a result of the birch damages, particularly in areas where birches do not recover and after the ground vegetation has reached the predamage stage.

## References

HÄMET-AHTI, L.: Zonation of the mountain birch forests in northern-most Fennoscandia. Ann. Botan. Soc. Vanamo **34:4**, 1–127 (1963).

HAUKIOJA, E., KOPONEN, S.: Birch herbivores and herbivory at Kevo. In: WIELGOLASKI, F. E. (Ed.): Fennoscandian tundra ecosystems, Part 2: Animals and systems analysis, pp. 181–188. Berlin-Heidelberg-New York: Springer 1975.

HUSTICH, I.: On the forest tundra and the northern treelines. Rep. Kevo Subarctic Res. Stat. **3**, 7–47 (1966).

KALELA, O.: On the fluctuations in the numbers of arctic and boreal small rodents as a problem of production biology. Ann. Acad. Sci. Fenn. A, IV Biologica **66**, 1–38 (1966).

KALLIO, P., KÄRENLAMPI, L.: A review of the stage reached in the Kevo IBP in 1970. In: HEAL, O. W. (Ed.): IBP Tundra Biome. Proc. of Kevo working meeting 1970, pp. 79–87. London: IBP Tundra Biome Steering Committee 1971.

KALLIO, P., LAINE, U., MÄKINEN, Y.: Vascular flora of Inari Lapland. 2. Pinaceae and Cupressaceae. Rep. Kevo Subarctic Res. Stat. **8**, 73–100 (1971).

KALLIO, P., LEHTONEN, J.: Birch forest damage caused by *Oporinia autumnata* (Bkh.) in 1965–1966 in Utsjoki, N-Finland. Rep. Kevo Subarctic Res. Stat. **10**, 55–69 (1973).

KALLIO, P., MÄKINEN, Y.: Vascular flora of Inari Lapland. 3. Salicaceae and Betulaceae. Rep. Kevo Subarctic Res. Stat. **12**, (1976).

KÄRENLAMPI, L.: Comparisons between the microclimates of the Kevo ecosystem study sites and the Kevo Meteorological Station. Rep. Kevo Subarctic Res. Stat. **9**, 50–65 (1972).

KÄRENLAMPI, L., KAUHANEN, H.: A direct gradient analysis of the vegetation of the surroundings of the Kevo Subarctic Station. Rep. Kevo Subarctic Res. Stat. **9**, 82–98 (1972).

KJELVIK, S., KÄRENLAMPI, L.: Plant biomass and primary production of Fennoscandian subarctic and subalpine forests and of alpine willow and heath ecosystems. In: WIELGOLASKI, F. E. (Ed.): Fennoscandian tundra ecosystems. Part 1: Plants and microorganisms, pp. 111–120. Berlin-Heidelberg-New York: Springer 1975.

MIKOLA, P.: Reflection of climatic fluctuation in the forestry practices of Northern Finland. Rep. Kevo Subarctic Res. Stat. **8**, 116–121 (1971).

SKUNCKE, F.: Renbetet, marklavarna och skogsbruket. Lappväsendet—Renforskningen, Meddelande **8**, 1–114 (1963).

SONESSON, M., WIELGOLASKI, F. E., KALLIO, P.: Description of Fennoscandian tundra ecosystems. In: WIELGOLASKI, F. E. (Ed.): Fennoscandian tundra ecosystems. Part 1: Plants and microorganisms, pp. 3–28. Berlin-Heidelberg-New York: Springer 1975.

TAST, J., KALELA, O.: Comparisons between rodent cycles and plant production in Finnish Lapland. Ann. Acad. Sci. Fenn. A, IV Biologica **186**, 1–14 (1971).

TENOW, O.: The outbreaks of *Oporinia autumnata* Bkh. and *Operophtera* spp. (Lep., Geometridae) in the Scandinavian mountain chain and northern Finland 1862–1968. Zool. Bidr. Uppsala, Suppl. **2**, 1–107 (1972).

VAARAMA, A., VALANNE, T.: Induced mutations and polyploidy in birch, *Betula* spp. Final Rep. Part V (Project No. E8—FS—47. Grant No. FG—Fi—133. Turku: Dept. Bot. Univ. Turku), 7C. On the biology and origin of *Betula tortuosa*, Ledeb: 63–97. Mimeogr. (1973).

# Birch Herbivores and Herbivory at Kevo

E. HAUKIOJA and S. KOPONEN

Foliage feeders may in extreme cases severely affect the production of a plant and in some cases kill the host (e.g. FRANKLIN, 1970). Even when they occur in moderate densities, their effects are far from negligible (RAFES, 1970).

In the neighborhood of the Kevo IBP areas, severe damage was observed in birch forests in 1965–1966 just before the IBP period (KALLIO and LEHTONEN, 1973, 1975). It was therefore natural that feeding upon the birch foliage was chosen exclusively as the main object in zoological studies in the Kevo IBP. In this paper results concerning herbivores and their affect in the IBP birch site and its immediate vicinity are described. As in ecosystem studies in general, some guessing of parameters was necessary to obtain figures for the whole area. To make later refinements possible, the bases used for estimating different figures have been given.

## Birch as a Host Plant

Birch *(Betula tortuosa)* is the dominant primary producer of the birch zone in northern Finland (KÄRENLAMPI and KAUHANEN, 1972) and this zone covers a large area in northernmost Fennoscandia (HÄMET-AHTI, 1963).

In the IBP birch site at Kevo, the height of the birch stand is about 4–5 m and birch represents 57% of the total above-ground biomass and a large proportion of the below-ground biomass. Biomass estimates are 11035 kg dw ha$^{-1}$ for the total birch biomass of which about 595 kg ha$^{-1}$ is formed of leaf biomass (KJEL-

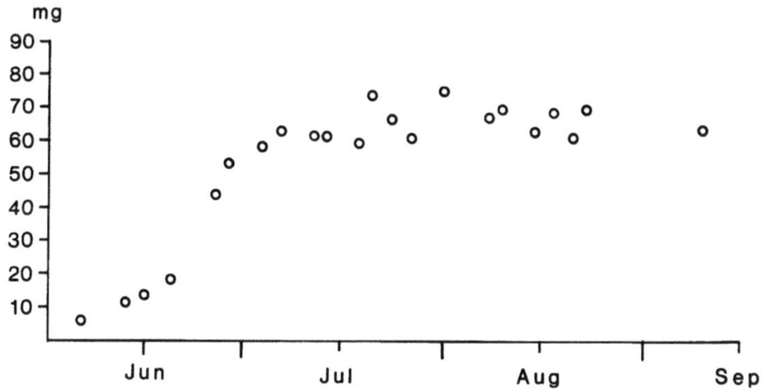


Fig. 1. Seasonal mean dry weights of birch leaves at Kevo in 1973

VIK and KÄRENLAMPI, 1975). Primary production by birch is estimated to be about 1700 kg $ha^{-1}$ $y^{-1}$ (KALLIO and KÄRENLAMPI, pers. comm.). Birch is relatively rich in nitrogen; especially the buds and young leaves contain a high proportion of protein (WIELGOLASKI et al., 1975). Birch leaves emerge in the first half of June at the IBP site and fall around the middle of September. The maximum leaf weight is attained after the middle of July (Fig. 1).

## Herbivores

Birch leaves and sap are the chief above-ground resources of herbivores. Vertebrates graze on leaves, buds, small twigs and shoots, but are not dependent on the birch as their sole food. Reindeer use it as a summer diet and voles also take considerable quantities during peak years. However, birch represents only a small proportion of their total annual energy intake when compared with invertebrate herbivores. Many or most of these are monophagous, the birch being their host plant and their specialization to a certain resource of the host is strict. Three main functional groups can be distinguished according to feeding habits in invertebrates: free-living leaf grazers (e.g. lepidopteran and hymenopteran larvae), free-living sap suckers (e.g. psyllids) and intra-leaf (in mines or galls) living forms (e.g. lepidopteran and dipteran larvae). Many species seem to share the same resource, leaves, as the only food at the same time of the season.

**Methods.** Herbivorous invertebrates were tested twice a week during the summers of 1972 and 1973 and less frequently in 1971 while only sporadic information is available from 1970 and 1974. The method used was to enclose a branch rapidly in a net, then cut it off, after which the branches were stored in plastic sacks, closed, and inspected later in the laboratory. Each leaf was removed separately and all animals found were picked off and weighed. The accuracy of this kind of hand-picking cannot be completely efficient and small animals are more often overlooked than large ones. The same person, however, did nearly all the picking work and his finding efficiency, when tested with individually marked saw-fly larvae, proved to be higher than 90%. The numbers found are dependent on weather factors, among other factors whose inportance is not known. Dry weights of animals were estimated from regression equations made from independent data.

For each sample both the dry weight of leaves and the fresh weight of the twigs and branches were determined. Animal numbers were converted to values per square unit by using the amount of birch leaves per unit area, the ratio leaf weight/branch weight in the sample, and the seasonal development of the leaves (Fig. 1).

Densities of vertebrate herbivores were estimated in the case of small rodents from decreases in the daily snap-trap catches. The density of reindeer is an estimate based on official figures reported by reindeer co-operatives.

**Main Herbivores.** As animals later are combined into broader categories (e.g. insects into orders) a short description of the most important groups and species within orders is given below. More detailed information of the invertebrate herbivore fauna of the mountain birch is given by KOPONEN (1973 b).

Homopterans predominated in the samples taken and they include the suborders Psyllina, Cicadina, Aphidina, and a few Coccina. The most abundant species was *Psylla hartigi* (see later). In the suborder Cicadina the species *Colladonus torneellus* and *Oncopsis flavicollis* were common. In Aphidina *Euceraphis punctipennis* was probably the only species.

The most important invertebrate leaf grazers were symphytan larvae. The number of saw-fly larvae living on the birch was rather high but one species *Dineura virididorsata* clearly dominated during its larval period, in July–September. Some symphytan larvae less common in numbers with a higher mean weight (e.g. *Pteronidea* spp. and *Amauronematus* spp.) dominated among the early summer species. Also a mining saw-fly, *Messa nana*, was common at Kevo.

The majority of the lepidopteran species were so-called mining micro-lepidopterans. The most abundant genera were *Eriocrania, Nepticula, Lithocolletis* and *Parornix*, and the most abundant species was *Eriocrania semipurpurella.* Among non-mining lepidopterans no species was abundant during the study period. However, some species with low numbers but relatively large larvae (e.g. *Polyploca flavicornis, Notodonta dromedarius,* and *Trichiura crataegi*) made some contribution to biomasses.

Three Curculionidae species (*Coeliodes rubicundus, C. nigritarsis,* and *Polydrosus ruficornis*) dominated among coleopterans. In other coleopteran families *Phyllodecta polaris* dominated among Chrysomelidae and *Corymbites costalis* among Elateridae. Gall-inhabiting Cecidomyiidae larvae (Dipt.) dominated in leaves just before leaf fall, important species include *Massalongia rubra.*

The reindeer is the only large herbivore with any appreciable effect on birch. Among small rodents at least *Clethrionomys rufocanus* and *C. rutilus* use the birch in their diet.

**Densities and Production.** For cyclically occurring rodents no normal-year value exists and Table 1 shows the estimated densities for the study period. Reindeer numbers in the whole commune of Utsjoki amount to approx. 20000 which makes about four animals per $km^2$; no knowledge of the variation between the study years is available.

Among invertebrate birch defoliators the geometrid moth *Oporinia autumnata* is known to have 9–10-year cycle (TENOW, 1972), but during the study period its population levels were very low. However, during its previous mass outbreak, in 1965–1966, it defoliated birch forests in areas of thousands of square kilometres around the Kevo area (KALLIO and LEHTONEN, 1973, 1975). Long-term oscillations in invertebrate densities may be more common than the present information indicates because, when long series of annual density determinations are lacking, they are noticed only in the case of some remarkable features, as in *Oporinia* damage. In short-term studies they resemble normal annual variation in numbers.

Table 1. Density (ind. $ha^{-1}$) estimates of *Clethrionomys* voles at Kevo during summers 1970–1974

| 1970 | 1971 | 1972 | 1973 | 1974 |
|---|---|---|---|---|
| 25 | <1 | <1 | 30 | >50 |

Table 2. Percentages of mined birch leaves at Kevo in 1971–1974. Beginnings smaller than 3 mm in diameter have been ignored

| | 1971 | 1972 | 1973 | 1974 |
|---|---|---|---|---|
| Leaves checked | 11222 | 25347 | 43243 | 2179 |
| Mined by | | | | |
| *Eriocrania* spp. | 0.94 | 0.03 | 1.20 | 0.41 |
| *Nepticula* spp. | 0.63 | 0.06 | 0.10 | — |

Mining lepidopterans offer an example of the magnitude of these annual fluctuations (Table 2), e.g. in *Eriocrania* spp. the increase from 1972 to 1973 was 40-fold (see also KOPONEN, 1973a, 1974). The occurrence of species within a family or genus may also vary annually. The leafhopper *Oncopsis flavicollis* was not found in the samples of 1971 and 1972, but in 1973 it represented about 70% of all leafhoppers. All *Coeliodes* (Col., Curculionidae) specimens in 1971 belonged to *C. rubicundus*, but in 1972, about 60% were representatives of *C. nigritarsis* and only 40% of *C. rubicundus*.

On the other hand, some groups (especially Psyllina) did not show any extreme variation in the proportion which they formed of the total biomass and numbers of all invertebrate herbivores (Table 3).

Table 4 gives the estimated densities of herbivorous invertebrates in the year 1973 according to the leaf-branch samples described above. Grouping of the dates is made according to the phenology of the leaf-growth. The first group includes emergence of buds and early growth of leaves. During this phase Psyllina nymphs dominated, especially in numbers but also in biomass. In the following period, before leaves attain their maximum weights, psyllids still dominate, but also larval symphytans are important. In late summer periods, up to the end of August, sawfly larvae, especially *Dineura viridorsata*, dominated.

In August and September, the gall-inhabiting Cecidomyiidae larvae quickly built up large numbers and also biomass. Their biomass was estimated in the

Table 3. Proportion of different insect orders among birch herbivore samples at Kevo in 1971–1973

| Group | % of numbers | | | | | | % of biomass | | | |
|---|---|---|---|---|---|---|---|---|---|---|
| | 1971 20 June–1 July | | 1972 22 June–13 July | | 1973 26 June–13 July | | 1972 22 June–13 July | | 1973 26 June–13 July | |
| Homoptera total | 59.3 | | 77.8 | | 85.6 | | 16.9 | | 42.5 | |
| Psyllina | | 54.0 | | 76.1 | | 81.4 | | 15.4 | | 33.3 |
| Cicadina | | 0.4 | | 1.5 | | 2.0 | | 1.5 | | 6.3 |
| Aphidina | | 4.9 | | 0.2 | | 2.2 | | 0.0 | | 2.9 |
| Hymenoptera larvae | 4.5 | | 12.8 | | 6.7 | | 64.1 | | 36.0 | |
| Lepidoptera larvae | 25.5 | | 2.0 | | 1.4 | | 13.2 | | 3.4 | |
| Coleoptera | 5.5 | | 0.6 | | 1.2 | | 4.2 | | 12.7 | |
| Thysanoptera | 0.6 | | 5.6 | | 2.6 | | 0.1 | | 0.0 | |
| Others | 4.6 | | 1.2 | | 2.5 | | 1.5 | | 5.4 | |
| *n* | 491 | | 539 | | 1206 | | | | | |

Table 4. Densities D (ind. $10^3$ $ha^{-1}$) and biomasses B, (g dw $ha^{-1}$) of insects among birch herbivore samples at Kevo in 1973

| | 5–19 June | | 26 June–13 July | | 18–31 July | | 7–22 August | |
|---|---|---|---|---|---|---|---|---|
| | D | B | D | B | D | B | D | B |
| Homoptera total | 1923 | 56.3 | 782 | 156.3 | 76 | 38.8 | 47 | 23.1 |
| Psyllina | 1888 | 49.5 | 748 | 122.4 | 44 | 6.1 | 6 | 1.4 |
| Cicadina | 25 | 2.9 | 16 | 24.0 | 8 | 20.3 | 11 | 13.7 |
| Aphidina | 10 | 3.8 | 18 | 9.8 | 24 | 12.4 | 29 | 8.0 |
| Hym., Symphyta total | 10 | 35.5 | 69 | 105.0 | 102 | 100.0 | 43 | 235.7 |
| *D. viridi dorsata* larvae | — | — | 49 | 2.1 | 94 | 52.8 | 41 | 207.3 |
| Coleoptera total | 22 | 21.1 | 10 | 42.4 | 12 | 23.0 | 49 | 51.6 |
| Curculionidae | 22 | 21.1 | 5 | 13.9 | 10 | 18.5 | 49 | 51.6 |
| Lepidoptera larvae tot. | 57 | 2.4 | 14 | 7.8 | 4 | 3.1 | 4 | 0.6 |
| Eriocraniidae | 56 | 2.4 | 10 | 4.5 | — | — | — | — |
| Thysanoptera | 39 | 0.6 | 25 | 0.2 | 4 | 0.0 | <1 | 0.0 |
| Hemiptera | 6 | 1.1 | 7 | 5.1 | < 1 | 0.0 | 1 | 3.6 |
| Total | 2058 | 116.9 | 906 | 316.8 | 199 | 164.9 | 145 | 314.6 |

following way: In September 1973 34.9% of the leaves inspected from Kevo contained Cecidomyiidae larvae, 8.1 individuals on an average in an infested leaf. Dry weight of a larva ready for pupation was 0.25 mg. According to the mean weight of a leaf, there was 735 leaves $m^{-2}$ which gives approx. 520 mg of Cecidomyiidae larvae $m^{-2}$. Thus the group with the highest biomass among invertebrate herbivores was not included in Table 4 at all.

As shown earlier (Tables 2 and 3), the year 1973 did not drastically deviate from the previous ones. The same species of saw-fly, *Dineura viridi dorsata*, dominated e.g. every year among saw-flies, and these were dominant over other mid-summer and late-summer larvae, excluding Cecidomyiidae. Some information from the year 1972 (KOPONEN, 1973b) indicated that Cecidomyiidae were also important in other years, although this was not revealed by the hand-picking method used.

The IBP birch site at Kevo has the same kind of invertebrate herbivore fauna as other sites studied in Finnish Lapland and Norwegian Finnmark (HAUKIOJA et al., 1973, and unpublished), the numbers of non-mining lepidopteran larvae, Cimbicidae (Hym.) larvae and Chrysomelidae (Col.), being perhaps lower than in other comparable sites. Knowing the numbers and mean weights of animals in successive catches makes it possible to calculate their net production, which, in turn, allows estimation of consumption after working out the ratio production/consumption. Practical difficulties are, however, obvious. First, the actual densities of most invertebrate herbivores are so low that there are difficulties in obtaining meaningful estimates of the trends in numbers. Furthermore, in common species between-sample variation is so large that smoothing density values before computing production is necessary (the practice used e.g. by WIEGERT, 1965). Secondly, computing production figures for a larger group (insect order, for instance) is possible only if the trends in population density, mean weights and growth rates are the same. In other cases, production values have to be calculated for each species in a sample and this has not been possible. Therefore, production values were calculated only for those groups investigated thoroughly. As they were selected for a detailed study according to their supposed importance, however, it is possible to get a general picture of the total production.

Most of the field and laboratory work was devoted to the study of the caterpillars of *Dineura virididorsata*, which was thought to be the species with the highest biomass (HAUKIOJA, 1973; HAUKIOJA and NIEMELÄ 1974a, 1974b). Net production by *Dineura* was estimated to be 75 and 26 mg dw $m^{-2}$ $yr^{-1}$ in 1972 and 1973, respectively (HAUKIOJA and ISO-IIVARI, in prep.). These values are near the maximum biomasses observed and indicate a low mortality rate during the larval stage. This is supported also by the observation that no parasited larvae were found among the hundreds of caterpillars raised.

Cecidomyiidae, with a very high momentary biomass, have also a production figure probably not much in excess of the maximum biomass, 520 mg $m^{-2}$.

For psyllids a net production value of 52 mg $m^{-2}$ $yr^{-1}$ was computed and for aphids the value arrived at was of the magnitude of 10 mg $m^{-2}$ $yr^{-1}$.

The above values are, with the exception of *Dineura*, from the year 1973 and thus depict only one season. They indicate that the total secondary production probably does not exceed 1 g $m^{-2}$ $yr^{-1}$ in the invertebrate herbivores of the birch.

The same kind of calculation is not meaningful for vertebrates as they are polyphagous and therefore data of birch production, leaf biomasses etc. cannot be used as a basis for comparison.

## Material Losses Due to Herbivory

Herbivore action causes losses of plant tissue due to consumption, waste of materials while feeding, and due to the reactions of the plant, e.g. deficiency in photosynthesis as a result of reduction in photosynthetic surface. Reactions of the plants to grazing are very difficult to evaluate, and in the case of birch herbivory only a fragmentary knowledge of this phenomenon is available (HAUKIOJA, 1975). When herbivory is studied as a part of an integrated ecological study, the sum of all the components above is the basic object of interest.

Consumption can often be measured in laboratory conditions only, and this was done at Kevo in some of the most important foliage-feeders (HAUKIOJA and NIEMELÄ, 1974a). It is possible to analyze the total removal of materials from plants (consumed + wasted) also from feeding signs in vegetation, but this method is tedious. If the goal is to determine the consumption, extra difficulties will arise as leaves partly consumed during the active growth period later expand their feeding areas, too, and special procedures are needed for estimating the original removal of materials. If, however, the goal is to determine losses in photosynthetic surface due to grazing, the task is easier, because these losses also include other influences in the plant than direct removal of material, for instance malformation of leaves. A possible method in this case is to estimate the percentage of leaves with feeding signs by defoliators (Table 5) and their mean weights. In 1973, leaves at Kevo containing eating marks by defoliating herbivores ("hole-eaters") were 8.1% lighter in September than untouched leaves (46.3 and 50.4 mg, respectively). About 63% of all leaves showed signs of herbivory, and it was

Table 5. Percentages of birch leaves with feeding signs by different insect groups at Kevo in September 1973. 8763 leaves checked

| | |
|---|---|
| Hole-eating Symphyta, Lepidoptera and Coleoptera | 62.9 |
| *Dineura virididorsata* (Hym.) | 38.3 |
| Miners | 1.9 |
| Cecidomyiidea (Diptera) | 34.9 |

calculated that the reduction in biomass was about 3 g dw $m^{-2}$. This estimate thus includes consumption, wasted materials and the possibly induced loss in leaf weight by the reactions of the plant caused by the action of non-mining lepidopteran and saw-fly larvae and herbivorous Coleoptera. Uncertainties in the computation include the assumption that these invertebrates have selected their diet randomly; evaluation of this is not yet available.

An evaluation of the removal of materials was also possible in *Dineura virididorsata*, whose feeding signs are characteristic, estimated at 0.5 g $m^{-2}$ $yr^{-1}$ in 1973. Unlike hole-eaters, this figure is largely caused by consumption, as *Dineura* do not waste materials to any considerable degree while eating, and their consumption occurs mainly in late summer, when leaves have attained their maximum weight. Inaccuracy in the estimate of removed plant biomass may derive particularly from using a mean loss per leaf of 2.2%, a figure taken from 1972 when density of the species was much higher. If consumption by *Dineura* is estimated from production figures, considerably lower intake is found in 1973, 0.23 g $m^{-2}$ $yr^{-1}$.

Consumption of materials by Cecidomyiidae was not measured, but using the maximum biomass as an estimate of the production and assuming a ratio of 1:9 for production and consumption (as in *Dineura*), a value of 4.7 g $m^{-2}$ $yr^{-1}$ is arrived at. In Homopteran the same kind of calculation gives an estimate of 0.5 g $m^{-2}$ $yr^{-1}$ consumed. These values do not include the possible reactions by the plants.

Losses of plant materials caused by other invertebrate herbivores are probably insignificant when compared with these groups mentioned and the sum of plant material lost due to invertebrate herbivory is about 9 g $m^{-2}$ $yr^{-1}$, or perhaps somewhat more as reactions of plants were incompletely recorded.

Among vertebrates, birch-leaf consumption by reindeer has been measured in a large number of plots and the mean was about 0.2 g $m^{-2}$ $yr^{-1}$ (HAUKIOJA and HEINO, 1974). This figure does not include losses caused by the reactions of the plants. Birch consumption by voles was very close to 0 in 1973.

As a summary, the material losses by different groups were as follows:

| | |
|---|---|
| Homoptera | 0.5 g $m^{-2}$ $yr^{-1}$ |
| "Hole-eaters" | 3.0 g $m^{-2}$ $yr^{-1}$ |
| *Dineura* | 0.5 g $m^{-2}$ $yr^{-1}$ |
| Cecidomyiidae | 4.7 g $m^{-2}$ $yr^{-1}$ |
| Reindeer | 0.2 g $m^{-2}$ $yr^{-1}$ |

The total losses were about 15% of the remaining leaf-biomass. Vertebrate herbivory upon birch was unimportant when compared with invertebrate herbivory.

## References

FRANKLIN, R. T.: Insect influences on the forest canopy. In: REICHLE, D. E. (Ed.): Analysis of temperate forest ecosystems, pp. 86–99. Berlin-Heidelberg-New York: Springer 1970.

HÄMET-AHTI, L.: Zonation of the mountain birch forests in northern-most Fennoscandia. Ann. Botan. Soc. Vanamo **34:4**, 1–127 (1963).

HAUKIOJA, E.: Weight development, consumption and egestion of *Dineura virididorsata* (Hym. Tenthredinidae) larvae. Rep. Kevo Subarctic Res. Stat. **10**, 9–13 (1973).

HAUKIOJA, E.: Importance criteria in birch defoliators. In: WIELGOLASKI, F. E. (Ed.): Fennoscandian tundra ecosystems, Part 2: Animals and systems analysis pp. 189–194. Berlin-Heidelberg-New York: Springer 1975.

HAUKIOJA, E., HEINO, J.: Birch consumption by reindeer *(Rangifer tarandus)* in Finnish Lapland. Rep. Kevo Subarctic Res. Stat. **11**, 22–25 (1974).

HAUKIOJA, E., KOPONEN, S., OJALA, H.: Local differences in birch leaf consumption by invertebrates in northern Norway and Finland. Rep. Kevo Subarctic Res. Stat. **10**, 29–33 (1973).

HAUKIOJA, E., NIEMELÄ, P.: Growth and energy requirements of the larvae of *Dineura virididorsata* (Retz.) (Hym., Tenthredinidae) and *Oporinia autumnata* (Bkh.) (Lep., Geometridae) feeding on birch. Ann. Zool. Fennici **11**, 207–211 (1974a).

HAUKIOJA, E., NIEMELÄ, P.: Applicability of laboratory measurements of consumption to field populations of *Dineura virididorsata* (Retz.) (Hym., Tenthredinidae) larvae. Ann. Zool. Fennici **11**, 297–299 (1974b).

KALLIO, P., LEHTONEN, J.: Birch forest damage caused by *Oporinia autumnata* (Bkh.) in 1965–66 in Utsjoki, N Finland. Rep. Kevo Subarctic Res. Stat. **10**, 55–69 (1973).

KALLIO, P., LEHTONEN, J.: On the ecocatastrophe of birch forests caused by *Oporinia autummata* (Bkk.) and the problem of reforestation. In: WIELGOLASKI, F. E. (Ed.): Fennoscandian tundra ecosystems. Part 2: Animals and systems analysis, pp. 174–180. Berlin-Heidelberg-New York: Springer 1975.

KÄRENLAMPI, L., KAUHANEN, H.: A direct gradient analysis of the vegetation of the surroundings of the Kevo Subarctic Station. Rep. Kevo Subarctic Res. Stat. **9**, 82–98 (1972).

KJELVIK, S., KÄRENLAMPI, L.: Plant biomass and primary production of Fennoscandian subarctic and subalpine forests and of alpine willow and heath ecosystems. In: WIELGOLASKI, F. E. (Ed.): Fennoscandian tundra ecosystems. Part 1: Plants and microorganisms, pp. 111–120. Berlin-Heidelberg-New York: Springer 1975.

KOPONEN, S.: On the mining insects of the mountain birch in northernmost Fennoscandia. Rep. Kevo Subarctic Res. Stat. **10**, 14–19 (1973a).

KOPONEN, S.: Herbivorous invertebrates of the mountain birch at Kevo, Finnish Lapland. Rep. Kevo Subarctic Res. Stat. **10**, 20–28 (1973b).

KOPONEN, S.: On the occurrence and ecology of *Eriocrania* spp. (Lep., Eriocraniidae) and other mining insects of the birch in northernmost Fennoscandia in 1973. Rep. Kevo Subarctic Res. Stat. **11**, 52–64 (1974).

RAFES, P. M.: Estimation of the effects of phytophagous insects on forest production. In: REICHLE, D. E. (Ed.): Analysis of temperate forest ecosystems, pp. 100–106. Berlin-Heidelberg-New York: Springer 1970.

TENOW, O.: The outbreaks of *Oporinia autumnata* Bkh. and *Operophthera* spp. (Lep., Geometridae) in the Scandinavian mountain chain and northern Finland 1862–1968. Zool. Bidrag, Uppsala Suppl. **2**, 1–107 (1972).

WIEGERT, R. G.: Energy dynamics of the grasshopper populations in old field and alfalfa field ecosystems. Oikos **16**, 161–176 (1965).

WIELGOLASKI, F. E., KJELVIK, S., KALLIO, P.: Mineral content of tundra and forest tundra plants in Fennoscandia. In: WIELGOLASKI, F. E. (Ed.): Fennoscandian tundra ecosystems. Part 1: Plants and microorganisms, pp. 316–332. Berlin-Heidelberg-New York: Springer 1975.

# Importance Criteria in Birch Defoliators

E. Haukioja

Different criteria may be used for the importance of various animal species in a community. The most simple of these is to use numbers, which however, totally neglect the size and activity of animals. Biomass figures provide a sounder basis for this kind of evaluation but here, in turn, the ratio size/assimilation is neglected. The most widely used parameters in IBP studies have been production and assimilation. Production is a possible basis for determining the relative importance of species. Assimilation is the amount of energy or matter absorbed from the alimentary canal and thus available for different physiological processes. In IBP studies the term has been used as a synonym for energy flow. Petrusewicz and MacFadyen (1970) have used the term energy flow to assess also the relative importance of different species or populations in an ecosystem. Mertens (1973) has criticized this and stresses the descriptive nature of this parameter. Mertens (1973) proposed the respired energy in relation to the total amount of energy captured by the autotrophs of the ecosystem, as a measure of the relative importance of a species in the functioning of the ecosystem. Respiration indicates how energy assimilated by plants escapes from the ecosystem, and the sum of autotrophic assimilation should equal the respired energy by all parts of the ecosystem if there is no run-off or accumulation of energy. The method therefore seems very applicable as it allows comparison between populations at different trophic levels.

Another possible way ist to study the effects of different species, and to use these criteria in the order of their proportional importance. Comparing different trophic levels is not possible in this kind of situation. The impact of a population on the preceding trophic level seems to be a logical measure of the effect. Mertens (1973) proposed for this the amount of materials removed from the preceding trophic level, i.e. the sum of materials consumed plus those wasted in the feeding process. The method is useful if preying or grazing by animals does not affect the prey or the host more than the actual removal of materials indicates. In predators this does not apply if the removal of a prey creates a vacancy for another individual. The same is true in herbivores, if plants react to grazing e.g. by changes in production. As an importance criterium of a species Hurlbert (1971) proposed the sum of the changes (over-all species) in productivity which occurs if the particular species is removed from the community. This method would also be useful for comparing the effects of more than one species, but in practice this kind of determination is quite impossible.

In the following different aspects are discussed for evaluating the proportional importance of some dominant defoliator species of the birch *(Betula tortuosa)* in the birch tundra of Finnish Lapland.

## Energy Budget Data

Larval populations of *Dineura virididorsata* (Hym., Tenthredinidae) and *Oporinia autumnata* (Lep., Geometridae) are first used as examples. These species were chosen for closer study because, in the case of *Dineura*, it has high biomass figures (HAUKIOJA and KOPONEN, 1975).and *Oporinia*, in turn, is a species which causes vast defoliation and kills birches during its mass outbreaks. The last of these was observed in the Kevo area in 1965–1966 (KALLIO and LEHTONEN, 1973, 1975). In the early seventies, however, larvae of *Oporinia* were almost entirely absent from the samples taken for determining the occurrence and biomass of birch herbivores at Kevo.

Figures characterizing energy budgets of the above two species originate from larval cultures kept at Kevo in 1972 and 1973. The methods used and results found are given in HAUKIOJA and NIEMELÄ (1974a) and the relevance of culture-raising experiments for field populations has been evaluated in the case of *Dineura* in HAUKIOJA and NIEMELÄ (1974b). According to these sources, the flow of materials through these species is distributed as follows:

| | *Dineura* | *Oporinia* |
|---|---|---|
| $NU/MR$ | ca. 0 | ? |
| $FU/C$ | 0.84 | 0.63 |
| $A/C$ | 0.16 | 0.37 |
| $P/A$ | 0.57 | 0.45 |
| $R/A$ | 0.43 | 0.55 |

where $MR$ = materials removed in the feeding process from *Betula* leaves, $NU$ = materials wasted in feeding, $FU$ = excreta, $C$ = consumption, $A$ = assimilation, $P$ = net production and $R$ = respiration. Assimilation was calculated as $C-FU$, and respiration as $A-P$; other parameters were measured. Net production here means the weight gain in the culture up to the maximum weight attained without taking moulted skins into account, and consequently respiration values are biased to the same extent.

When compared with references in the literature, the efficiency in *Oporinia* closely resembles that mentioned for defoliating larvae (SMITH, 1972; WINTER, 1971) but in *Dineura* $A/C$ is lower and $P/R$ higher. The low assimilation efficiency (and high defecation rate) of ingested food is probably explained by a seasonal trend in the digestibility of the food, birch leaves. This is because *Dineura* occurs late in the summer and the nutritive value of the leaves (especially protein content) decreases with the advancing season (WIELGOLASKI et al., 1975). *Oporinia*, an early season species, and *Dineura* belong near the extreme ends of this gradient. The high production/respiration ratio in *Dineura* is probably explained by the slow growth of this species, making a more economic use of energy possible.

Table 1. Estimated energy budget factors for *Dineura virididorsata* and *Oporinia autumnata* populations at Petsikko Fell in 1973. Leaf biomass energy and primary production concern the birch

| | *Dineura* larvae | | | *Oporinia* larvae | | |
|---|---|---|---|---|---|---|
| | kcal $m^{-2}$ $yr^{-1}$ | % leaf biom. energy | % prim. prod. | kcal $m^{-2}$ $yr^{-1}$ | % leaf biom. energy | % prim. prod. |
| *P* | 0.91 | 0.306 | 0.107 | 0.08 | 0.027 | 0.009 |
| *A* | 1.59 | 0.534 | 0.187 | 0.18 | 0.061 | 0.021 |
| *R* | 0.68 | 0.229 | 0.080 | 0.10 | 0.034 | 0.012 |
| *C* | 9.95 | 3.345 | 1.171 | 0.49 | 0.165 | 0.058 |
| *FU* | 8.36 | 2.810 | 0.984 | 0.31 | 0.104 | 0.036 |

In the calculations given later, the energetic value of all substances has been assumed to be 5 cal $(mg\ dw)^{-1}$, an assumption which probably is not more than ten percent inaccurate. The calculations were made from computed field production estimates based on weekly samples. The values were given for 100 g dried leaves and transferred to values per $m^2$ by applying the figure given by KÄRENLAMPI (KJELVIK and KÄRENLAMPI, 1975) for *Betula* leaves in the IBP birch site, 595 kg $ha^{-1}$. The area for which the present calculations were made, is situated about 50 km south of the IBP birch site, but has much the same kind of characteristics, probably also for leaf biomass. This area was the only one during 1971–1973 with both *Dineura* and *Oporinia* in considerable numbers at the same place. The primary production value of the birch used in the calculations (1,700 kg $ha^{-1}$ $yr^{-1}$) is a very rough estimate (KALLIO and KÄRENLAMPI, pers. comm.).

The different herbivore energy budget parameters given in Table 1 represent a very low percentage of the energy content in the birch leaf-biomass, from about 3% for consumption in *Dineura* to about 0.25% for respiration. The energy in relation to primary production of birch is about one third of the value on a leaf biomass basis, e.g. 0,08% for respiration to primary production, the importance parameter recommended by MERTENS (1973).

In *Oporinia* the corresponding figure was 0.012%. The ratio of these importance criteria between species is 0.15. A different figure (0.05) is found if the ratio is calculated from consumption. Thus different criteria in energy budget data may offer different results; even the order of the most important species may be reversed.

The data support the view that, in energy terms, birch defoliators are negligible factors in their community. This expression is, however, applicable to the present level of populations only. In *Dineura* large-scale defoliations are not known, but *Oporinia* is potentially capable of defoliating an area similar to that studied. In that case, assuming a consumption of the total leaf production, 595 kg $ha^{-1}$, and assuming the same ratios for energy budget parameters as with low population densities, *Oporinia* population would respire about 7% of the normal-year primary production and a much larger proportion from the primary production of defoliated trees. *Oporinia* is thus an important potential factor in the ecosystem.

## Importance of the Effects of Defoliators

Large-scale effects, e.g. defoliations, are to a large degree dependent on the numbers of herbivores. However, a defoliator alone does not determine its effects; the way the host reacts to grazing may be even more important—and more complicated.

Hosts have evidently adapted themselves to the normal presence of low-population densities of herbivores. RAFES (1970) mentions that only foliage losses above 75% by phytophagous insects usually cause significant increment losses in the plants, which may result in death of the host. Factors affecting the survival of birches after heavy defoliation were studied at Kevo and in some years, after favorable growth seasons, birches were able to recover from artificial defoliation. One reason why the *Oporinia* damage in 1965–1966 caused such destruction among birches was probably to be found in the unfavorable seasons with low temperatures preceding defoliation (KALLIO and LEHTONEN, 1973, 1975).

The time of defoliation is also important in determining the effects of the host. Loss in photosynthetic surface is evidently more fatal early in the season than just before the annual leaf-fall. The time of defoliation affects the accumulation of storage materials in the tree, but is very difficult to quantify.

An experiment was carried out at Kevo in order to simulate the grazing situation by caterpillars such as e.g. *Oporinia*, at different times of the season. During the spring growth period small pieces were removed from leaves, leaving a control leaf of the same size from a nearby twig. The removed leaf piece was weighed and the weight of the "grazed" leaf and its control were also taken after the maximum weight of leaves had been attained. Results (Table 2) indicate that even relatively small "eating" during the early growth period of the leaves causes larger losses of photosynthetic surface than the mere removal of materials indicates. After leaves were nearly full-grown, they were capable of compensating for the loss of materials. In leaves "grazed" in June the final weight of the grazed leaf + the weight of the removed piece was heavier than the control leaf in five cases while the opposite was true in 22 cases. In leaves grazed in July, but before they had completed their growth, the corresponding figures were 16 and 14. The difference is statistically significant ($p<0{,}025$, $X^2=5.98$, $f=7$).

It was also noticed in a field situation that leaves can compensate for the effects of material removal, at least to a certain degree, at a late stage of their development. HAUKIOJA (1974) has shown that in a mining micro-lepidopteran

Table 2. Results of an experiment concerning artificial "consumption" of birch leaves. Weights are dry weights in mg

| Date of "consumption" | Mean material taken (A) | % of leaf weight | Date of checking | Mean weight of "consumed" leaves (B) | Mean weight of control leaves (C) | (A+B)−C |
|---|---|---|---|---|---|---|
| June 12 | 1.92 | ca. 15 | July 25 | 44.7 | 59.2 | −12.58 |
| July 12 | 11.27 | ca. 20 | July 26 | 65.1 | 73.6 | + 2.77 |

genus, *Eriocrania*, consumption during a life-time was of the order of 3 mg, calculated from production and an assumed ratio of consumption to production. However, weights of leaves mined by an early season species were on an average 10–60 mg lower than predicted according to a regression equation for weight/width data of unmined leaves from the same place. Thus reactions of the plant to grazing, which happened while leaves were still growing, resulted in larger losses of green tissue than from the consumption directly (in the present case particularly due to a curving of mined leaves). In a miner species occurring later in summer some kind of compensation in weight occurred in the unmined half of the leaf.

## Discussion

Criteria of importance depend on the goal of the study. If a number of species are compared, belonging to different groups, the respiration-production energy method proposed by MERTENS (1973) seems very good. In energy-budget data, some other parameters may also be used as a basis for comparison. Production has a special position among energy-budget factors according to the objectives of the Tundra Biome programme of the IBP (ANON, 1972). Furthermore, production usually acts as a source of food for another population or populations and thus establishes a link between them. Assimilation of a population is a less obvious goal in an ecosystem study, but the situation is different if the population is studied without reference to other parts of the ecosystem.

In studies concerning transfer of matter and energy, it is logical to regard as important those elements in an energy budget which have the largest quantities of matter and energy. Thus a population may be understood as a unit channeling energy along different pathways. Among the energy-budget factors of a herbivore population, material removed from plants has the highest order, with the largest energy or matter content.

The effects of herbivore populations are necessarily not in the same ratio as their proportional importance e.g. according to the respiration/primary production ratio. This is because the respiration/consumption ratio is rather variable, depending e.g. on the age of the animal and on the quality of the diet. There is probably no simple, feasible way to describe the proportional importance of the impact of a species in an ecosystem e.g. in the way outlined by HURLBERT (1971). This is perhaps why the effects of herbivores have received very little attention in spite of the fact that the total effects of herbivory may largely depend on reactions occurring in plants as a consequence of eating (e.g. VARLEY, 1967).

## References

ANON: Strategy of the tundra biome. IBP Norden, **8**, 3–11 (1972).

HAUKIOJA, E.: Measuring consumption in *Eriocrania* (Eriocraniidae, Lep.) miners with reference to interaction between the leaf and the miner. Rep. Kevo Subarctic Res. Stat. **11**, 16–21 (1974).

HAUKIOJA, E., KOPONEN, S.: Birch herbivores and herbivory at Kevo. In: WIELGOLASKI, F. E. (Ed.): Fennoscandian tundra ecosystems. Part 2: Animals and systems analysis pp. 181–188. Berlin-Heidelberg-New York: Springer 1975.

HAUKIOJA, E., NIEMELÄ, P.: Growth and energy requirements of the larvae of *Dineura viridi-dorsata* (Retz.) (Hym., Tenthredinidae) and *Oporinia autumnata* (Bkh.) (Lep., Geometridae) feeding on birch Ann. Zool. Fennici **11**, 207–211 (1974a).

HAUKIOJA, E., NIEMELÄ, P.: Applicability of laboratory measurements of consumption to field populations of *Dineura virididorsata* (Retz.) (Hymenoptera, Tenthredinidae) larvae. Ann. Zool. Fennici **11**, 297–299 (1974b).

HURLBERT, S. H.: The nonconcept of species diversity: A critique and alternative parameters. Ecology **52**, 577–586 (1971).

KALLIO, P., LEHTONEN, J.: Birch forest damage caused by *Oporinia autumnata* (Bkh.) in 1965–66 in Utsjoki, N Finland. Rep. Kevo Subarctic Res. Stat. **10**, 55–69 (1973).

KALLIO, P., LEHTONEN, J.: On the ecocatastrophe of birch forests caused by *Oporinia autumnata* (Bkh.) and the problem of reforestation. In: WIELGOLASKI, F. E. (Ed.): Fennoscandian tundra ecosystems. Part 2: Animals and systems analysis, pp. 174–180. Berlin-Heidelberg-New York: Springer 1975.

KJELVIK, S., KÄRENLAMPI, L.: Plant biomass and primary production of Fennoscandian subarctic and subalpine forests and of alpine willow and heath ecosystems. In: WIELGOLASKI, F. E. (Ed.): Fennoscandian tundra ecosystems. Part 1: Plants and microorganisms, pp. 111–120. Berlin-Heidelberg-New York: Springer 1975.

MERTENS, A. L.: The use and misuse of the "energy flow" concept. In: KENDEIGH, S. C., PINOWSKI, J. (Eds.): Productivity, population dynamics and systematics of granivorous Birds, pp. 407–410. Warszawa: PWN 1973.

PETRUSEWICZ, K., MACFADYEN, A.: Productivity of terrestrial animals (principles and methods) IBP-handbook No. 13. Oxford-Edinburgh: Blackwell 1970.

RAFES, P. M.: Estimation of the effects of phytophagous insects on forest production. In: REICHLE, D. E. (Ed.): Analysis of temperate forest ecosystems, pp. 100–106. Berlin-Heidelberg-New York: Springer 1970.

SMITH, P. H.: The energy relations of defoliating insects in a hazel coppice. J. Anim. Ecol. **41**, 567–587 (1972).

VARLEY, G. C.: The effects of grazing by animals on plant productivity. In: PETRUSEWICZ, K. (Ed.): Secondary productivity of terrestrial ecosystems (Principles and methods), Vol. 2, pp. 773–778. Warszawa-Krakow: PWN 1967.

WIELGOLASKI, F. E., KJELVIK, S., KALLIO, P.: Mineral content of tundra and forest tundra plants in Fennoscandia. In: WIELGOLASKI, F. E. (Ed.): Fennoscandian tundra ecosystems. Part 1: Plants and microorganisms, pp. 316–332. Berlin-Heidelberg-New York: Springer 1975.

WINTER, K.: Studies in the productivity of Lepidoptera populations. In: ELLENBERG, H. (Ed.): Integrated experimental ecology. Methods and results of ecosystem research in the German Solling project, pp. 94–99. Berlin-Heidelberg-New York 1971.

# Wild Reindeer Food Habits and Range Use at Hardangervidda

E. GAARE and T. SKOGLAND

The management of wild or tame reindeer must be partly based on household studies of the pastures. This is also true for other ungulates, and studies on feeding behavior have been numerous. A review of the literature may be found in GOLLEY and BUECHNER (1968) or PETRUSEWICZ and MACFADYEN (1970).

The work presented here is part of the IBP-organized study of the Hardangervidda ecosystem. The team studied reindeer diet in 1970 and habitat utilization patterns during 1971–1973. The animals have been localized to plant communities and the time spent in each has been measured. These data are thus available for a total synthesis. Detailed progress reports have been published: GAARE et al. (1970), GAARE and SKOGLAND (1971), SKOGLAND (1974, 1975).

The botanical nomenclature follows LID (1963) for higher plants, DAHL and KROG (1973) for lichens and NYHOLM (1956–69) for mosses. Names of vegetation types have been given according to the recommendation of DAHL et al. (1971).

## The Study Area

The alpine Hardangervidda plateau ranging from 1000–1500 m altitude with a total non-forest ground range of ca. 7500 km$^2$ has carried a wild reindeer herd since prehistoric times. The habitat shows a gradient from oceanic climate in the SW to a more continental climate in the NE, with an annual precipitation ranging from nearly 2000 mm to 500 mm.

As this creates very different snow depths and vegetation periods, the influence on vegetation distribution and composition is considerable. The low alpine region ranges from ca 1100–1450 m in the NE and 850–1200 in the SW. Within this region the western part is dominated by grass heaths and dwarf shrub heaths in which mosses play an important role. In the continental NE dwarf shrub heaths dominate with fruticose lichens as an important element. The greater topographic variation in the western and the south-western part of the range creates local continental pockets exploited by the reindeer during late winter migration. Migration distances vary from 100–150 km from early winter areas to calving grounds. The altitudinal variation from 1000–1500 m is exploited by the reindeer in the western part of the range during the early and late growing season.

## The Reindeer Population

The sizes of the deer were estimated by comparative analysis of mandibular lengths and dressed weights (REIMERS pers. comm.) to be in the category of the heavily-exploited Snøhetta-range herd (GAARE, 1968) where lichen heath over-grazing was particularly evident. This evidence led to a plan for reducing the Hardangervidda herd to under 10000 deer in 1970.

The population has undergone great fluctuations in size with an estimated low of ca. 1500 around 1930 (THAMBS-LYCHE, 1941) to an estimated high of ca. 32000 in the middle 1960s (KRAFFT, 1970 and pers. comm.). During the second world war and early 1950s there was a gradual influx of at least 3000 domestic reindeer into the herd. The last domestic reindeer company abandoned its function in 1956 due to conflict and mixing with wild deer. Estimates of total population size based on aerial censuses (KRAFFT, pers. comm.) during the present study period 1970–73 have varied from a summer total of 14700 in 1970 to 7600 in 1971 to 10000 in 1972. This represents a total standing biomass of 552 t in the autumn 1972. The figure is based upon dressed weight $\times 2$ from the hunt in 1971 and sex and age structure in September 1972. The hunt may harvest about 130 t of this under stabilized conditions. However, no hunting was allowed during the 1971 and 1972 autumn seasons. Ground counts of herd compositions were carried out throughout the study period. Early summer calf to female ratios were found to vary between 35% and 54% in the period 1970–73. A first-month mean calf death rate of 15% (SKOGLAND, 1974) was found in all 4 years. October (rut) herd compositions were found to be 13–20% calves, 2–10% yearlings, 10–11% $2^1/_2$-year-old, 2.5–9% $3^1/_2$-year-old and older bucks, and about 50% females older than $2^1/_2$ years for 1971 and 1972. During those years hunting prohibition left herd growth undisturbed, with an annual growth rate of 20%.

## Materials and Methods

Food habits and plant preferences of reindeer were studied during 1970 by botanical analyses of culled reindeer rumen content samples and ingested plants collected in esophagus fistula bags. Rumen samples were collected in the field, preserved in FAA and screened mechanically and by running water in a 150-micrometer sieve mesh in the laboratory. The sample was then spread on a $30 \times 70$ cm dish and 400 point identification readings were carried out through a $10–40 \times$ pancreatic binocular microscope (GOLLEY et al. 1968).

Due to accidents only one $1^1/_2$-year-old buck fitted with an esophageal fistula (EF) was available. (Concerning fistula techniques see SKJENNEBERG et al. (1971). This buck and a control buck of the same age were brought to the study area where the main wild herd was known to graze at that time. Between sampling periods they stayed at a lowland farm some 250 km away. EF samples were collected in nylon bags attached to the neck of the deer, and plants were collected during grazing for periods of 20 min. Samples were preserved in 70% alcohol and analyzed botanically in the same manner as rumen samples. Standard chemical analyses of rumen and EF samples were made.

Plant coverage in grazed areas was estimated in 33 × 100 cm plots. Habitat utilization patterns of free-ranging undisturbed groups of deer were studied during 1971–73, following group movements with 10–40 X field-glasses and classifying animal usage of slope, aspect and phytosiological plant alliances (SKOGLAND, 1975). Food consumption rates were estimated by weighting EF bag-collected ingesta per time unit (GAARE and SKOGLAND, 1973), and by use of comparative literature data for related ungulates.

## Results

**Food Habits.** Tables 1 and 2 show the results from the EF (esophageal fistula) and rumen samples. The average reading is shown together with its 80% confidence limits. In addition Fig. 1 gives a survey of the rumen samples data through the year. The uncertainty in plant species identification in the point readings of both types of samples made grouping necessary. For clarity the graph shows the most prevalent grouped species. It is worth mentioning however that the total number of species determined in 81 rumen samples was 55. This was distributed over 7 sample periods through the year. In addition the 52 EF samples in 5 sample periods gave 14 more species.

In the winter period the animals utilized a broad spectrum of available species. In contrast, they can afford to be more selective during the summer period. Species like *Cetraria islandica* and *Cladonia mitis* in the winter and *Deschampsia caespitosa* and *D. flexuosa* in the summer are often selected out of proportion to their standing biomass. The same is true for epilithic lichens in April. The selection pattern is much the same as shown by SKJENNEBERG et al. (1971) from a coastal range in Northern Norway.

Table 1. Reindeer diet shown as a percent score of 400 points read from esophageal fistula samples. An 80% confidence interval is given

| Month | January | April | June | July | October |
|---|---|---|---|---|---|
| No. of samples | 8 | 12 | 10 | 12 | 10 |
| | $\bar{x}$ 80% c.i. | $\bar{x}$ 80% c.i. | $\bar{x}$ 80% c.i. | $\bar{x}$ 80% c.i. | $\bar{x}$ 80% c.i. |
| *Betula nana* | 0.5 ± 0.2 | 0 | 0 | 11 ± 6.2 | 0 |
| *Salix* spp. | 0.3 ± 0.1 | 0 | 46 ± 7.4 | 9.6 ± 2.5 | 0 |
| Other woody plants | 4.9 ± 1.8 | 9.1 ± 1.5 | 0.3 ± 0.1 | 0.1 ± 0.0 | 1.5 ± 0.6 |
| *Carex/Eriophorum* | 0 | 2.4 ± 0.4 | 8.5 ± 1.8 | 5.6 ± 1.7 | 0.3 ± 0.2 |
| Other graminids | 0.0 ± 0.0 | 1.7 ± 0.3 | 8.7 ± 1.0 | 16 ± 5.5 | 4.2 ± 1.6 |
| Forbs | 0 | 0 | 14 ± 7.0 | 41 ± 6.9 | 1.4 ± 0.5 |
| Mosses | 0.8 ± 0.5 | 4.4 ± 0.4 | 2.8 ± 0.5 | 0.3 ± 0.1 | 0.7 ± 0.3 |
| Mushrooms | 0 | 0 | 0 | 0 | 3.8 ± 3.1 |
| *Cetraria* spp. | 19 ± 2.5 | 46 ± 2.0 | 8.6 ± 1.9 | 10 ± 2.7 | 43 ± 6.0 |
| Cladina group | 65 ± 3.4 | 9.3 ± 1.1 | 3.7 ± 1.1 | 3.6 ± 1.5 | 42 ± 5.6 |
| Other epigeic lichens | 4.0 ± 0.9 | 9.0 ± 1.4 | 4.8 ± 2.4 | 0.6 ± 0.2 | 1.8 ± 0.4 |
| Epilithic lichens | 0 | 7.1 ± 1.1 | 0 | 0 | 0 |
| Litter | 6.3 ± 1.0 | 11 ± 1.1 | 4.8 ± 0.8 | 1.0 ± 0.5 | 2.1 ± 0.4 |

Table 2. Reindeer diet shown as a percent score of 400 points read from rumen content samples. An 80% confidence interval is given

| Month | January | | April | | June | | July | | September | | October | | December | |
|---|---|---|---|---|---|---|---|---|---|---|---|---|---|---|
| No. of samples | 9 | | 12 | | 8 | | 12 | | 14 | | 13 | | 13 | |
| | $\bar{x}$ | 80% c. i. | $\bar{x}$ | 80% c. i. | $\bar{x}$ | 80% c. i. | $\bar{x}$ | 80% c. i. | $\bar{x}$ | 80% c. i. | $\bar{x}$ | 80% c. i. | $\bar{x}$ | 80% c. i. |
| *Salix* spp. | 0 | | 0 | | 17 | ±3.5 | 11 | ±0.9 | 3.0 | ±0.6 | 0 | | 0 | |
| Other woody plants | 8.9 | ±1.6 | 22 | ±2.4 | 8.0 | ±2.9 | 2.3 | ±1.6 | 7.5 | ±1.2 | 3.9 | ±1.0 | 12 | ±1.5 |
| *Carex*/*Eriophorum* | 8.9 | ±2.0 | 7.8 | ±0.9 | 6.1 | ±0.9 | 17 | ±2.9 | 17 | ±1.5 | 19 | ±2.5 | 8.3 | ±1.1 |
| Other graminids | 19 | ±2.2 | 14 | ±2.7 | 19 | ±2.7 | 32 | ±2.4 | 30 | ±2.1 | 19 | ±3.0 | 18 | ±1.1 |
| Forbs | 0 | | 0 | | 0.1 | | 7.6 | ±1.0 | 1.1 | ±0.3 | 0 | | 0 | |
| Mosses | 2.5 | ±0.6 | 12 | ±1.1 | 6.1 | ±1.0 | 3.7 | ±0.5 | 5.3 | ±0.7 | 2.1 | ±0.3 | 3.2 | ±0.4 |
| Mushrooms | 0 | | 0 | | 0 | | 0 | | 0.5 | ±0.2 | 0 | | 0 | |
| *Cetraria* spp. | 9.1 | ±1.4 | 5.6 | ±1.2 | 12 | ±1.6 | 6.9 | ±0.9 | 5.8 | ±0.9 | 8.0 | ±0.7 | 15 | ±2.0 |
| Cladina group | 44 | ±4.6 | 18 | ±2.0 | 14 | ±2.6 | 16 | ±2.6 | 24 | ±1.3 | 40 | ±3.1 | 26 | ±1.8 |
| Other epigeic lichens | 2.7 | ±0.5 | 4.1 | ±1.2 | 4.6 | ±1.0 | 2.2 | ±0.4 | 4.1 | ±0.8 | 2.3 | ±0.5 | 12 | ±0.9 |
| Epilithic lichens | 0.1 | ±0.1 | 6.4 | ±1.4 | 0.6 | ±0.4 | 0 | | 0.1 | ±0.1 | 1.0 | ±0.5 | 0.2 | ±0.1 |
| Litter | 3.3 | ±0.4 | 9.4 | ±0.6 | 9.8 | ±1.9 | 0 | | 0 | | 4.9 | ±0.4 | 4.9 | ±0.6 |

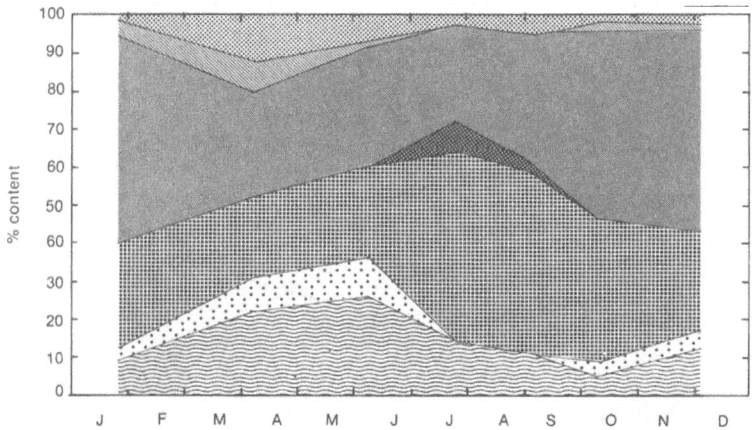


Fig. 1. Reindeer diet, Hardangervidda 1970. Based on 82 rumen content samples. From top to bottom the percentage score of the following plant groups is shown: Mosses, Epilithic lichens, Epigeic lichens, Forbs, Graminids, Litter and Woody plants

Most of the woody plants eaten are leaves, but in the winter period only the evergreens fall into this group. Leaves of *Betula nana* and other shrubs are part of the litter from October to early June. Most of the litter group consists of dead leaves from shrubs. Together with mosses they form part of the winter diet which is unavoidably ingested during lichen grazing, so that litter is scarce in the summer diet from July to September.

The EF and rumen samples show some distinct differences. Lichens show the highest content in the rumen samples during the summer while the EF samples show the highest lichen content in the winter period. Graminids show consistently the highest score in the rumen samples, especially in the summer. Forbs score highest in the EF samples.

A difference is also evident in chemical composition, shown in Fig. 2. From early June to December rumen analysis shows a slightly increasing sugar and nitrogen content, and a slightly decreasing fiber content. Clipped plant samples showed in the same period an even N content of about 3% till August, then a marked drop in December to about 1%. Plant clippings were taken in 1971. The fiber content of clipped samples was 27–28% from June to August, and dropped to 21% in December. This pattern compares well with the fiber content of rumen samples.

The variation between samples from the same period is larger in the botanical than in the chemical analyses and consistently greater in the EF samples than in the rumen samples.

**Habitat Utilization Pattern.** During the annual cycle of movement, the deer roam a variety of seasonal range types and plant alliances. Most vegetation receives some use, as also indicated by the rumen and EF samples. The reindeer pattern of feeding is very extensive: they nibble at the ground as they move, aggregate in spots of feeding interest and disperse again when food is abundant.

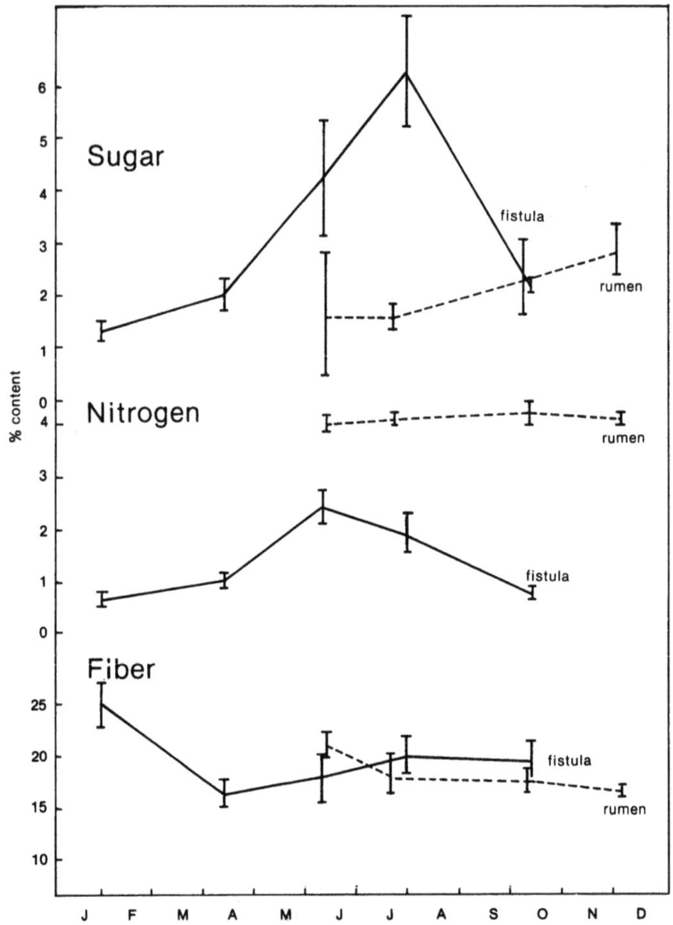


Fig. 2. Percentage composition of fistula and rumen content samples. A 95% confidence interval is indicated

Distances moved each day vary with season and external factors such as weather and harassment by insects and predators. By continuous observation a certain annual range-use pattern was established. Figure 3 shows the part of Hardangervidda frequented seasonally by at least 60% of the population, i.e. 4000–6000 deer, and Table 3 shows the mean percentage of the year the deer spent in various plant alliances and vegetation types. The eastern winter lichen heaths were entirely of the Arctostaphylio-Cetrarion nivalis type (Table 3). Grass-dominated associations of the Phyllodoco-Vaccinion myrtilli were important particulary during the spring season, and during the summer.

Small scattered groups, particularly male bands, could be found outside the hatched area especially near glaciers and other inaccessible western mountains, but the home range usage pattern of the majority of the herd was regular within the seasons over the 4-year study period. Important alliances were Nardo-Caricion bigelowii, Lactucion alpina and Caricion canescentis-nigrae.

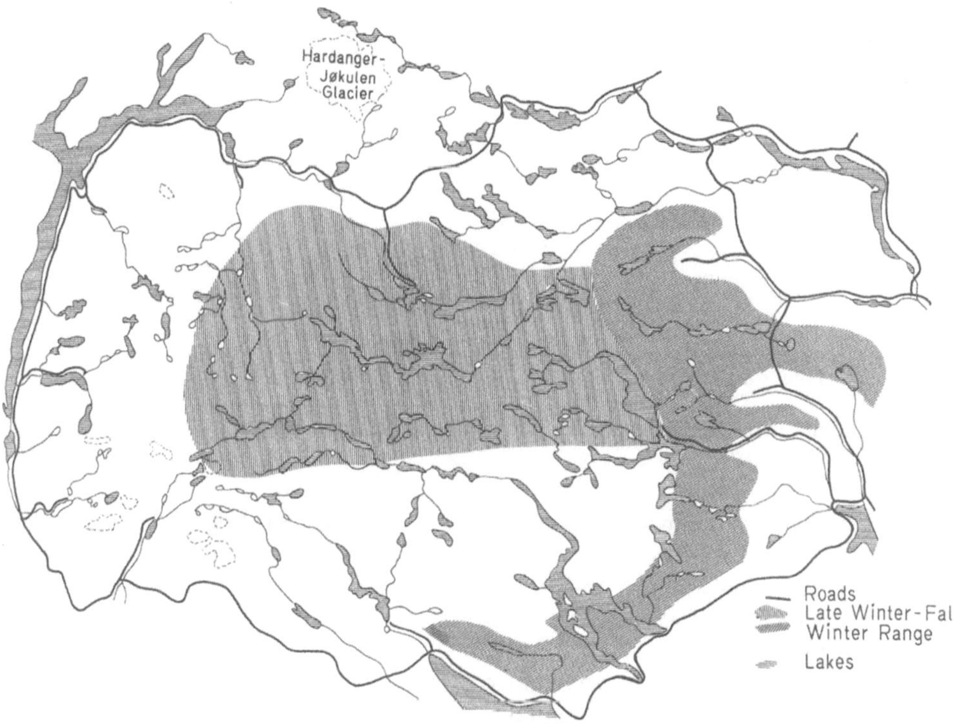


Fig. 3. Map showing main reindeer home range 1970–1973 at Hardangervidda

Table 3. Reindeer grazing time (percentage of the year), and estimated amounts consumed, from different vegetation types and alliances. Grazing percentages are means for the 1971–73 period. Annual consumption by the total reindeer herd in tons dry matter (DM)

| | % of year | | Consumed tons DM |
|---|---|---|---|
| Chionophobous lichen heath (total) | | 43.1 | |
| Arctostaphylio-Cetrarion nivalis | 43.1 | | 3129 |
| Graminid or forb-rich associations of moderately chionophobous alliances (total) | | 15.9 | |
| Kobresio-Dryadion | 1.4 | | 102 |
| Juncion trifidi scandinavicum | 2.1 | | 152 |
| Potentillo-Polygonion vivipari | 1.1 | | 80 |
| Phyllodoco-Vaccinion myrtilli | 11.3 | | 820 |
| Chionophilous graminid or forb-rich heaths and meadows (total) | | 19.5 | |
| Nardo-Caricion bigelowii | 15.7 | | 1140 |
| Ranunculo-Oxyrion digynae | 3.8 | | 276 |
| Chionophilous tall herb meadow (total) | | 11.6 | |
| Lactucion alpinae | 11.6 | | 842 |
| Mires (total) | | 7.4 | |
| Caricion canescentis-nigrae and Salicion myrsinitis | 6.5 | | 472 |
| Oxycocco-Empetrion hermaphroditi | 0.9 | | 65 |
| Snow beds (total) | | 2.5 | |
| Cassiopo-Salicion herbacea and Polarion | 2.5 | | 182 |

Table 4. Observed home range size and animal density, means of 1970–1973. The number of the observed deer was 4000–6000

| Season | Months | Size $km^2$ | Density deer $km^{-2}$ |
|---|---|---|---|
| Winter | Nov–Mar | 1270 | 3–5 |
| Spring | Mar–May | 610 | 7–10 |
| Calving | May-June | 340 | 12–18 |
| Early summer | June | 340 | 12–18 |
| Summer [a] | July | 320 | 13–19 |
| Total for 9 months | | 2900 | |

[a] Late summer and fall are excluded. During this mushroom season dispersal complicates density calculations.

Table 4 shows the mean estimated seasonal and annual home range size and deer densities. Lowest densities were found during winter with 3–5 deer $km^{-2}$ and highest during summer, 13–19. The total range exploited by 4000–6000 deer was ca. 2900 $km^2$ per year, i.e. 38% of total available habitat, giving a density of 1.4–2.1 reindeer $m^{-2}$.

Within the winter range an estimated 147–220 $m^2$ lichen mat were available per deer per day. The mean annual density observed lies within the expected range based on literature inferences from other studies (KELSALL, 1968). However, monthly densities were up to $18.5 \pm 0.77$ determined for 9 months of the year for ca. 6000 deer. Densities up to 33 deer $km^{-2}$ were found in July during insect harassment in all 4 observation years.

The small variation between months indicates that the same number of deer move in a clockwise manner through the seasonal biotopes. During winter a mean of only 25% of the landscape is available for grazing due to the snowpack (data based on transect lines). This means that feeding densities of up to 75 deer $km^{-2}$ during the winter can be expected. This figure compares well with observed mean group sizes of 50–100 deer. There is a 50% chance of observing this group size at any time of the year (SKOGLAND, 1974). This evidence indicates that density might be affected both by vegetation and other external factors like predation and parasites with a subsequent influence on social organization.

**Food Consumption.** Estimated intake rates have been calculated from the food entry rate into EF bags as dry matter $min^{-1}$ (GAARE and SKOGLAND, 1973; WHITE, pers. comm.). The data indicate that females have slow growth at first pregnancy, thus mostly following male growth to up 1½ years. Table 5 shows estimated annual consumption and live weight productivity during 1972. An annual cumulative herd intake of 7260 tons dry matter (DM) was estimated for the home range area, based on a population of ca. 10000 animals. This means a daily herd intake of ca. 20 t DM. The mean food intake in kg dry matter of plants per kg live weight gain of reindeer was ca. 400 kg. Mean live meat weight was ca. 60 kg $km^{-2}$. Estimated annual consumption from different vegetation types, given in Table 3, shows that ca. 3100 tons DM were taken from the most impor-

Table 5. Estimated plant consumption (dry matter) and live weight productivity in the reindeer herd during 1972. The calculation is based on the censused herd size of ca. 10000

| Sex and age | Live weight increase kg $yr^{-1}$ | Food intake kg $deer^{-1} yr^{-1}$ | Herd % | Herd consumption tons $yr^{-1}$ | Live weight production tons $yr^{-1}$ |
|---|---|---|---|---|---|
| Calf | 2 | 250 | 21 | 522 | 48 |
| Yearling ♀♂ | 22 | 450 | 10 | 450 | 44 |
| Male 2 yr | 15 | 900 | 11 | 1008 | 16 |
| Male 3 yr + | 14 | 1350 | 9 | 1269 | 13 |
| Female 2 yr + | 12 | 827 | 48 | 4010 | 58 |
| | | | | 7260 | 180 |

tant community, the lichen-rich Arctostaphylio-Cetrarion nivalis alliance, during a year.

Data for the effects of trampling on lichen mats and material removed but not eaten are not available from this study. A pilot test from a different mountain area in Røros, Norway, with a considerably greater standing biomass of lichen, gave a figure for material removed but not eaten during the snow season of 103 g DM $m^{-2}$ ($n = 6$ SD = 71). Compared to estimated consumption of 10 g DM $m^{-2}$ this gave a spillage factor of 10.

## Discussion

Data are presented both from EF and rumen content samples. The clear difference between them is undoubtedly due to differences in digestibility and variable resistance to mechanical breakage of different food plants. In addition there may be differences caused by the general nutritional status of wild animals and the penned artificially-fed EF animal. It is believed however that rumen content samples give a more reliable quantitative picture of the diet than EF samples, but that EF sampling from preselected areas could give a good indication of individual plant preferences.

Hoehne et al. (1967) found that salivary contamination and preparation influenced the chemical composition of EF samples. Crude protein and carbohydrates were lower in EF samples than amounts fed, but in most cases crude fiber and in all cases lignin were not affected by EF sampling. In this study deviations in fiber content in the EF, rumen and clipped samples were not found, but the nitrogen level and the sugar level of the rumen contents were less fluctuating than in clipped and EF samples. The microbial activity and buffering capacity of the rumen system could explain the smaller seasonal variation in the nitrogen level of the rumen contents.

The differences between years (1971–1973) in the reindeer use of plant alliances were small. Snow conditions early in winter (December) influencing digging, initiation of snow-melt in the spring (May), and the weather conditions during the summer season (July–August) influenced use of plant alliances both geographically and altitudinally. During the winter only the lichen heath Arctostaphylio-Cetrarion nivalis alliance is normally available for grazing on the windblown ridges. Variable snow pack and spring melt-off then expose other heath communities lower down the slopes, first Phyllodoco-Vaccinion myrtilli in the low alpine zone and Juncion trifidi scandinavicum in the middle alpine zone and subsequently communities like Nardo-Caricion bigelowii and Lactucion alpinae in the low alpine zone and the Potentillo-Polygonion viviparum and Ranunculo-Oxyrion digynae alliances in the middle alpine zone. All the 4 years of this study had different snow pack, snow-melt, temperature and precipitation conditions. The reindeer consistently used the same seasonal areas, but dispersed altitudinally within the area during grazing according to vegetational zonation, slope and aspect with the most favorable conditions at any time. Only during August was there a marked geographical dispersion based on two strategies: either to follow the spring progression altitudinally up toward the glaciers on snow bed communities, Cassiopo-Salicion herbacea and Polarion, or to move toward the lower, more continental winter range where mushrooms could be fruiting in August in the Phyllodoco-Vaccinion myrtilli and Lactucion alpinae alliances. The ecology of mushrooms is poorly understood, but humid and warm weather seems to favor fruiting in mid-August in low alpine regions.

The high spillage factor of 10 found in a pilot study of material removed during winter lichen grazing by digging in snow is considerable. The reindeer at Hardangervidda spend at least 80% of their active time during December-February digging in snow to obtain food, while from March–May they graze mostly bare-blown and sun-exposed ridges free of snow. During grazing on snow-free ground the deer would not waste lichens to the same extent simply because pawing is unnecessary. A high standing biomass of lichen heath would presumably also lower the amount of material removed but not eaten during snow pawing. It seems reasonable to assume, however, that the Hardangervidda reindeer in about 20% of their annual time spent in the Arctostaphylio-Cetrarion nivalis alliance remove up to 15000 tons due to pawing in the snow.

Field observations have shown that most of the early winter snow digging is from the *Betula nana–Cladonia alpestris* association in the lower part of the Arctostaphylio-Cetrarion nivalis alliance and that the largest part of material removed is *Cladonia alpestris*. WHITE (pers. comm.) has shown that the *in vitro* digestibility of *C. alpestris* is only 16% compared to 77% for *Cetraria cucullata*. The rumen content samples have shown a very low proportion of *C. alpestris* in the diet compared to other *Cladonia* and *Cetraria* spp. This seems to indicate that reindeer have a very low preference for *C. alpestris* and that they mainly select *C. mitis* in associations of *Betula nana–C. alpestris*. In late winter they confine most of their grazing to the *Empetrum hermaphroditum–Cetraria nivalis* association with *C. nivalis*, *C. cucullata* and *C. mitis* free of snow.

It is concluded that the considerable waste of *C. alpestris* is largely due to the active search for preferred lichen species under the snow cover.

## References

DAHL, E., KALLIOLA, R., MARKER, E., PERSSON, Å.: Nordisk vegetationsklassificering för kartleggning. Fjällvegetation. IBP Norden **7**, 3–12 (1971).

DAHL, E., KROG, H.: Macrolichens of Denmark, Finland, Norway and Sweden. Oslo-Bergen-Tromsø: Universitetsforlaget 1973.

GAARE, E.: A preliminary report on winter nutrition of wild reindeer in the Southern Scandes, Norway. Symp. Zool. Soc. London **21**, 109–115 (1968).

GAARE, E., SKOGLAND, T.: Villreinens næringsvaner. Hardangervidda: Juli–Desember 1970. Trondheim: Norsk IBP, Statens viltundersøkelser 1971 (mimeogr.).

GAARE, E., SKOGLAND, T.: Reinprosjektet på Hardangervidda 1972. In: IBP i Norge. Årsrapport 1972, pp. 257–268. Oslo: Norwegian National IBP Committee 1973 (mimeogr.).

GAARE, E., SKOGLAND, T., THOMSON, B. R.: Villreinens næringsvaner og adferd. Hardangervidda: Januar–Juni 1970. Trondheim: Norsk IBP, Statens viltundersøkelser 1970 (mimeogr.).

GOLLEY, F. B., BUECHNER, H. C.: A practical guide to the study of the productivity of large herbivores. IBP Handbook No. 7. Oxford-Edinburgh: Blackwell 1968.

HOEHNE, O. E., CLANTON, D. C., STREETER, C. L.: Chemical changes in esophageal fistula samples caused by salivary contamination and sample preparation. J. Anim. Sci. **26**, 628–631 (1967).

KELSALL, J. P.: The migratory barren-ground Caribou of Canada. Ottawa: Dept. of Indian Affairs and Northern Dev. Canad. Wildlife Service 1968.

KRAFFT, A.: Villreintelling på Hardangervidda. Jakt, Fiske, Friluftsliv. **99**, 8–10 (1970).

LID, J.: Norsk og svensk flora. Oslo: Det norske samlaget 1963.

NYHOLM, E.: Illustrated moss flora of Fennoscandia. Stockholm: Natural Science Research Council 1954–1969.

PETRUSEWICZ, K., MACFADYEN, A.: Productivity of terrestrial animals. Principles and Methods. IBP Handbook No. 13. Oxford-Edingburgh: Blackwell 1970.

SKJENNEBERG, S., FJELLHEIM, P., GAARE, E., LENVIK, D.: Reinens beiteopptak. Forsøk med spiserørsfistel utført i området Hinnøy-Gratangen. Statens Reinforsøk: Melding Nr. 2, 1971 (mimeogr.).

SKOGLAND, T.: Villreinens habitatatferd. Hardangervidda 1970–1973. Trondheim: Norsk IBP, Direktoratet for vilt og ferskvannsfisk, Viltforskningen, Trondheim 1974 (mimeogr.).

SKOGLAND, T.: Wild reindeer range use and selectivity in Southern Norway. In: LUICK, J. R. (Ed.): Proc. First Int. Reindeer/Caribou Symp. Fairbanks: University of Alaska 1975.

THAMBS-LYCHE, H.: Utbyttet av villreinjakta i tida 1920–1938. Statens viltundersøkelser. Medd. **11** (1941).

# Reindeer Activity on Hardangervidda

E. Gaare, B. R. Thomson, and O. Kjos-Hanssen

The purpose of this activity study was to promote a better understanding of behavior and movement patterns of wild reindeer, and to provide quantitative data on activity, linked with a simultaneously conducted study of food habits and range use. The results of these studies have been published and discussed in progress reports by Gaare et al. (1970), Gaare and Skogland (1971), Thomson (1971, 1973, 1975), and Skogland (1974).

With the activity study we have attempted to measure the time that herds devote to various activities throughout the day and year. Such data can then be related to feed-intake data and to seasonal energy expenditure.

Also by comparing the activity of reindeer on different mountain ranges, a tool for range condition comparison may be expected. A general resumé of Hardangervidda reindeer population structure and social habits is given in Østbye et al. (1975), and food habits and range use by Gaare and Skogland (1975).

## Methods

Activity is taken to mean one of the normal behavioral types, and recorded were grazing, lying, standing, walking, trotting, and running. These are the major activities, and a rhythmic shift between them can be expected within a 24-hrs period. The number of individuals in a herd engaged in the different activity types was investigated at $^1/_4$-hr intervals by means of a simple hand counter. A spotting-scope with pancratic optics (X20–X45) and ordinary 10 × 50 binoculars aided in this. Most of the observations were made on wild reindeer during the light period, but in addition the activity of a few domesticated male reindeer, which were brought out in the range, were also recorded.

Lying, trotting and running were readily distinguished. Standing and walking were recorded even if often part of the general grazing activity. Grazing in this study was taken to mean when an animal was biting, chewing, slowly moving forward with head low, or digging in the snow. Seasonally frequent behavior patterns e.g. antler sparring by males, bush thrashing, and suckling by calves, were recorded as "other" activities. A few weather data were recorded, temperature and wind speed most regularly.

The data was collected on the Hardangervidda range during 1970–1973 in trips lasting from 5–30 days in a variety of weather conditions. Supplementary data was collected on trips to other mountain ranges further north in southern Norway, in particular Snøhetta, North Ottadalen and Forelhogna.

Each field trip on foot, ski or snow-scooter was planned with practical help from local inhabitants, who also provided information about large reindeer concentrations. In the later years we could rely more on our own experience as the animals proved to be quite regular in their seasonal range-use pattern from one year to the next.

## Results and Discussion

**Alternation of Active and Rest Periods.** Reindeer activity is characterized by a more or less regular alternation between periods of activity and rest. Typically, after a few hours of grazing and moving, the animals lie down to ruminate and rest for a similar period. This rhythm holds true for individual animals, groups and herds of all sizes. Reindeer in a herd of hundreds or even thousands show a marked tendency to synchronize their activity, so that within a quarter-hour a herd may have switched from all grazing and moving to all lying (Fig. 1).

In winter (December–March) herd active periods averaged 2.6 hrs and lying periods 2.0 hrs, while in summer (July–September) active periods averaged 2.0 hrs and lying periods 1.1 hrs. All these differences were statistically significant. The endogenous regulation mechanics controlling periodicity in ruminants are not yet fully understood, but it is generally thought that it is connected with the rate of rumen depletion, and concentration of digesta products in the rumen and blood (reviewed in BAILE and FORBES, 1974). The lower digestibility of feed in winter can account for the reindeer's longer lying periods (for rumination), while the longer active period in winter is at least partly accounted for by the extra time spent on digging and foraging in the snow for feed.

A further basic difference between winter and summer activity is that, in early winter particularly, the alternating periods are linked to a 24 hrs cycle, whereas for the rest of the year the alternation appears at random with no predictable link with time. Thus the winter season is characterized by a 2–3-hrs grazing period in morning and again in the afternoon, with a regular 2–2.5-hrs lying period in the middle of the day (Fig. 1). As the winter proceeds and day length increases, two day-time rest periods occur, but with increasing irregularity, and the coordination between herds becomes less striking. Examples of daily activity patterns over the year are shown in Fig. 2.

The time that reindeer spent on various activities over the year from November 1969 to July 1971 is shown in Table 1. Grazing takes up 35–50% of a reindeer's time, more in winter than in summer, lying normally 20–40%, while the rest of the time is largely taken up by walking or standing.

**Sex and Age Differences.** The description above holds true for an average mature animal. However, in the pre- and post-calving periods, mother females on Hardangervidda devoted 54% of their time to grazing, well above average. Independently, SKOGLAND (1974) found that the feeding intensity of lactating females

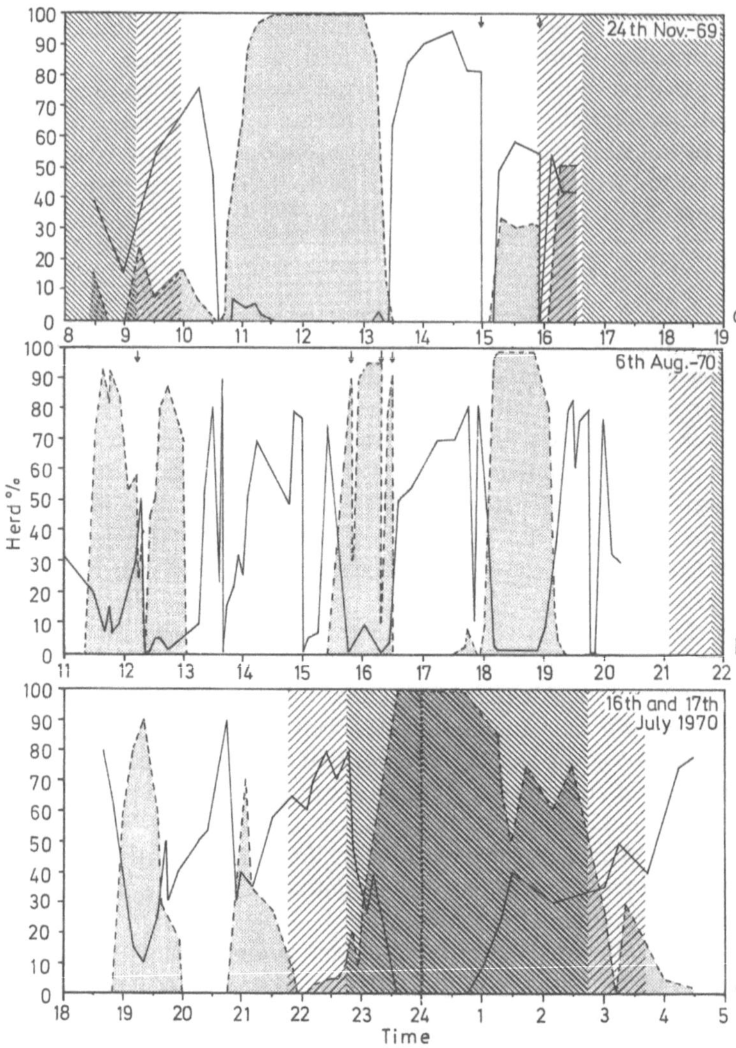


Fig. 1 a–c. Percent of reindeer herd showing grazing (solid line) and lying (grey) activity. Twilight and darkness are hatched. Human disturbance indicated with arrows. (a)—typical winter pattern, (b)—typical summer pattern, (c)—night activity in summer

was significantly high at this time. By contrast, on the North Ottadalen range, some 160 km further North, mother females at this time spent only 36% of their time grazing, with much time spent in idle standing and walking (Fig. 3). In climate and environment these ranges are very comparable, but reindeer density for several decades has been substantially lower in Ottadalen than on Hardangervidda, less than 1 animal $km^{-2}$ gross range compared with 2–3 animals $km^{-2}$. As a result, the quality of the pasture is very different. Large areas of Hardangervidda winter range now hold a biomass of lichens 200–400 g $m^{-2}$ or less, while Ottada-

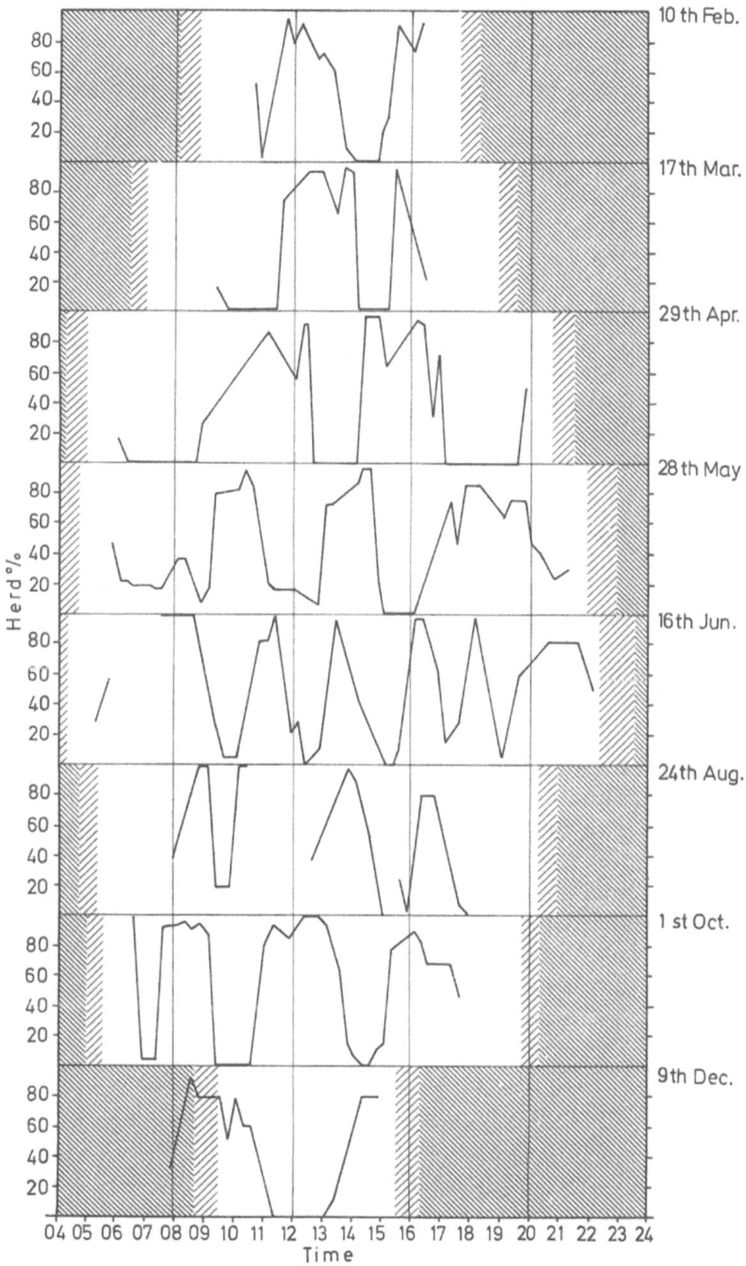


Fig. 2. Grazing activity throughout 1972. Redrawn after SKOGLAND (1974)

len seldom supports less than 800 g m$^{-2}$ and often more. The difference in animal condition between the ranges has been demonstrated (REIMERS, 1972) and this would seem to be clearly reflected in the different grazing times, and also in significant differences in nursing behavior. On Ottadalen, the mother seldom

Table 1. Mean percentage time spent by reindeer on different activities each month

| Month | Grazing | Lying | Standing | Walking | Trotting and running | Other | Nos. quarter hour counts |
|---|---|---|---|---|---|---|---|
| 1969 November[a] | 41 | 38 | 11 | 9 | 1 | <1 | 141 |
| 1970 January | 46 | 30 | 11 | 12 | 1 | <1 | 114 |
| Feb.-March[b] | 44 | 36 | 9 | 10 | 1 | <1 | 228 |
| April | 48 | 40 | 2 | 9 | < 1 | <0.1 | 338 |
| May ♀[c] | 36 | 42 | 10[d] | 10 | 2 | <0.1 | 233 |
| May calf[c] | 7 | 63 | 13 | 10 | 3 | 4[e] | 145 |
| July | 36 | 28 | 3 | 18 | 15 | <0.1 | 176 |
| August | 42 | 21 | 8 | 18 | 11 | <0.1 | 253 |
| August with flies | 37 | 10 | 15 | 22 | 17 | <0.1 | 110 |
| August no flies | 49 | 30 | 3 | 14 | 3 | <0.1 | 99 |
| September hunt | 30 | 17 | 3 | 19 | 31 | <0.1 | 32 |
| October | (43[f]) | 25 | (10[f]) | (14[f]) | (6[f]) | (3[f]) | 155 |
| December | 45 | 33 | 8 | 11 | 2 | 1 | 95 |
| 1971 January | 40 | 44 | 4 | 9 | 1 | 2 | 124 |
| March | 39 | 37 | 8 | 14 | 1 | 1 | 305 |
| May ♀ | 54 | 37 | 4[d] | 4 | <0.1 | <0.1 | 36 |
| May calf | 7 | 66 | 9 | 12 | 2 | 4[e] | 36 |
| July | 36 | 24 | 11 | 18 | 10 | <0.1 | 160 |

[a] Snøhetta, [b] Forelhogna, [c] North Ottadalen, [d] Includes 4% nursing, [e] Suckling, [f] The % varies depending on the stage of rut.

refuses to nurse her calf from the normal side position, and she more often allows suckling to continue until the calf is satisfied and terminates (THOMSON, 1973). This contrasts with the mothers on Hardangervidda who utilize more time on foraging.

The male reindeer, while in hard antler and particularly during the rut, also shows a changed activity pattern. As the rut approaches, males devote less time to grazing and more to antler sparring, sexual displays, approaches to females and wild chasing (THOMSON, 1971; SKOGLAND, 1974). At the peak of their libido, mature males stop grazing altogether. By the end of the rut, males are clearly at an energetic low. This is confirmed by SKJENNEBERG and SLAGSVOLD (1968) and BASKIN (1970). On Hardangervidda, snow and weather conditions are normally such that a re-formation of fat reserves is possible to December.

For the calf, behavior patterns are acquired rapidly. Before a reindeer is half a day old, it is not only lying, standing, walking, running and suckling, but has made its first tentative efforts to graze. Calves up to 10-days-old spend two-thirds of their time in lying, and on an average 7% in grazing (Fig. 3). One-week-old calves have a rumen content volume of 0.9–1.2 l so that the grazing is already contributing to their energy balance. Suckling is of course the main energy source at this time. Data from Hardangervidda and for caribou calves in Alaska (WHITE et al., 1975) allow us to calculate that the total suckling time is reduced in the following manner:

| | | | | | | |
|---|---|---|---|---|---|---|
| Approximate age of calf (weeks) | 1 | 2 | 4 | 8 | 11 | 16 |
| Calculated suckling time (min per 24hrs period) | 77 | 24 | 14 | 4 | 3 | 2.4 |

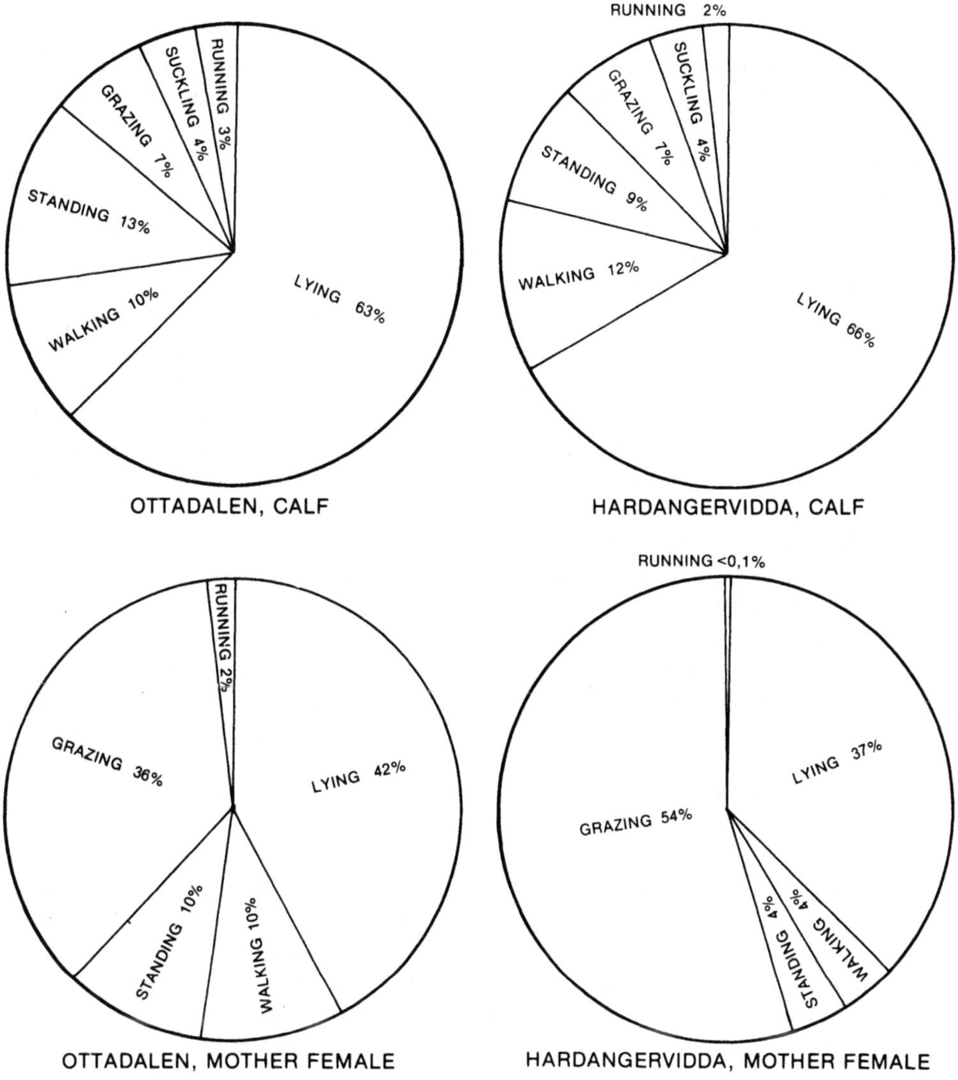


Fig. 3. Activity times of reindeer mothers and their 1–10-day old calves in two mountain areas of southern Norway

By the age of 6–8 weeks the calves' grazing and overall activity pattern fully matches that of the adults, although there is still an observable tendency for them to lie more. Weaning is a difficult term to define for reindeer, suckling can still be seen with decreasing frequency through the autumn, winter and even into the next spring and summer.

**Environmental Influences.** The controlling features of the activity pattern, namely the short-term rhythm of alternating active and rest periods and the 24-hrs rhythm apparent in winter, are common to all Cervidae populations. It became unique to reindeer only through the changes and constraints imposed by the reindeer's specific reproductive biology and social behavior patterns. The reindeer's activity and social organization generally are further strongly influenced by the environment, in this case by the conditions of Hardangervidda. The nature of this habitat, the quality and availability of the food supply, snow cover and weather, other animals and man's influence all shape an activity pattern unique to the reindeer population on Hardangervidda. Some of the main external influences on activity are considered below. GAARE and SKOGLAND (1975) have described the reindeer's seasonal use of plant communities on Hardangervidda.

As far as air temperature alone is concerned, this is well within reindeer tolerance limits on Hardangervidda, as elsewhere in the Norwegian mountains. Temperatures of $-30°$ C on still days had no observable effect on reindeer activity. ZHIGUNOV (1968) reports little or no effect on metabolism of temperatures down to $-60°$ C, a temperature not experienced in Norway. However, when cold temperatures are combined with wind, then reindeer were seen to respond. In winter with strong winds, Beaufort Force 5, grazing was progressively disrupted. The deer would tend to clump together, some standing with their heads facing low into the wind, others lying with their backs to the wind and heads tucked round compactly into their bodies. With winds of Force 7–8, near gale, and gale, some wild running bursts would occur and herds would often split up as a result. This behavior would appear to be consistent with attempts to maintain body heat. Strong winds of Force 8 combined with temperatures around $-10°$ C is the strongest wind-chill conditions normally experienced on Hardangervidda. According to SIPLE and PASSEL (1945) this corresponds to a heat loss in humans of up to 1450 kcal $m^{-2}$ $h^{-1}$.

Other weather features, such as precipitation, cloud cover, etc. affect the activity little or not at all. In mist there is a tendency for reindeer to group closer together.

In summer, high temperatures and low wind speeds are associated with flying insects which cause a profound change in reindeer activity. As ambient temperature exceeds 7° C on still days, reindeer can be seen reacting to insect harassment by shaking, tail wagging, twitching and stamping legs. If the temperature exceeds 12–13° C then harassment becomes increasingly severe, with grazing disrupted as the herd move quicker with outbursts of running. At temperatures of up to and beyond 20° C animals seek out, and may remain for hours, on insect-relief areas such as hill tops, snow patches if available, or mill together in a marsh or in the shallow water at a lake-side.

On Hardangervidda, the following Diptera are thought to be important: Oestridae warble and nostril flies, Culicidae mosquitos, Tabanidae horseflies and Simuliidae blackflies. Harassment by these insects is a regular part of reindeer life in summer. In the 77 days from mid-June to the end of August, 46–47 days were "insect days" causing mild to severe harassment in 1970 and 1971. This harassment results in a considerable loss of grazing and lying time and an increase in the time spent standing, walking, trotting, and running (Table 1). In warm summers, the total energy loss may therefore be large, a combined effect of reduced feed intake and increased energy expenditure. Complementary studies of caribou on Alaska arctic tundra found that insect harassment was the major influence on summer activity, movement and social structure and that similar thresholds existed with air temperature and wind speed (WHITE et al., 1975).

Compared with insects, other animals do not have a prolonged influence on activity. Reindeer become alert and often run off at the presence of potential predators and scavengers, notably the golden eagle *(Aquila chrysaëtos)*, raven *(Corvus corax)*, red fox *(Vulpes vulpes)* and common gull *(Larus canus)*. Usually, once the disturbance has passed, the herd shows a remarkably consistent return to its former activities (Fig. 1). The deer are also put to flight by hunters, hill-walkers, loud aircraft and snow-scooters. The reaction to different animals and man vary very much; while man usually causes the most violent reaction, that to gulls may be negligible. The activity of a group or herd was temporarily disrupted every 4 hrs on an average during the day. Of 190 recorded disturbances 45% were from hunters, hikers or aircraft, and 55% from other animals. Of the latter ravens accounted for 63%, gulls for 14%, eagles for 9%, foxes for 9% and others for 6%.

Since the wolf became extinct in the Hardangervidda area some 50 years ago, man is the major predator and influence on the population size and structure. About 84% of the annual harvest on a biomass basis is due to hunting (ØSTBYE et al., 1975). The population was allowed to grow too large up to the 1960s so that a reduction hunt became necessary between 1965–1970. In 1970 the level of disturbance during this reduction hunt was registered. 4000 hunters shot 4500 animals legally, while illegal hunting was reported to be high. During the permitted hunting hours, 08.00 to 18.00, reindeer groups were alarmed every 20 min on an average, resulting in a substantial change in normal activity (Table 1). Half the reindeers' day-time was taken up by moving, mostly trotting and running in alarm.

Although a normally regulated hunt may result in somewhat less disturbance than the 1970 reduction hunt, there is no doubt that the normal hunting period in the first 3 weeks of September represents to the reindeer an energy cost at least equal to the worst days of insect harassment.

**Energy Budget.** The observed activity pattern and the effect of external influences on the seasonal activity, allow preliminary calculations of the energy budget for reindeer on Hardangervidda.

The relative cost of the various activity types are presented in a number of papers, MOEN (1973) gives an updated review. Other data and estimates are taken from a current development of a simulation model for reindeer published in a Workshop Report by Statens viltundersøkelser (1974). The activity pattern for 1970 is used, but the population is assumed stabilized by hunting. Its annual variation is shown on Fig. 4, together with the energy cost for maintenance for the whole population. As population structure varies in sex and age, as well as in size, through the year, demands for an average animal cannot be deduced from this figure.

The sum of the year, 12400 Gcal (gigacalories), requires a gross energy intake of 23700 Gcal. The latter is about 80% of the calculated total intake based on grazing-time data (GAARE and SKOGLAND, 1975; ØSTBYE et al., 1975).

The reindeer conserve their energy resources in winter by spending a minimal time on trotting and running, whereas in summer with a herd's hurried style of grazing and frequent harassment by flies, it is a significant and major activity. Insect harassment as well as the hunt is shown to be costly in terms of energy and

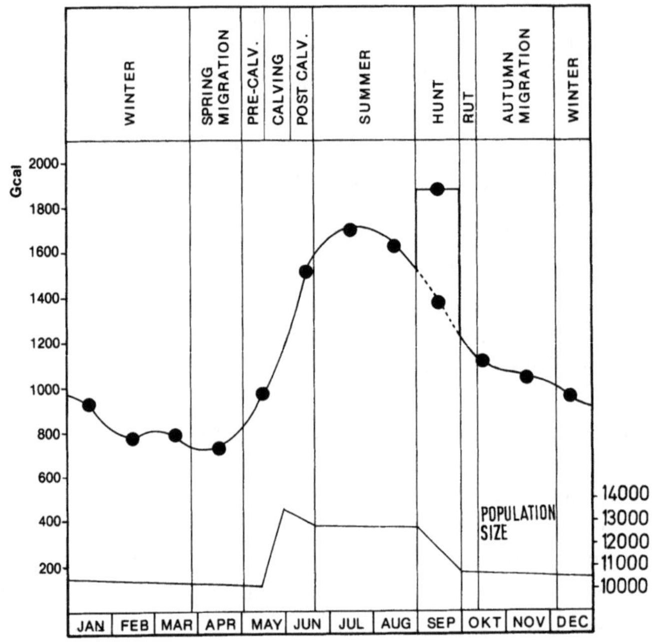


Fig. 4. Hardangervidda annual reindeer population size and maintenance energy in gigacalories (Gcal)

also to have an important effect on the food intake. If we estimate a September grazing intake value without hunt by interpolation, the population intake drops 600 Gcal (from 3875 to 3275 Gcal) during a reduction hunt, and the requirement rises about 500 Gcal (Fig. 4). The total effect of the hunt then amounts to 8–9% of the annual maintenance requirement. Even if the animals may compensate lost intake more than is assumed, it seems justified to try to find ways of diminishing this effect.

*Acknowledgements.* The studies were financed in equal parts by Norwegian IBP (sections PT-UM, Tundra Biome Program) and the Norwegian Directorate for Wildlife and Freshwater Fish. The authors wish to thank T. SKOGLAND for valuable assistance in field work and discussions.

## References

BAILE, C. A., FORBES, J. M.: Control of feed intake and regulation of energy balance in ruminants. Physiol. Rev. **54**, 160–214 (1974).

BASKIN, L. M.: Reindeer. Their ecology and behaviour. Moscow: "Nauka", AS USSR, Institute of evolutionary morphology and ecology of animals 1970. Canadian Wildl. Serv. Transl. 1971.

GAARE, E., SKOGLAND, T.: Villreinens næringsvaner. Hardangervidda: Juli–Desember 1970. Trondheim: Norsk IBP, Statens viltundersøkelser 1971 (mimeogr.).

GAARE, E., SKOGLAND, T.: Wild reindeer food habits and range use at Hardangervidda, Norway. In: WIELGOLASKI, F. E. (Ed.): Fennoscandian tundra ecosystems. Part 2: Animals and systems analysis, pp. 195–205. Berlin-Heidelberg-New York: Springer 1975.

GAARE, E., SKOGLAND, T., THOMSON, B. R.: Villreinens næringsvaner og adferd. Hardangervidda: Januar–Juni 1970. Trondheim: Norsk IBP, Statens viltundersøkelser 1970 (mimeogr.).
MOEN, A. N.: Wildlife Ecology. San Francisco: W. H. Follmann and Co. 1973.
ØSTBYE, E. (Ed.), BERG, A., BLEHR, O., ESPELAND, M., GAARE, E., HAGEN, A., HESJEDAL, O., HÅGVAR, S., KJELVIK, S., LIEN, L., MYSTERUD, J., SANDHAUG, A., SKAR, H.-J., SKARTVEIT, A., SKRE, O., SKOGLAND, T., SOLHØY, T., STENSETH, N. C., WIELGOLASKI, F. E.: Hardangervidda. In: ROSSWALL, T., HEAL, O. W. (Eds.): Structure and function of tundra ecosystems, pp. 225–264. Ecol. Bull. **20**. Stockholm: Swed. Nat. Sci. Res. Council 1975.
REIMERS, E.: Growth in domestic and wild reindeer in Norway. J. Wildl. Management **36**, 612–619 (1972).
SIPLE, P. A., PASSEL, C. F.: Dry atmospheric coding in subfreezing temperatures. Proc. Am. Phil. Soc. **89**, 177–199 (1945).
SKJENNEBERG, S., SLAGSVOLD, L.: Reindriften og dens naturgrunnlag. Oslo-Bergen-Tromsø: Universitetsforlaget 1968.
SKOGLAND, T.: Villreinens habitatatferd. Hardangervidda 1970–73. Trondheim: Norsk IBP, Direktoratet for vilt og ferskvannsfisk, Viltforskningen 1974 (mimeogr.).
Statens viltundersøkelser, Direktoratet for jakt, viltstell og ferskvannsfiske and Norwegian IBP (PT/UM): *Rangifer* workshop in Trondheim, 12–17th of November 1973. 1974 (mimeogr.).
THOMSON, B. R.: Wild reindeer activity, Hardangervidda. Trondheim: Norwegian IBP-Statens viltundersøkelser 1971; 1973 (mimeogr.).
THOMSON, B. R.: Leadership in wild reindeer in Norway. In: LUICK, J. R. (Ed.): Proc. First Int. Reindeer/Caribou Symp. Fairbanks: University of Alaska 1975.
WHITE, R. G., THOMSON, B. R., SKOGLAND, T., PERSON, S. J., RUSSELL, D. E., HOLLEMAN, D. F., LUICK, J. R.: Ecology of caribou at Prudhoe Bay, Alaska. In: BROWN, J. (Ed.): Ecological investigations of the Tundra Biome in the Prudhoe Bay Region, Alaska. Biol. Pap. Univ. Alaska, Spec. Rep. **2**, 151–201 (1975).
ZHIGUNOV, P. S. (Ed.): Reindeer husbandry. Jerusalem: Israel Programme for Scientific Translations, Washington D.C.: U.S. Dept. of Interior 1968.

# Grazing by Sheep

F. E. Wielgolaski

## Introduction and Methods

The present investigations were carried out within the Norwegian IBP project at two meadow exclosures (about 1 ha) in the upper low-alpine region of Hardangervidda. A review of various abiotic and biotic data obtained at the sites is given by Sonesson et al. (1975). Plant biomass and primary production at the meadows are discussed by Wielgolaski (1975b). Herbage intake by sheep was found to influence the primary production strongly and also the ratios top/root and dead/live above-ground plant parts at the sites (Wielgolaski, 1976).

Changes in vegetation structure found by sheep grazing in the exclosures at Hardangervidda (Wielgolaski, 1975a) indicate a selective food intake by the animals. This is also well known from several earlier investigations on sheep (Ivins, 1952; Hunter, 1962; Arnold, 1964; Selsjord, 1966; Rice et al., 1971; Baadshaug, 1974) as on other large vertebrates, e.g. reindeer (Gaare and Skogland, 1975). The selection of plant species by sheep on the Hardangervidda dry and wet meadows, under strong grazing for three years (four years at the dry meadow), is discussed in the present paper. Generally one ewe and two lambs of the variety *Dalasau* were placed within the exclosures each year at the beginning of July and removed in the first half of September (for two years only one lamb was placed on the dry meadow).

About 50 vascular plant species were found on the dry meadow investigated, and about 35 species on the wet meadow (Lye, 1972). Only few species, however, had a relatively homogenously distributed biomass within the area (Wielgolaski and Kjelvik, 1975b). Standard errors in biomass and intake by sheep were very high for species infrequently found at the sites (often more than 50%). Therefore, only the grazing of the main species was studied separately in the present investigations.

The intake by sheep was found by comparison of cut above-ground biomass of these main plant species and other species taken together at 10 grazed and ungrazed plots of the same community by the wire-cage method outlined by Milner and Hughes (1968). The approximate annual total intake is given by $(B_2-W_2)+(B_3-W_3)$ where $B_2$ = average above-ground biomass at time $t_2$ (one month after start of grazing) inside the 10 cages (0.5 × 1.0 m), $W_2$ = biomass left after grazing at 10 plots outside the cages at the same time, $B_3$ = biomass inside

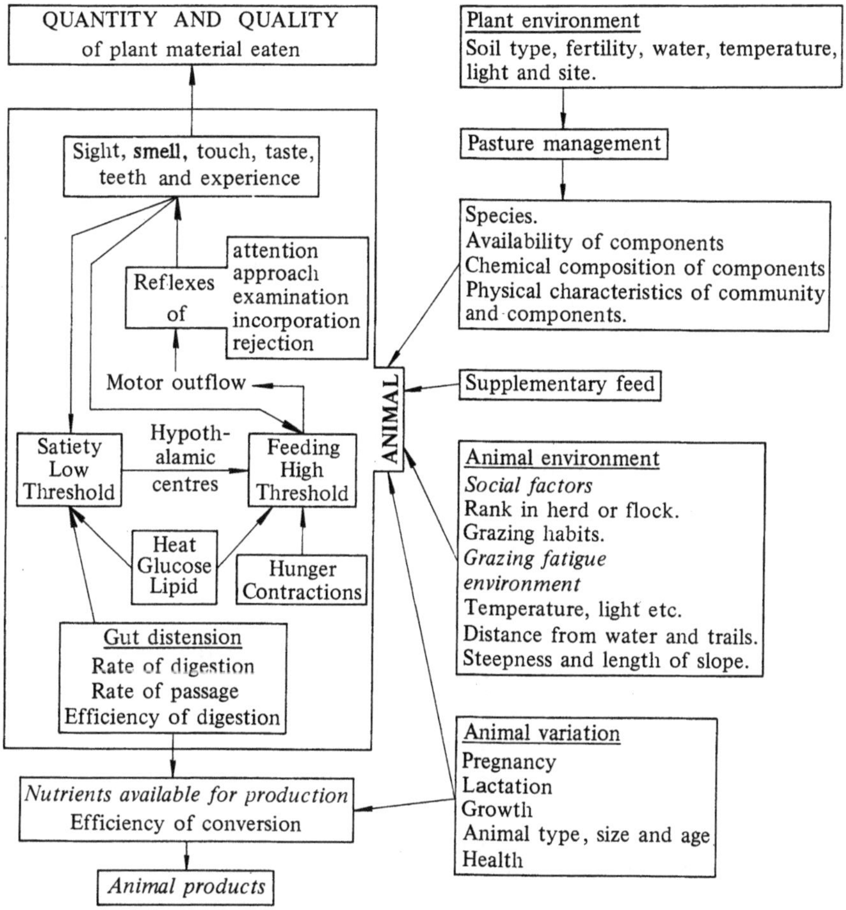


Fig. 1. The complexity of soil-plant-animal relationships (ARNOLD, 1964)

and $W_3$ = biomass outside the cages at time $t_3$ (in the end of the grazing period = end of growing season).

There are many reasons for selective herbage intake, and the soil-plant-animal relationship is very complex, as mentioned by HUGHES et al. (1964) and seen in Fig. 1 from ARNOLD (1964); often toxidity of plants to animals is also involved (MCKEY, 1974). However, in the present study an attempt is made to explain some of the variations found in the preferred intake, by variation in chemical composition of the plants.

Investigations on forage quality and value of alpine pasture plants are earlier carried out in Norway by several workers (review by BAADSHAUG, 1974). Still more data are available on chemical analyses of lowland pasture plants in many countries (see e.g. ARNOLD, 1964, and in Norway HOMB, 1952). Usually different minerals, total ash, crude protein (total N × 6.25), ether extract (raw fat), crude fiber and N-free extract are analyzed. CRAMPTON and MAYNARD (1938) proposed lignin, cellulose and "other carbohydrates" for analysis instead of crude fiber and N-free extract. In the present investigation, no ether extractions were carried out, but minerals, total ash and total nitrogen were analyzed (WIELGOLASKI et al., 1975; KJELVIK and WIELGOLASKI, 1975) and so were lignin and holocellulose (SKRE et al., 1975), while "other carbohydrates" were subdivided in analyses of storage polysaccharides (including hemicelluloses), oligo- and monosaccharides (SKRE, 1975). Often the energy value of herbage is said to be of greater significance to the grazing animal than the chemical composition, since the voluntary intake of food is linked with digestability (BLAXTER et al., 1961; HUGHES et al., 1964). Therefore also the caloric values of the plants in the present investigation (WIELGOLASKI and KJELVIK, 1975a; LUNDE et al., 1975) were compared with sheep intake.

The chemical analyses were mostly based on two replicates. Ash, mineral elements and total nitrogen of the plants were determined as described by WIELGOLASKI and KJELVIK (1973), the carbohydrate fractions as described by SKRE et al. (1975). The energy content was determined in an adiabatic bomb calorimeter by the method described by LIETH (1968).

## Plant Intake and Sheep Growth

Average intake of above-ground vascular plants by the sheep through all seasons is given in Table 1 for the dry meadow and in Table 2 for the wet meadow, while variation in consumption for the different years is presented in another paper (WIELGOLASKI, 1976). The average total grazing was somewhat higher at the dry than at the wet meadow (on a daily average basis about $1.0\,g\,m^{-2}$ and $0.7\,g\,m^{-2}$, respectively). The difference was partly caused by the different plant species found at the wet and dry meadows, partly by a slightly smaller exclosure at the last site. At both sites, however, the intake was higher than observed in Great Britain under similar grazing pressure (RAWES and WELCH, 1964) during approximately the same period of growing season (July and August).

The fresh weight increase of the lambs was found on an average to be 100–$150\,g\,d^{-1}$ on the dry meadow during the two-month grazing period and 150–$200\,g\,d^{-1}$ on the wet meadow (KAURI et al., personal communication 1971). Particularly the increase in lamb weight on the dry meadow was somewhat lower than found for the same sheep variety (180–$250\,g\,d^{-1}$) in other Norwegian mountain grazing studies (SANDBERG, 1960; SELSJORD, 1964), as well as lower than summer values (180–$320\,g\,d^{-1}$) reported from Iceland (OLAFSSON, 1973). This, together with approximately no weight increase of the ewes during summer in the present study, and the high percentage of plant material grazed compared with the British investigations, all indicate that the grazing pressure at Hardangervidda was very high compared to the area of the exclosures, particularly on the dry meadow. This is also indicated by the relatively strong lichen grazing on the dry meadow in the last part of the season (from mid-August); an average for four years shows above $5\,g\,m^{-2}$ grazed, or about 46% of the lichen biomass (in July no lichen grazing was observed).

The grazing intensity usually varies strongly in different plant communities. SELSJORD (1966) found that in the Norwegian low alpine zone, sheep under free

Table 1. Dry meadow average sheep grazing 1969–1972. Sum dry weight g $m^{-2}$ per season in percentage of total ungrazed above-ground biomass per season. Average chemical content in percentage of dry weight of green material in the seasons. Mineral and crude protein values are based on data from WIELGOLASKI et al. (1975) and KJELVIK and WIELGOLASKI (1975). Saccharide content is based on SKRE (1975), lignin and holocellulose values on SKRE et al. (1975) and average energy content (calories) per g dw on data from WIELGOLASKI and KJELVIK (1975a) and LUNDE et al. (1975)

| Species | Grazing | | P | K | Ca | Ash | Crude Protein | Saccharides | | Holo-cell. | Lignin | Cal. |
|---|---|---|---|---|---|---|---|---|---|---|---|---|
| | g $m^{-2}$ | % | | | | | | Ethanol soluble | Storage | | | |
| *Anthoxanthum alpinum* | 4.3 | 63.2 | 0.15 | 1.2 | 0.5 | 5.1 | 11.3 | 18 | 12 | | | 4388 |
| *Carex bigelowii* (+*C. vaginata*) | 7.7 | 44.8 | 0.20 | 1.4 | 0.7 | 6.3 | 21.3 | 15 | 16 | | | 4462 |
| *Deschampsia flexuosa* | 9.3 | 46.5 | 0.13 | 1.0 | 0.7 | 4.6 | 10.0 | 16 | 11 | 39 | 13 | 4514 |
| *Poa alpina* | 6.7 | 62.7 | 0.13 | 1.1 | 0.3 | 3.5 | 10.6 | 22 | 11 | | | 4334 |
| Other monocotyledons | 1.6 | 43.9 | 0.15 | 1.4 | 0.4 | 6.0 | 15.6 | 16 | 8 | | | — |
| *Astragalus alpinus* | 1.5 | 58.0 | 0.25 | 1.7 | 1.5 | 8.1 | 25.0 | 5 | 6 | | | 4402 |
| *Polygonum viviparum* | 4.8 | 44.9 | 0.30 | 1.5 | 0.9 | 6.1 | 19.4 | 8 | 6 | | | 4641 |
| *Ranunculus acris* | 2.1 | 51.1 | 0.19 | 2.3 | 1.3 | 9.4 | 13.8 | 12 | 9 | 27 | 19 | 4174 |
| *Thalictrum alpinum* | 1.3 | 48.1 | 0.21 | 2.0 | 1.4 | 7.8 | 18.1 | 11 | 7 | | | 4269 |
| Other eutrophic forbs | 2.5 | 54.7 | 0.19 | 1.6 | 1.6 | 8.7 | 13.1 | 18 | 10 | | | 4346 |
| Other forbs | 8.8 | 44.3 | 0.17 | 1.6 | 1.4 | 8.2 | 13.8 | 7 | 9 | | | 4308 |
| Sums and averages | 50.6 | 51.1 | | | | | | | | | | |
| *Salix herbacea* | 1.6 | 27.5 | 0.18 | 1.1 | 1.2 | 5.3 | 11.9 | 10 | 7 | | | 4679 |
| *Salix reticulata* | 2.6 | 26.1 | 0.17 | 0.8 | 2.2 | 6.9 | 11.3 | 10 | 7 | 27 | 24 | 4712 |
| *Vaccinium vitis-idaea* | 0.4 | 17.8 | 0.13 | 0.4 | 0.8 | 3.3 | 9.4 | 15 | 6 | | | 4895 |

grazing preferred plants of the alliance Lactucion alpinae (and within this, grazed particularly in willow thickets dominated by *Deschampsia caespitosa* in the field layer). Slightly lower grazing intensity was, in the same study, found in plant communities which can be classified as Ranunculeto-Oxyrion digynae according to DAHL et al. (1971). The dry meadow belongs mostly to the Alchemillion glabrae chionophilum suballiance of this alliance (WIELGOLASKI, 1975a). This means that the species composition at the site is probably favorable to sheep. The wet meadow belongs to the alliance Caricion canescentis-nigrae according to DAHL et al. (1971). These mire communities were found to be less preferred by sheep in free grazing (SELSJORD, 1966). The lower weight increase of the lambs at the dry than the wet meadow in spite of more favorable species composition, may have several explanations. One reason may be the higher grazing pressure on the dry meadow, forcing the animals to consume also less palatable species and other plant parts, particularly late in the season. This effect may be increased by drought at the dry meadow.

*Deschampsia flexuosa* was generally less heavily grazed in percentage of available biomass than species which may be classified as "broad-leaved grasses" (*Anthoxanthum alpinum, Poa alpina, Phleum commutatum*, and to some degree *Festuca rubra*) as seen in Tables 1–3. This is in agreement with British studies (HUNTER, 1962; RAWES and WELCH, 1964). On the basis of other Norwegian mountain sheep grazing studies, BAADSHAUG (1974) reports *Deschampsia flexuosa* to be of great value for grazing in alpine pastures. This is due to the commonness of the species in Fennoscandia, also in the alpine tundra region. On a biomass basis, the species was also the most strongly grazed on the dry meadow (Table 1) in the present investigations. Even if the *Carex* spp. were found to be only moderately grazed at the Hardangervidda sites (Tables 1–3) on a percentage basis in agreement with e.g. the British studies mentioned, the high percentage cover of the species caused them to be amongst the main components of the sheep forage at the sites. This was also found for *Carex* spp. in the Rocky Mountains (PAULSEN, 1960) in spite of only about 10% utilization. VIGERUST (1936) reported both *C. nigra* and *C. bigelowii* to be important forage plants in Norwegian mountains.

Mesotrophic-eutrophic forb species were relatively strongly grazed at the Hardangervidda sites, particularly *Astragalus alpinus* at the dry meadow and *Ranunculus acris* at both the dry and wet sites (Tables 1–3). Species within the same genera were also some of the most preferred species in the American studies mentioned, and SELSJORD (1966) also reports *Ranunculus acris* to be important. This species is, however, found in other studies to be rejected by domestic animals (ULVESLI, 1958) and to cause digestive disturbances and poisoning, particularly on dry pastures, in cattle (SVANBERG and ÅBERG, 1955). The woody plants at the Hardangervidda sites were only moderately grazed, but *Salix* spp. showed some importance, particularly *Salix lapponum* and related species at the wet meadow (Table 2), due to relatively high and easily available biomass (WIELGOLASKI, 1975b).

The availability of preferred plant species will strongly influence the intake (ARNOLD, 1964). Small species beneath less grazed species will be overlooked by the animals. This may have been the case particularly for *Salix herbacea* at the wet

Table 2. Wet meadow average sheep grazing 1969–1971. For further explanations, see Table 1

| Species | Grazing | | P | K | Ca | Ash | Crude Protein | Saccharides | | Holo-cell. | Lignin | Cal. |
|---|---|---|---|---|---|---|---|---|---|---|---|---|
| | $g\,m^{-2}$ | % | | | | | | Ethanol soluble | Storage | | | |
| *Carex nigra* | 18.2 | 15.2 | 0.27 | 1.0 | 0.5 | 4.8 | 18.8 | 8 | 16 | | | 4437 |
| *Festuca rubra* | 3.4 | 21.1 | 0.21 | 1.0 | 0.5 | 5.1 | 10.0 | 15 | 10 | 41 | 15 | 4554 |
| *Phleum commutatum* | 1.1 | 59.0 | — | — | — | — | — | — | — | | | — |
| Other monocotyledons | 2.4 | 36.5 | 0.21 | 0.9 | 0.6 | 4.7 | 13.8 | — | — | | | 4392 |
| *Leontodon autumnalis* | 0.4 | 30.9 | 0.23 | 1.0 | 1.4 | 6.7 | 12.5 | — | — | | | 4202 |
| *Polygonum viviparum* | 1.3 | 34.2 | 0.32 | 1.1 | 0.9 | 5.5 | 12.5 | 10 | 7 | | | 4435 |
| *Ranunculus acris* | 0.8 | 62.8 | 0.25 | 1.2 | 1.8 | 9.7 | 12.5 | 15 | 12 | 23 | 19 | 4149 |
| Other herb. dicotyled. | 0.9 | 24.0 | 0.27 | 0.6 | 1.9 | 7.1 | 13.8 | — | — | | | 4246 |
| *Equisetum palustre* | 0.2 | 45.0 | 0.31 | 1.4 | 1.6 | 8.0 | 16.3 | — | — | | | 4100 |
| Sums and averages | 28.7 | 36.5 | | | | | | | | | | |
| *Salix herbacea* | 1.0 | 12.8 | 0.24 | 0.7 | 1.3 | 5.1 | 16.3 | 14 | 7 | | | 4604 |
| *Salix lapponum* et al. | 5.1 | 23.0 | 0.26 | 0.7 | 1.2 | 4.4 | 15.0 | 16 | 8 | 21 | 26 | 4692 |

meadow, a species often grazed by sheep (KELLGREN, 1924; VIGERUST, 1949; SELSJORD, 1966). Probably this is also one reason for the generally relatively low intake at the same site of *Festuca rubra*, a species at the upper growth limit at the wet meadow and therefore reduced viability compared to *Carex nigra* (WIELGOLASKI, 1975c). The sheep diet is selected almost exclusively from green compartments (ARNOLD, 1964). The presence of old and dry material has an important effect on utilization (HOMB, 1952; MILTON, 1953) and on species selection. *Carex nigra* is found to have a very high percentage of standing dead material (WIELGOLASKI and KJELVIK, 1975b). This may be one reason for the relatively low intake of the species at the wet meadow.

## Selective Grazing Related to Chemical Plant Composition

The selective intake by sheep of different plant species may be caused either by a direct avoidance of, or a preference for some species due to their qualitative chemical composition. Avoidance may be caused by specific compounds in the plants, e.g. various alkaloids, phenolics and other plant substances, which may be toxic at least to some animals not adjusted to them (FREELAND and JANZEN, 1974; LEVIN, 1971; MCKEY, 1974). Sheep will usually avoid species or plant parts with high amounts of several chemicals (MCLEAN, 1970), often due to the specific taste of these compounds, e.g. the almost universally bitter taste of alkaloids (MCKEY, 1974).

Some workers have found that the phosphate content of plant species, particularly selected by sheep was higher than in other species (PLICE, 1952; ARNOLD, 1964). In the present investigations no clear P-influence was found on grazing. HOMB (1952) reports highest phosphorus content in years with highest precipitation, and in the present investigations there were tendencies to higher amounts in green plant parts at the wet than at the dry meadow (Tables 1 and 2), while the grazing percentages were lower. Positive correlation between grazing preference and the content of potassium and calcium was reported by HUNTER (1962), DAVIES et al. (1959) and HUGHES et al. (1964), and a high total mineral content as found in Norwegian mountain pasture plants (SOLBERG, 1964) has also been reported to be preferable for grazing (ARNOLD, 1964). There was probably a weak indication to highest K, Ca and ash content in the most heavily grazed forb species in the present investigation, but not in the monocotyledonous preferred species; this may be because of intercorrelations between ash and Si-content. In *Astragalus* lowest potassium amount (KJELVIK and WIELGOLASKI, 1975) was found in the plants in the year with lowest intake of the species (WIELGOLASKI, 1976), but the findings were probably not significant.

No significant positive correlation or probably even a weak negative relationship was found between the caloric values of different species at Hardangervidda and grazing preference, (Tables 1 and 2). This indicates that the high energy constituents of the plants e.g. fat, are of minor importance to selection by sheep of plant species compared to carbohydrates, for example. The results do not directly fit the hypothesis that voluntary intake of fodder is related to its apparent digestible energy (COOK et al., 1956; BLAXTER et al., 1961). The higher caloric values in green than in non-green above-ground live material (WIELGOLASKI and KJELVIK, 1975a), however, may partly explain the generally high intake of leaves (ØDELIEN, 1951; OLSEN, 1973) and the high fodder value of mountain plants (OLSEN, 1973). Sheep seem to have some preference for plant species with a high crude-protein content in the present investigation, in agreement with several other studies (HOMB, 1952; ARNOLD, 1964; SOLBERG, 1964; HERNES, 1972; VALBERG and BØ, 1972; BAADSHAUG, 1974). This was particularly true for the legume *Astragalus alpinus*, but also in *Carex* spp. heavy grazing was found (WIELGOLASKI, 1976) in the years with highest nitrogen content, which was probably caused by a longer vegetative growth period in these years (SKRE et al., 1975).

There seems, however, to be a negative correlation between the crude protein content and the ethanol-soluble sugar (Tables 1 and 2), as earlier found by e.g. HOMB (1952). Generally, the ethanol-soluble sugars (sum of mono- and oligosaccharides) are clearly positively correlated with the plant species most preferred by sheep. This is in agreement with PLICE (1952), and is another reason for the high nutritive value of highland fodder since the sugar content is found to be higher in mountain than in lower areas (SOLBERG, 1964). Particularly oligosaccharides are higher in plants at high incoming radiation (MYHR and SÆBØ, 1969) compared to shaded plants. SELSJORD (1968) also found highest content of sugar (and of N-free extracts) in *Deschampsia flexuosa* at sun-exposed plots. He found this species to be more heavily grazed in mountain areas than in forests, and explains this by the increased sugar content. The nitrogen content, however, is found to be considerably higher in shaded plants than in the same species at the same developmental stage on exposed plots (SELSJORD, 1968; MYHR and SÆBØ, 1969). This indicates that the sugar content is more important than crude protein for the selection of plant species by sheep. Similar conclusions can be drawn from the present investigation, but a low sugar content can to some degree be compensated for by a high nitrogen content (as in *Astragalus alpinus*, Table 1). This is also reported by TERRY and TILLEY (1964). The importance of easily soluble carbohydrates in animal nutrition has been reported by several workers during the last years (e.g. KIVIMÄE, 1959; WITT, 1970).

On the other hand, the intake can be negatively influenced by a high lignin content as in shrubs (Tables 1 and 2), even if the sugar content is relatively high. The lowest lignin content in the present investigation was found in the heavily grazed eutrophic grass species at the sites. The negative influence of lignin in digestibility of fodder has been stressed earlier by several authors, e.g. CRAMPTON and MAYNARD (1938), HOMB (1952), ARNOLD (1964), and is probably one of the main reasons for the better digestibility of young material (HOMB, 1952), as the lignin content increases during the growing season (SKRE et al., 1975). Lignin is included in crude fiber analyses and therefore often a negative influence has been found between grazing and crude fiber content of the species (HUNTER, 1962). The crude fiber content is, however, lower in less-grazed, shaded plants than in sun-exposed (MYHR and SÆBØ, 1969). Holocellulose is also included in the crude fiber content, but is generally known to have a higher digestibility than lignin, and shows higher content in young than in older green material (SKRE et al., 1975). In the present investigation (Tables 1 and 2), the holocellulose content seems to be of minor importance to the selection of species for sheep grazing.

Table 3. Sheep grazing in g (dry weight) $m^{-2}$ and in percentage of ungrazed above-ground biomass for some plant species from the beginning of July to the beginning of August, and further to the first part of September. Mono-, oligo- and storage polysaccharide content of green material (% of dw) during the same periods (based on SKRE, 1975)

| Species | Grazing g $m^{-2}$ | | Grazing % | | Monosacch. | | Oligosacch. | | Storage polysacch. | |
|---|---|---|---|---|---|---|---|---|---|---|
| | early summer | late summer | early summer | late summer | early summer | late summer | early summer | late summer | early summer | late summer |
| Dry meadow | | | | | | | | | | |
| *Anthoxanthum alpinum* | 1.26 | 5.35 | 20.8 | 68.2 | 3.9 | 10.3 | 3.1 | 13.1 | 19.3 | 8.4 |
| *Carex bigelowii* (+ *C. vaginata*) | 3.43 | 8.07 | 26.6 | 38.3 | 1.8 | 4.5 | 7.5 | 17.3 | 15.9 | 16.5 |
| *Deschampsia flexuosa* | 2.36 | 8.71 | 15.3 | 32.8 | 3.8 | 5.4 | 3.0 | 12.0 | 16.4 | 9.2 |
| *Poa alpina* | 8.43 | 8.33 | 57.9 | 64.9 | 3.7 | 7.2 | 16.6 | 16.3 | 15.8 | 8.4 |
| *Astragalus alpinus* | 0.56 | 0.20 | 50.0 | 53.1 | 2.5 | 1.6 | 0.1 | 2.4 | 6.8 | 6.1 |
| *Polygonum viviparum* | 1.35 | 3.13 | 21.9 | 40.9 | 3.5 | 7.7 | 0.3 | 4.8 | 7.6 | 5.0 |
| *Ranunculus acris* | 2.44 | 0.29 | 59.5 | 14.3 | 7.8 | 5.7 | 4.8 | 3.9 | 10.8 | 9.7 |
| *Thalictrum alpinum* | 0.91 | 1.18 | 32.2 | 31.5 | 2.4 | 3.3 | 4.9 | 10.3 | 11.9 | 4.8 |
| Wet meadow | | | | | | | | | | |
| *Carex nigra* | 6.82 | 14.35 | 6.9 | 25.0 | 2.7 | 3.2 | 5.5 | 7.1 | 16.8 | 16.5 |
| *Festuca rubra* | 2.06 | 4.00 | 22.2 | 45.6 | 6.0 | 3.7 | 4.7 | 16.8 | 13.0 | 8.0 |
| *Polygonum viviparum* | 1.56 | 0.34 | 42.9 | 49.7 | 3.4 | 5.3 | 3.8 | 5.4 | 10.4 | 4.5 |
| *Ranunculus acris* | 0.77 | 0.14 | 57.7 | 74.1 | 9.1 | 6.9 | 9.1 | 5.3 | 30.6 | 7.3 |

A more detailed comparison of species selection by sheep with saccharide content to various parts of the growing season is given in Table 3 for some species (one year of study only). There was a very clear positive correlation between the percentage of material grazed and both the mono- and oligosaccharide content particularly at the dry meadow, while the amount of carbohydrates analyzed as storage polysaccharide was of minor direct influence. Low sugar content and thus low grazing intensity in July was found in *Anthoxanthum alpinum*. Higher sugar content in August–September combined with moderate lignin content caused strong grazing of this species and of *Poa alpina* on the dry meadow, and to some degree also of *Festuca rubra* on the wet meadow in late season. The opposite was found for *Ranunculus acris* on the dry meadow with highest sugar content and grazing in July. The lignin content probably increased more during the season in this species on the dry meadow than on the wet meadow where a high intake was found in *Ranunculus acris* particularly in late season. A higher crude protein content was also observed in this species late in the season, which may also have influenced the selection by sheep, but the low weight intake of the species on the wet meadow may have caused misinterpretations.

In conclusion it can be stated that the selective grazing by sheep at Hardangervidda is mostly occasioned by the content of monosaccharides and oligosaccharides in the green plant parts of species easily available to the animals. The lignin content produced a strong negative modifying effect (lignin generally increasing during the season and therefore decreasing the digestibility of the older herbage material). The crude protein content, and to some degree some minerals such as K and Ca, had a positive modifying effect on the palatability of the plant species to the sheep, but more important was the percentage of green leaf material compared to non-green above-ground plant organs.

## References


ARNOLD, G. W.: Factors within plant associations affecting the behaviour and performance of grazing animals. In: CRISP, D. J. (Ed.): Grazing in marine and terrestrial environments, pp. 133–154. Symp. Brit. Ecol. Soc. 1962. Oxford: Blackwell 1964.

BAADSHAUG, O. H.: Jordbruksmessig utnytting av fjelltraktene. Forskn. Fors. Landbr. **25**, Suppl. 1., 1–53 (1974).

BLAXTER, K. L., WAINMAN, F. W., WILSON, R. S.: The regulation of food intake by sheep. Anim. Prod. **3**, 51–61 (1961).

COOK, C. W., STODDART, L. A., HARRIS, L. E.: Comparative nutritive value and palatability of some introduced and native forage plants for spring and summer grazing. Bull. 385 Utah Agric. Exp. Sta. 1956.

CRAMPTON, E. W., MAYNARD, L. A.: The relation of cellulose and lignin content to the nutritive value of animal feeds. J. Nutr. **15**, 383–395 (1938).

DAHL, E., KALLIOLA, R., MARKER, E., PERSSON, A.: Nordisk vegetations-klassificering för kartläggning. IBP Norden **7**, 3–12 (1971).

DAVIES, R. O., JONES, D. T. H., MILTON, W. E.: Factors influencing the composition and value of herbage from fescue and *Molinia* areas. J. Agric. Sci. **53**, 268–285 (1959).

FREELAND, W.J., JANZEN, D.H.: Strategies in herbivory by mammals: The role of plant secondary compounds. Am. Naturalist **108**, 269–289 (1974).

GAARE, E., SKOGLAND, T.: Wild reindeer food habitats and range use at Hardangervidda, Norway. In: WIELGOLASKI, F. E. (Ed.): Fennoscandian tundra ecosystems, Part 2: Animal and systems analysis, pp. 195–205. Berlin-Heidelberg-New York: Springer 1975.

HERNES, O.: Forsøk med en og flere gangers slått, og høstetidspunktet for første slått (Engl. summ.) Forskn. Fors. Landbr. **23**, 435–445 (1972).

HOMB, T.: Kjemisk sammensetning og fordøyelighet av engvekster (Engl. summ.) N.L.H. Foringsfors. 71 beretn. 1952.

HUGHES, R.E., MILNER, C., DALE, J.: Selectivity in grazing. In: CRISP, D.J. (Ed.): Grazing in marine and terrestrial environments, pp. 189–202. Symp. Brit. Ecol. Soc. 1962. Oxford: Blackwell 1964.

HUNTER, R.F.: Hill sheep and their pasture: A study of sheep-grazing in South-east Scotland. J. Ecol. **50**, 651–680 (1962).

IVINS, J.D.: The relative palatability of herbage plants. J. Brit. Grassland Soc. **7**, 43–54 (1952).

KELLGREN, A.G.: *Salix herbacea*. En säregen men betydelsesfull betesväxt. Sv. Betes-Och Vallför. Årsskr. 1924, 122–124.

KIVIMÄE, A.: Chemical composition and digestibility of some grassland crops. Acta Agr. Scand. Suppl. **5**, 1–142 (1959).

KJELVIK, S., WIELGOLASKI, F.E.: Mineral content in plants at Hardangervidda, Norway. In: VIK, R. (Ed.): IBP in Norway. Methods and results. Sections PT-UM Grazing project Hardangervidda. Botanical investigations, pp. 89–158. Oslo: Norwegian Nat. IBP Comm. 1975.

LEVIN, D.A.: Plant phenolics: an ecological perspective. Am. Naturalist **105**, 157–181 (1971).

LIETH, H.: The measurement of calorific values of biological material and the determination of ecological efficiency. In: ECKARDT, F.E. (Ed.): Functioning of terrestrial ecosystems at the primary production level, pp. 233–242. Paris: UNESCO 1968.

LUNDE, E.M., KJELVIK, S., WIELGOLASKI, F.E.: Caloric values on plant material from Hardangervidda, Norway. In: VIK, R. (Ed.): IBP in Norway. Methods and results. Sections PT-UM Grazing project Hardangervidda. Botanical investigations, pp. 237–257. Oslo: Norwegian Nat. IBP Comm. 1975.

LYE, K.A.: Vegetation of selected localities for IBP on investigation in Hardangervidda, southern Norway. IBP Norden **8**, 101–111 (1972).

MCKEY, D.: Adaptive patterns in alkaloid physiology. Am. Naturalist **108**, 305–320 (1974).

MCLEAN, E.K.: The toxic actions of pyrrolizidine *(Senecio)* alkaloids. Pharmacol. Rev. **22**, 429–463 (1970).

MILNER, C., HUGHES, R.E.: Methods for the measurement of the primary production of grassland. IBP Handbook No. 6. Oxford: Blackwell 1968.

MILTON, W.E.J.: The palatability of herbage on underdeveloped grasslands in West Central Wales. Emp. J. Exp. Agric. **21**, 116–122 (1953).

MYHR, K., SÆBØ, S.: Verknaden av skygging på vekst, utvikling og kjemisk samansetning hos nokre grasarter (Engl. summ.) Forskn. Fors. Landbr. **20**, 297–315 (1969).

ØDELIEN, M.: Bladprosenten hos timotei og dens betydning for høyets fórverdi. Forskn. Fors. Landbr. **2**, 52–62 (1951).

OLAFSSON, G.: Fórgrunnlaget for sauehold under nordeuropeiske forhold. N.L.H. Fóringfors. Stensiltrykk nr. 20, 1973.

OLSEN, E.: Undersøkelser av forholdet mellom blad og stengel i gras høstet til forskjellige tidspunkt og på to høgdetrinn (Engl. summ.) Forskn. Fors. Landbr. **24**, 73–88 (1973).

PAULSEN, H.A. JR.: Plant cover and forage use of alpine sheep ranges in the central Rocky Mountains. Iowa St. J. Sci. **34**, 731–748 (1960).

PLICE, M.J.: Sugar versus the intuitive choice of foods by livestock. J. Range Mgmt. **5**, 69–74 (1952).

RAWES, M., WELCH, D.: Studies on sheep grazing in the Northern Pennines. J. Brit. Grassland Soc. **19**, 403–411 (1964).

RICE, R.W., CUNDY, D.R., WEYERTS, P.R.: Botanical and chemical composition of esophageal and rumen fistula samples of sheep. J. Range Mgmt. **24**, 121–125 (1971).

SANDBERG, M.: Avdråttskontroll på skogs- og fjellbeite (Engl. summ.). Forskn. Fors. Landbr. **11**, 311–325 (1960).

SELSJORD, I.: Vektkontroll av sau på fjellbeiter i Sør-Norge (Engl. summ.). Forskn. Fors. Landbr. **15**, 371–404 (1964).

SELSJORD, I.: Vegetasjons- og beitegranskinger i fjellet (Engl. summ.). Forskn. Fors. Landbr. **17**, 325–381 (1966).

SELSJORD, I.: Kjemiske analyser av beiteplanter (Engl. summ.). Forskn. Fors. Landbr. **19**, 1–7 (1968).

SKRE, O.: Methods and results of carbohydrate measurement on Norwegian alpine plants. In: VIK, R. (Ed.): IBP in Norway. Methods and results. Sections PT-UM Grazing project Hardangervidda. Botanical investigations, pp. 195–235. Oslo: Norwegian Nat. IBP Comm. 1975.

SKRE, O., BERG, A., WIELGOLASKI, F. E.: Organic compounds in alpine plants. In: WIELGOLASKI, F.E. (Ed.): Fennoscandian tundra ecosystems. Part 1: Plants and microorganisms, pp. 339–350. Berlin-Heidelberg-New York: Springer 1975.

SKRE, O., KJELVIK, S., BERG, A., WIELGOLASKI, F. E.: Nitrogen content in Norwegian alpine plants. In: VIK, R. (Ed.): IBP in Norway. Methods and results. Sections PT-UM Grazing project Hardangervidda. Botanical investigations, pp. 159–194. Oslo: Norwegian Nat. IBP Comm. 1975.

SOLBERG, P.: Dyrking av eng i fjellet, sammenlignet med dalen, og orienterende analyser av jord- og planteprøver (Eng. summ.). Forskn. Fors. Landbr. **15**, 45–87 (1964).

SONESSON, M., WIELGOLASKI, F. E., KALLIO, P.: Description of Fennoscandian tundra ecosystems. In: WIELGOLASKI, F. E. (Ed.): Fennoscandian tundra ecosystems, Part 1: Plants and microorganisms, pp. 3–28. Berlin-Heidelberg-New York: Springer 1975.

SVANBERG, O., ÅBERG, E.: On environmental conditions for Lymphadenosis in dairy cattle in Sweden. Kungl. Lantbrukshögskolans ann. **22**, 17–24 (1955).

TERRY, R. A., TILLEY, J. M. A.: The digestibility of the leaves and stems of perennial ryegrass, cocksfoot, tall fescue, lucerne and sainfoin, as measured by an *in vitro* procedure. J. Brit. Grassland Soc. **19**, 363–372 (1964).

ULVESLI, O.: Sammensetningen av fórverdien av høy fra soleieng (Engl. summ.). Meld. N.L.H. **37**, no 5, 1–20 (1958).

VALBERG, E., BØ, S.: Forsøk med slåttetid og gjødsling på eng i Nord-Norge 1958–1965 (Engl. summ.). Forskn. Fors. Landbr. **23**, 405–434 (1972).

VIGERUST, Y.: Starrvekstene *(Carex)* og deres betydning i fjellbeitene. Årb. Beitebr. Norge 1934–1935 *XII*, 42–64 (1936).

VIGERUST, Y.: Fjellbeitene i Sikilsdalen. Årb. Beitebr. Norge 1946–47 *XVIII*, 18–184 (1949).

WIELGOLASKI, F. E.: Comparison of plant structure on grazed and ungrazed tundra meadows. In: WIELGOLASKI, F. E. (Ed.): Fennoscandian tundra ecosystems. Part 1: Plants and microorganisms, pp. 86–93. Berlin-Heidelberg-New York: Springer 1975a.

WIELGOLASKI, F. E.: Primary productivity of alpine meadow communities. In: WIELGOLASKI, F.E. (Ed.): Fennoscandian tundra ecosystems. Part 1: Plants and microorganisms, pp. 121–128. Berlin-Heidelberg-New York: Springer 1975b.

WIELGOLASKI, F. E.: Phenology and population density of Norwegian tundra plants. In: VIK, R. (Ed.): IBP in Norway. Methods and results. Sections PT-UM Grazing project Hardangervidda. Botanical investigations, pp. 323–367. Oslo: Norwegian Nat. IBP Comm. 1975c.

WIELGOLASKI, F. E.: The effect of herbage intake by sheep on primary production, ratios top-root and dead-live above-ground parts. (Hardangervidda, Norway). Polish Ecol. Studies **1** (in press) 1976.

WIELGOLASKI, F. E., KJELVIK, S.: Mineral elements and energy of plants at Hardangervidda, Norway. In: BLISS, L. C., WIELGOLASKI, F. E. (Eds.): Primary production and production processes, Tundra Biome, pp. 231–238. Edmonton-Oslo: IBP Tundra Biome Steering Committee. 1973.

WIELGOLASKI, F. E., KJELVIK, S.: Energy content and use of solar radiation of Fennoscandian tundra plants. In: WIELGOLASKI, F. E., (Ed.): Fennoscandian tundra ecosystems, Part 1: Plants and microorganisms, pp. 201–207. Berlin-Heidelberg-New York: Springer 1975a.

WIELGOLASKI, F. E., KJELVIK, S.: Plant biomass at the Norwegian IBP sites at Hardangervidda 1969–1972. In: VIK, R. (Ed.): IBP in Norway. Methods and results. Sections PT-UM Grazing project Hardangervidda. Botanical investigations, pp. 1–88. Oslo: Norwegian Nat. IBP Comm. 1975b.

WIELGOLASKI, F. E., KJELVIK, S., KALLIO, P.: Mineral content of tundra and forest tundra plants in Fennoscandia. In: WIELGOLASKI, F. E. (Ed.): Fennoscandian tundra ecosystems. Part 1: Plants and microorganisms, pp. 316–332. Berlin-Heidelberg-New York: Springer 1975.

WITT, N.: Undersøgelser over græsmarkplanternes sukkerindhold. Tidsskr. Pl. Avl. **70**, 498–504 (1970).

# 2. Conservation and Recreation in Tundra Ecosystems

# Stability of Tundra Ecosystems in Fennoscandia

E. Dahl

## Stability of Vegetation Pattern

The vegetation pattern in Fennoscandian tundra areas undisturbed by human interference or violent fluctuations in animal populations remains very constant, by reason of the strong environmental control of the vegetation pattern. Probably nowhere in the world do environmental conditions change more abruptly over short distances than in alpine tundra areas (Dahl, 1975). Microclimatic differences caused by differences in slope, exposure and degree and duration of snow cover in winter are very large; hence within a distance of 20 m, mean annual soil temperatures may vary as much as 4° C or about the difference between Oslo and Paris. Drainage conditions and soil fertility can also vary very strongly within short distances. All these variables exert a strong influence on the composition of vegetation, which is strongly adapted to the specific, environmental conditions. Normally environmental conditions do not vary drastically from one year to another compared with environmental differences in neighboring localities. Therefore, as long as topographic conditions, soil properties and climate remain constant, the vegetation pattern also remains constant.

The constancy of vegetation affects not only the dominant species, but the subordinate species as well. Nordhagen (1956 and personal communication) re-visited sites analyzed by him 30 years earlier in the Sylene area and found even subordinate species present within almost the same square meter. Similar results apply also to Rondane (Dahl, 1956 and later, unpublished observations). In sites where rare plants were found by botanists 200 years ago, the same species may be found by the botanists of today. On the other hand, fluctuations of altitudinal limits of pine in Northern Fennoscandia and birch in the Sylene area (Nordhagen, 1956) emphasize their sensitivity to climatic fluctuations, especially to changes in summer temperature.

Soil erosion is relatively unimportant within the Fennoscandian tundra areas. Erosional features resulting from trampling by man or animals, tracks left by vehicles or as an undesired result of road building are rarely seen, although establishment of new trails across sloping mires may cause some peat erosion (Karlsson, 1974). The parent soils are predominantly glacigenic deposits left by the glaciers of the Pleistocene glaciations. Such soils are far less vulnerable to soil erosion than residual soils, produced by weathering of parent rocks during long periods, commonly found in areas unglaciated during the Pleistocene glaciations.

Probably the strongest geomorphic process affecting the vegetation pattern is solifluction. Downslope displacements of sticks left in the soil of solifluction areas have often been observed. HÖGBOM (1914) believed that erosion due to solifluction and other effects of frost action could cause considerable geomorphic changes in short periods, but later observations by RAPP (1960), for example, have not confirmed this. An indication of stability is the common occurrence of large lichen specimens on the rocks. Such specimens may be hundreds, and probably even thousands of years old, and during the life span of the lichens, the rocks could not have been turned without destroying them. Cairns built by topographers or marking boundaries between properties hundreds of years ago still remain intact.

Fire is not considered an important factor in Fennoscandian tundra areas. The pine forests of northern Fennoscandia normally go through a fire cycle (HUSE, 1965; SARVAS, 1959), but the subalpine birch forests and the low alpine heaths and shrubs do not burn easily. It does sometimes happen that a luxuriant vegetation of reindeer lichen (*Cladonia stellaris* = *C. alpestris*) catches fire, and burns vigorously.

Nevertheless, it is probable that a slow succession takes place due to leaching of nutrients from the soils. In almost all Fennoscandian tundra areas, precipitation is in excess of evaporation, and on stable soils podsols develop. Since the leached soils support a vegetation of predominantly xeromorphic species, while soils rich in nutrients support a more mesomorphic vegetation, the succession proceeds in a xeric direction (DAHL, 1956). But this process is too slow to be observed directly.

## Changes Due to Fluctuations in Animal Populations

The vegetation pattern is thus intrinsically stable. But fluctuations in consumer populations nevertheless induce vegetation fluctuations.

The small rodents, and especially the lemming populations undergo cyclical changes. In years of high population densities, almost all new shoots of mosses such as *Pleurozium schreberi* may be eaten. In places with some, but not too deep, snow cover, the vegetation can be seriously affected. WIELGOLASKI (1975, and pers. comm.) observed a strong decline of abundance of *Salix herbacea* at Hardangervidda during peak rodent years, but in years with low rodent populations, the vegetation recovers.

In northern Fennoscandia, birch forests have been seriously damaged by mass attacks of caterpillars of the geometrid moth *Oporinia autumnata* (see KALLIO and LEHTONEN, 1973, 1975, for references). The destruction of the birch trees also has consequences for other ecosystem components, thus according to SILVOLA (1967) bird populations in defoliated areas decreased to about 50% in 1966, while the populations of birds in closely situated undamaged birch forests increased.

Changes in reindeer populations also affect other components of the ecosystems. With the introduction of firearms and their subsequent improvement, man's ability to decimate the reindeer populations increased, and numbers fell to low levels. At the same time, populations of predator species like wolf and wolverines decreased also, due to hunting. During the years with low reindeer populations, grazing pressure on the lichen vegetation, which is the most important source of winter food, was light, and hence a considerable biomass of lichens accumulated. With the introduction of severe restrictions on reindeer felling, the reindeer populations increased rapidly. But as long as an accumulated biomass of lichens remained, shortage of food in winter did not limit population growth. Thus the reindeer populations became larger than the corresponding carrying capacity of the winter grazing ranges. During the 1950s the lichen heaths of the Snøhetta area in the Dovre mountains in southern Norway were rapidly destroyed due to overgrazing. A decline in late winter condition and average shooting weight of the animals followed, and a decimation of the reindeer population was necessary to permit the lichen vegetation to recover. A similar development took place in the Hardangervidda area in the 1960s (GAARE and SKOGLAND, 1975).

## Changes Due to the Effects of Animal Husbandry

The most important use of tundra areas by man has been as grazing grounds for domestic animals. Especially in southern Norway most farms in the valleys had outfarms (Norwegian sæter) in the mountains, where cattle and goats were kept and milked during the summer months. These animals grazed near the outfarms, cows being kept inside overnight. Sheep, oxen and horses could graze more freely over larger areas. The number of these animals was limited by shortage of the food needed to keep them alive during winter, and in general the tundra areas could sustain larger populations in summer than those actually using the grazing grounds.

The most serious effect of animal husbandry on the tundra ecosystems is observed near the outfarms. The outfarms are all located in areas with relatively productive parent soils, and at or somewhat below the timber line. Wood is needed for heating the outfarms and for production of cheese, especially goat cheese. This led to a destruction of the forests near the outfarms which today, characteristically, are situated in a treeless area, while forests extend to higher levels at some distance from the farms. Due to the grazing pressure, especially the original tall-herb meadows were destroyed and replaced by a more grassy vegetation, but relics, such as *Aconitum septentrionale*, commonly remain of the original vegetation.

Sometimes juniper was used for fuel in addition to birch. Using a horse and a rope, the juniper shrubs were dragged out of the ground and transported to the outfarms. Under the juniper shrubs, species of relatively high grazing value such as *Deschampsia flexuosa* could survive, but with the removal of the juniper, grazing pressure increased. In oceanic areas, this gave competitive advantage to *Nardus stricta*, which is unpalatable, and secondary *Nardus* heaths developed (RESVOLL HOLMSEN, 1932). In more continental areas of southern Norway mountain regions like Sikilsdalen (NORDHAGEN, 1943) and Rondane (DAHL, 1956) *Festuca ovina* heaths develop under grazing pressure after the juniper is removed. *Festuca ovina* heaths are of relatively high value for grazing.

In recent times outfarming has declined. This commonly results in a dense regrowth of young birches and reestablishment of birch forests. Due to the slow growth of the juniper, the effects of its removal, however, last very long.

In areas with dense tourist populations, some damage occurs. Along the trails, trampling results in the destruction of vegetation, the lichen heaths are especially sensitive to trampling. But in general, tourists do not commonly stray far away from the trails and the vegetation remains intact. The effect of pollution on lakes and rivers can be serious.

## Sensitivity of Tundra Ecosystems to the Effect of Atmospheric Pollution

Much of the atmospheric pollution comes from the combustion of fossil fuels. If the fossil fuels contain sulphur, such as coal and oil, the sulphur is first oxidized to $SO_2$. Which within a few hours or days is oxidized to $SO_3$. With water $SO_3$ forms sulphuric acid. Roughly half of the sulphur oxides (including sulphuric acid) are removed from the atmosphere by precipitation, the other half is deposited directly on leaves and other surfaces (dry deposition). The amount of dry deposition depends on the vegetation, thus a greater quantity is deposited on forest vegetation, than on low vegetation.

Lichens are known to be very sensitive to air pollution, $SO_2$ in the air is especially toxic. But most Fennoscandian tundra areas are so far from industrial centers that most of the $SO_2$ has been oxidized by the time of its arrival to $SO_3$, which is far less toxic to lichens. Although the disappearance of some macrolichens from the highlands of Scotland may be related to air pollution, no similar effects have yet been documented from Fennoscandia.

When a strong mineral acid such as sulphuric acid is added to the ecosystem from the atmosphere, and precipitation is in excess of evaporation, the acid percolates through the soil layers. If the drainage water contains fewer hydrogen ions than is added from the atmosphere, the difference between hydrogen ions added and removed must correspond to a loss of mobile cations. Thus an acidification of the atmosphere leads to an increased leaching of cations from the ecosystems (DAHL and SKRE, 1971; SKRE, 1972; Sweden's Case Study, 1971). Available data permit a semiquantitative assessment of this effect.

Unfortunately, little information is available about the amount of sulphuric acid added to the tundra ecosystems. The best data cover a period of 7 months at Fillefjell, about 100 km north of Hardangervidda (SCHJOLDAGER, 1974). During this period the amount of hydrogen ions added was 8.8 me $m^{-2}$, calculated from measurements of pH and amount of precipitation. This is probably an underestimate since precipitation samplers collect less than the real amount of precipitation in the winter months. During the same period the amount of hydrogen ions precipitated at the station Kise, the nearest in an international network of stations measuring chemical composition of precipitation and rainwater, 150 km to the east, was 5.6 me $m^{-2}$. The average amount of hydrogen ions added each year in the period 1970–1973 at Kise was 12.2 me $m^{-2}$ $yr^{-1}$. Thus the amount of hydrogen ions in precipitation at Fillefjell can be assessed at 20 me $m^{-2}$ $yr^{-1}$. In addition comes the dry deposition, assessed at 10 me $m^{-2}$ $yr^{-1}$, marking a total of 30 me $m^{-2}$ $yr^{-1}$. This seems reasonable judged from the general pattern of acid deposition. The acid deposition at Hardangervidda is likely to be slightly in excess of that at Fillefjell.

When rainwater percolates through the soil, it leaves the ecosystem at a pH corresponding to that of the upper part of the C layer which is usually about 5.5. At this pH the mobile ions are Na, K, Ca, and Mg (and perhaps Mn), while Fe and Al are immobile. Assuming a precipitation of 1000 mm, of which 300 mm evaporates, the 700 mm percolation water removes 2.2 me $H^+$ $m^{-2}$ at pH 5.5. This is a small quantity compared with the hydrogen ions added.

HINNERI et al. (1975) give data on bioelement-pools, covering the IBP sites at Hardangervidda in southern Fennoscandia and at Abisko and Kevo in northern Fennoscandia. Their data include extractable amounts of Ca, Mg and K in the soils down to 30 or 35 cm, measured in g $m^{-2}$. Recalculating the weights to equivalents, the total, mobile cation pool of these elements taken together is calculated and given below, measured in me $m^{-2}$.

| | | |
|---|---|---|
| Kevo: | Pine forest | 895 |
| | Birch forest | 1815 |
| | Low alpine heath | 500 |
| Abisko, Stordalen: | Oligotrophic mire | 3640 |
| Hardangervidda: | Birch forest | 1510 |
| | Lichen heath | 795 |
| | Dry meadow | 24150 |
| | Snow bed | 1510 |
| | Wet meadow | 17100 |

The analyses do not include Na and Mn, but usually these elements are present in subordinate amounts compared with the ions analyzed.

Comparing the calculated annual loss of cations due to acid deposition on the ecosystems with the bioelement pools, it seems likely that at least some of the ecosystems lose appreciable parts of their cation capitals each year due to acid deposition. The calculated annual loss of cations is more than 3% of the cation capital in the lichen heath. Thus one must expect a rapid depletion of the bioelement pools in oligotrophic, nutrient-poor ecosystems. In general the slow natural process of leaching of nutrients from the soils, must be speeded up enormously by the acid deposition in southern Fennoscandia. This must be expected to have consequences both for ecosystem composition and productivity.

This calculation is probably an underrating of the total effect of acid precipitation on the ecosystems, since it does not take into account the effect of the acidification on the bicarbonate content of the precipitation.

The data of KALLIO and VEUM (1975) show that the acidification of precipitation at Kevo has not nearly reached the level of southern Norway.

The studies of RÜHLING and TYLER (1973) show that the pattern of heavy metal deposition, as measured by the amounts of heavy metals adsorbed on mosses, follows a pattern similar to that of acid deposition. Thus the tundra areas in southern Norway must have received far greater amounts of heavy metals than those of northern Fennoscandia. The observation by KALLIO and WIELGOLASKI (pers. comm.) that under otherwise comparable conditions, nitrogen fixation in tundra ecosystems in southern Norway is lower than in Kevo, is possibly an effect of atmospheric pollution. This question merits closer study.

## References


DAHL, E.: Rondane. Mountain vegetation in South Norway and its relation to the environment. Oslo. Skr. Norske Vid. Akad. Oslo I. Mat.-Nat. Kl. **3**, 1956.

DAHL, E.: Flora and plant sociology in Fennoscandian tundra areas. In: WIELGOLASKI, F. E. (Ed.): Fennoscandian tundra ecosystems. Part 1: Plants and microorganisms, pp. 62–67. Berlin-Heidelberg-New York: Springer 1975.

DAHL, E., SKRE, O.: En undersøkelse over virkningen av sur nedbør på produktiviteten i landbruket. Nordforsk Miljøvårdssekr. Publ. **1971: 1**, 27–40 (1971).

GAARE, E., SKOGLAND, T.: Wild reindeer food habitats and range use at Hardangervidda. In: WIELGOLASKI, F. E. (Ed.): Fennoscandian tundra ecosystems. Part 2: Animals and systems analysis, pp. 195–205. Berlin-Heidelberg-New York: Springer 1975.

HINNERI, S., SONESSON, M., VEUM, A. K.: Soils of Fennoscandian IBP tundra ecosystems. In: WIELGOLASKI, F. E. (Ed.): Fennoscandian tundra ecosystems. Part 1: Plants and microorganisms, pp. 31–40. Berlin-Heidelberg-New York: Springer 1975.

HÖGBOM, B.: Über die geologische Bedeutung des Frostes. Bull. Geol. Inst. Uppsala **12**, 1914.

HUSE, S.: Strukturformer hos urskogbestand i Øvre Pasvik (Strukturformen von Urwaldbeständen in Övre Pasvik). Meldinger Norges Landbrukshøgskole **44** (31), 1965.

KALLIO, P., LEHTONEN, J.: Birch forest damage caused by *Oporinia autumnata* (Bkh.) in 1965–1966 in Utsjoki, N. Finland. Rep. Kevo Subarctic Res. Sta. **10**, 55–69 (1973).

KALLIO, P., LEHTONEN, J.: On the ecocatastrophe of birch forests caused by *Oporinia autumnata* (Bkh.) and the problem of reforestation. In: WIELGOLASKI, F. E. (Ed.): Fennoscandian tundra ecosystems. Part 2: Animals and systems analysis, pp. 174–180. Berlin-Heidelberg-New York: Springer 1975.

KALLIO, P., VEUM, A.: Analysis of precipitation at Fennoscandian tundra sites. In: WIELGOLASKI, F. E. (Ed.): Fennoscandian tundra ecosystems. Part 1: Plants and microorganisms, pp. 333–338. Berlin-Heidelberg-New York: Springer 1975.

KARLSSON, L.: Turistslitaget i Nedre Rapadalen, Sareks Nationalpark. Medd. Växtbiol. Inst. Uppsala **7**, 1974.

NORDHAGEN, R.: Sikilsdalen og Norges fjellbeiter. Berg. Mus. Skr. **22**, 1–607 (1943).

NORDHAGEN, R.: Vegetasjonsforskyvninger i naturparken ved Sylene i perioden 1920–1954 som bevis på klimatiske endringer; en fotografisk dokumentasjon. Norske Vid. Akad. Oslo Årbok **1955**, 21–22 (1956).

RAPP, A.: Recent development of mountain slopes in Kärkevagge and surroundings, Northern Scandinavia. Geogr. Ann. **42**, 65–200 (1960).

RESVOLL HOLMSEN, H.: Om planteveksten i grensetrakter mellom Hallingdal og Valdres. Norske Vid. Akad. Oslo Skr. I Mat.-Nat. Kl. **9**, 1932.

RÜHLING, A., TYLER, G.: Heavy metal deposition in Scandinavia. Water, Air and Soil Pollution **2**, 445–455 (1973).

SARVAS, R.: Der nordische Urwald. Schweizerische Z. Forstwes. **110**, 124–135 (1959).

SCHJOLDAGER, J.: Svovelforurensninger i luft og nedbør. Døgnmålinger 1973. Norsk Inst. for Luftforskning. Teknisk notat **82/74**, 1974.

SILVOLA, T.: Tunturittmarin juokkoesiintymisen vaikurusesta linnustoon utsjoella 1964–1966. Ornis Fennica **44**, 65–67 (1967).

SKRE, O.: Sur nedbør. Årsaker og verknader. Oslo: Norges Naturvernforbund. 1972.

Sweden's Case Study for the United Nations conference on the human environment. Air pollution across national boundaries. The impact on the environment of sulphur in air and precipitation. Stockholm: Nordstedt och Söner 1971.

WIELGOLASKI, F. E.: Comparison of plant structure on grazed and ungrazed tundra meadows. In: WIELGOLASKI, F. E. (Ed.): Fennoscandian tundra ecosystems. Part 1: Plants and microorganisms, pp. 86–93. Berlin-Heidelberg-New York: Springer 1975.

# Remote Sensing of Vegetation in Conservation of Tundra Landscapes

O. HESJEDAL and J. LARSSON

The term "remote sensing" comprises remote registration of objects as well as data processing. Remote sensing is an interdisciplinary tool that may be used in all geo-sciences, and furthermore, may serve in many situations as a link between research and planning for conservation. Experts of various professions may use remote sensing in visualizing situations so that the recordings become more easily accessible for planners and conservationists.

In this paper, only a limited part of the field will be treated, namely airplane recordings with film cameras and satellite recording with multispectral scanners within the spectral band range 400–1100 nm; visible and near infrared radiation (Fig. 1).

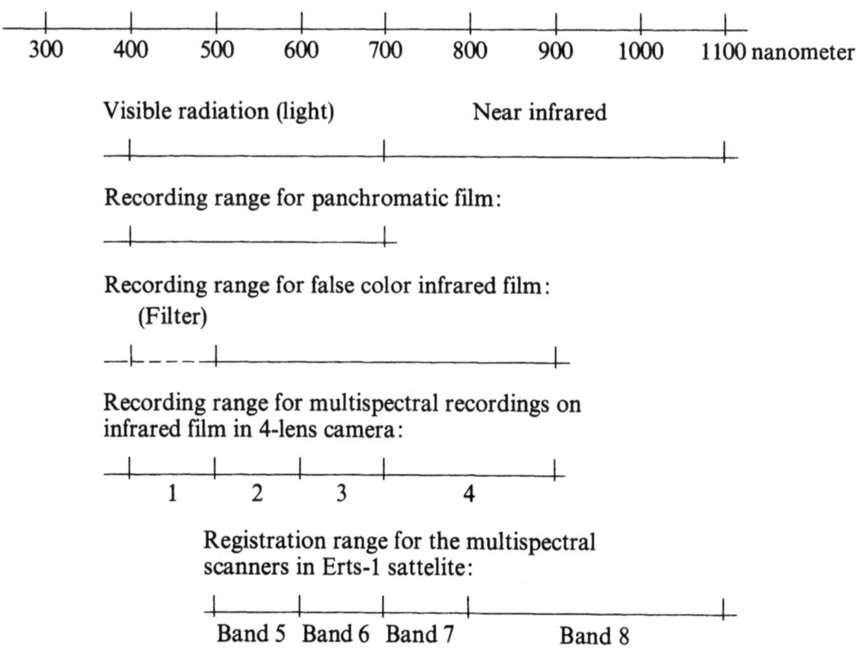


Fig. 1. Spectral band range 400–1100 nm in the electromagnetic spectrum; visible and near infrared radiation

A number of Fennoscandian projects which include remote sensing of vegetation in mountain areas, are related to conservation. A brief survey of some of them, and the experiences gained, are given.

## Methods and Material

Determination of boundary lines between different vegetation units is made possible by features appearing on air photos. Density, perhaps also color tone is most important for small-scale photos. For larger-scale photos, texture and pattern features will help to simplify interpretation, and in stereoscopic studies of air photos, topography and geomorphology are of great value. The experienced photo interpreter will intuitively integrate these parameters. For automatic interpretation, only one of these parameters can be used, i.e. density. Separation of units is done regardless of what the grey tone represents in the field. On the other hand, technical separation of grey tone intervals is far more accurate than visual separation.

There are two principal methods in use for air photo application in data on vegetation: (a) photo interpretation alone, possibly combined with field checks according to a point-method control and (b) all information received from the photo is directly applied in delineation of the units while working in the field. For (a), an interpretation key may be elaborated for photo density and structure. For (b), which applies to detailed recording of ecological or environmental features, field control of every unit delineated will be necessary. For delineation of vegetation units, the mapper must, while working with the stereoscopic photo model, be sufficiently experienced to separate units by integration of all the information he may obtain from the stereo model.

Panchromatic photos for stereoscopic interpretation have long played a major part in Fennoscandian mountain area mapping. This applies particularly to range-land inventories (EINEVOLL, 1958, 1968), while vegetation inventories have also been successfully performed in the last few years (HESJEDAL, 1973a; LARSSON, 1974). Preliminary method studies for mountain vegetation interpretation on different film types, including false-color infrared film, have been done by IHSE (1974) and SCHÜRHOLZ and LARSSON (1973). Agfacontour film for photographic separation of grey tone intervals has been tested in a mountain vegetation mapping project (SCHÜRHOLZ et al., 1972).

Multispectral scanner recordings from sattelite (Erts-1), are available for Fennoscandian tundra areas; however, a large part of the area is snow-covered and has a high degree of cloudiness. Film resolution on the Erts photos is approx. 80 m; in cases where the contrast is good, smaller units may be detected. However, scale and resolution appear too small for most purposes related to conservation of Fennoscandian tundra landscapes. The method seems promising, however, and with improved resolution capacity and more suitable time for recording, small-scale multispectral recordings in the future may prove very valuable as a tool in the conservation of tundra landscapes.

Multispectral recordings from a plane-mounted 4-lens camera are also available in Fennoscandia as also an additive color viewer for multispectral photo interpretation. These tools are expected to prove valuable in remote sensing for conservation purposes, and will be tested in mountain areas.

For photo interpretation alone, false-color infrared film and multispectral recordings appear superior to the traditional film types. In more detailed inventories, where field control of every unit is provided, our view is that stereo models of panchromatic photos (paper prints) is still the best alternative, regarded from a cost/benefit point of view. The material may be gathered at a reasonable cost, and is better suited for field use than most other types of photos.

In some situations, photo flights for conservation purposes alone will not be economically feasible. Panchromatic photos exist, however, of nearly all of Fennoscandia, material cost will thus be limited to the price of such copies as are needed. In all inventories where the information is to be subsequently transferred to official maps, earlier flight photos will be economically advantageous to use. Points and data necessary for map construction are found on the photo series from which the base map was made.

## Results

In 1970, vegetation mapping using black and white panchromatic air photos, was started in the mountain region of Valdres in southern Norway (HESJEDAL, 1973b). On the basis of this inventory, 8 map sheets were produced, each covering approx. 30 $km^2$, scale 1:10000.

Plant communities and combinations of communities comprising a mapping unit are termed vegetation types. These are divided into subtypes where necessary. The vegetation map features occurrence and distribution of the vegetation types. Due to the strong relationship between vegetation and habitat, the map will provide information on a number of environmental factors and variations in both. The plant community reflects ecosystem characteristics.

The mapper performs the inventory, using a stereo model in the field work and delineating each vegetation type directly on the photos, according to a combination of photo interpretation and field control. All vegetation units having an areal coverage of 0.2 ha or more are mapped.

Generally, this type of vegetation map may be applied for certain major puposes: (a) evaluation of biomass production, (b) survey of areas for conservation, (c) survey of areas that from a conservation point of view may be developed for constructional purposes and (d) education. In Valdres, the area mapped is subject to utilization for recreational purposes. Based on the information given by the vegetation maps, it will be possible to locate construction of recreation facilities, especially cabin building, to areas where no conflict will arise with respect to various conservation intentions. High productive areas, valuable wildlife areas, localities of special botanical interest, etc., will appear from information given by the map. Low-yielding areas with sufficient sand or gravel deposits for sewage water infiltration in the ground may be located on the basis of the information appearing, and selection of suitable contruction areas will be possible. The vegetation maps of the Valdres region are part of the material being used to make a recommendation on which areas should be used for agriculture, forestry, outdoor life and conservation.

At the Hardangervidda mountain plateau in southern Norway, air photos of several plots (black and white panchromatic film and false-color infrared film) mostly to the scale 1:10000 have been used in combination with field studies of some plots to achieve measurements for vegetation type mapping to the scale 1:10000 and for vegetation type distribution at Hardangervidda as a whole (Hesjedal, 1975). Correlated with primary production estimates (Kjelvik and Kärenlampi, 1975; Wielgolaski, 1975), it might be possible to calculate roughly the total annual primary production of the plateau. Another air photo project was carried out in the same region and at similar mapping scale (Hesjedal, 1974) to collect information about areal qualities in a region that may be subject to water-power regulations. Areal coverage of each mapped vegetation unit was recorded on an automatic reading planimeter. Tables for nutrient values, water conditions, snow cover and recreation values were elaborated. Records from the various potentially dammed areas were compared in order to establish a more secure basis for evaluation of which alternative should be preferred from the conservation point of view. Based on this material the previously prepared list of conservation value of the various areas in question, was changed.

In order to meet the existing demand for more generalized surveys of the occurrence and distribution of vegetation types over larger areas, a simplification and slight altering of the system may be made (Larsson, 1974). The mapping units in this survey system, recorded on panchromatic photos, scale 1:30000–40000 and reproduced on map scale 1:50000, are intended to coincide with major photo features, in order to faciliate recognition and delineation of vegetation types. Photo interpretation plays a major part; field control includes reconnaissance trips and checks of areas not easily interpreted. In this way, larger areas may be mapped at lower cost; however, accuracy and degree of detail will have to be reduced, as a result of reduced map scale. For many tundra areas, detailed mapping is not feasible and vegetation maps, scale 1:50000, might prove suitable as a source of information for conservation.

Kodak Aerochrome film has been employed extensively in the last few years for ecological surveys and vegetation inventories. The film type was tested in 3 different areas at Hardangervidda (Schürholz and Larsson 1973). Photo plots were sampled, interpreted and described with respect to color image. Consequently, the plots were controlled in the field, where a number of data were recorded for each plot. Certain vegetation types can be easily distinguished and identified by differences in photo color. This is particularly true of types that differ in vegetation density and vigor, and of types that differ in main hydrological features.

On false-color infrared film, eroded areas by reindeer winter overgrazing occurring in several Fennoscandian mountain regions, appear quite clearly in ice-blue color, entirely different from vegetated areas. Overgrazed areas may thus easily be detected. This should make remote sensing an important tool for documentation of the present state of wild reindeer winter grazing area, and also range use (cf. Gaare and Skogland, 1975). Furthermore, the false color infrared film can be used for documentation of other vegetation damages, i.e. severe insect infestations, immision damages and other ecocatastrophes (cf. Kallio and Lehtonen, 1975).

Densitometry techniques using Agfacontour film have been tested in Fennoscandian mountain areas (SCHÜRHOLZ et al., 1972). The techniques included a photographic separation of grey-tone intervals on originals, the equidensities being produced on Agfacontour film. Subsequently, the intervals were reproduced in colors that had various combination alternatives. In photographic separation such as this, areas appearing with equal grey tone on the photo will be separated regardless of area qualities. The method proved to be less suitable for detailed vegetation inventories.

Information obtained through vegetation inventories using remote sensing techniques ought to be integrated with other surveys for better understanding in evaluating ecosystem qualities. Preliminary tests performed in observing wildlife species directly related to mapped vegetation units indicate clearly that certain vegetation types are of greater value than others to different species of wildlife. It is hoped that further collection of such data, and data-analyzing in co-operation with zoologists and wildlife investigators, might aid us in the effort to assess vegetation type „biotope values" for some of the most important species of wildlife.

Evaluation of natural areas relies on adequate information about areal qualities. In all photography of vegetated areas, remission from vegetation is the prime factor to be recorded. Information may be interpreted directly from differences in tone and texture; however, a classification including ecological characteristics needs information about the actual vegetation, and this has to be obtained in the field. It is our experience that comprehensive field controls have been to some degree neglected in photo interpretation projects, and we have therefore stressed this point.

Vegetation is an important component of any ecosystem, and a practical tool for delineation of different areal units. A large number of environmental factors are reflected in the vegetative cover, and vegetation classification will also represent an integrated classification of the environment. Crucial data in evaluating conservation of natural areas are qualitative assessments of vegetation values, indications as to nutrient conditions, hydrology, snow cover, values for various wildlife species and populations etc. Detailed information demands thorough field work, whereas rough surveys are made more easily using remote sensing techniques. Using remote sensing as an inventory tool, much of the work may be performed by a technically educated staff, specially trained in vegetation mapping, so that fully qualified personel are left with control and detail studies.

## References

EINEVOLL, O.: Luftbilete ved fjellbeitegransking. Tidsskr. Norske Landbruk. **65**, 101–133 (1958).

EINEVOLL, O.: Photographic interpretation in the registering of reindeer grazing. Norsk Tidsskr. Jordskifte Landmåling **1**, 91–99 (1968).

GAARE, E., SKOGLAND, T.: Wild reindeer food habitats and range use at Hardangervidda. In: WIELGOLASKI, F. E. (Ed.): Fennoscandian tundra ecosystems. Part 2: Animals and systems analysis, pp. 195–205. Berlin-Heidelberg-New York: Springer 1975.

HESJEDAL, O.: Vegetasjonskartlegging. ISBN 82-557-0010-2. Ås: Landbruksbokhandelen 1973 a.

HESJEDAL, O.: Vegetasjonskartlegging i Øystre Slidre, Valdres. In: Inventering av arealressurser. Ås: Jorddirektoratet, avdeling for jordregistrering 1973b.

HESJEDAL, O.: Vegetasjonskartlegging av potensielle magasinområder for Dagali kraftverk. Ås: Jorddirektoratet, avd. for jordregistrering 1974.

HESJEDAL, O.: Vegetation mapping at Hardangervidda. In: WIELGOLASKI, F. E. (Ed.): Fennoscandian tundra ecosystems, Part 1: Plants and microorganisms, pp. 74–81. Berlin-Heidelberg-New York: Springer 1975.

IHSE, M.: Översiktlig kartering av fjellvegetation i flygbilder. En innledande metodstudie. Stockholm: Nat. Geo. Inst., Rapport **17**, 1–83 (1974).

KALLIO, P., LEHTONEN, J.: On the ecocatastrophe of birch forests caused by *Oporinia autumnata* (Bkh.) and the problem of reforestation. In: WIELGOLASKI, F. E. (Ed.): Fennoscandian tundra ecosystems Part 2: Animals and systems analysis, pp. 174–180. Berlin-Heidelberg-New York: Springer 1975.

KJELVIK, S., KÄRENLAMPI, L.: Plant biomass and primary production of Fennoscandian subarctic and subalpine forests, and of alpine willow and heath ecosystems. In: WIELGOLASKI, F.E. (Ed.): Fennoscandian tundra ecosystems. Part 1: Plants and microorganisms, pp. 111–120. Berlin-Heidelberg-New York: Springer 1975.

LARSSON, J.: Vegetasjonskartlegging, m. 1:50000. In: Inventering av arealressurser. Ås: Jorddirektoratet, avdeling for jordregistrering 1974.

SCHÜRHOLZ, G., EINEVOLL, O., HESJEDAL, O., LARSSON, J.: Die Multiple-Use-Bedeutung einer Vegetationskarte, dargestellt für Norwegen, verbunden mit einer kritischen Betrachtung von Äquidensiten als Hilfsmittel in der mechanischen Luftbildaufbereitung. Landschaft und Stadt, **1**, 1–11 (1972).

SCHÜRHOLZ, G., LARSSON, J.: Falschfarbenbilder für die Vegetationskartierung. Allgemeine Forst- und Jagdzeitung **5/6**, 111–117 (1973).

WIELGOLASKI, F. E.: Primary productivity of alpine meadow communities. In: WIELGOLASKI, F.E. (Ed.): Fennoscandian tundra ecosystems, Part 1: Plants and microorganisms, pp. 121–128. Berlin-Heidelberg-New York: Springer 1975.

# 3. Models for Integration and Prediction

# Principles in the Use of Wide-Scale Models on Tundra Data

F. E. WIELGOLASKI

## The Aim of Models in Ecology

A model is an abstraction of the real world attempting to describe a phenomenon. Models may be strictly descriptive, an example being the introductory Chapter of part 1 of the Fennoscandian tundra ecosystem Volumes (SONESSON et al., 1975), or may try, by words, to imply more functional relationships within the ecosystem, as in the final Chapter of the present Volume (WIELGOLASKI, 1975a). Mathematical symbols may in many cases be used for description of ecosystems, and formal notations of the interactions in the system may be expressed by equations. Such models are normally developed from strictly conceptual to more quantitative forms, but are always imperfect and abstract representations of the real world. They may, however, be very valuable tools for increasing our understanding of the functioning of ecosystems and to see where our available data is most inadequate (cf. SANDHAUG et al., 1975 in the present section). Mathematical models may also be used for the prediction of influences of change in certain parameters on the functioning of the system, e.g. by repeated computations and curve fitting, (cf. LIESTØL et al., 1975 in the present section). Sometimes mathematical model building is said to be equal to systems analysis and systems ecology. Strictly speaking, however, systems analysis cover not only the mathematical description of the system (the model) and other mathematical handling of data, but also observations of different components by measurement of several parameters, as well as analyses to find the most important variables for the functioning of the system.

## Types of Mathematical Models

A mathematical model may be static or dynamic. The static model shows the relationship between factors in the system at a given point (or from calculated average values). A typical example of static models are regression models as given e.g. by KJELVIK et al. (1975). Regression models, sometimes stepwise multiple regressions, have been used in biology for many years. They may be of great value in ecology, for instance in singling out the environmental factor or factors of greatest importance to biological events. However, they do not express anything about the kinetics or dynamics of the system. Today, mathematical modeling mostly means dynamic modeling by simulation and/or optimization techniques. Simulation is the use of mathematical models to reproduce quantitatively some aspects of the real world, as exemplified by the models described in the present section. By optimization, attempts are made to maximize (or minimize) one or several important parameters of the system under given conditions (RAU, 1970; SWARTZMAN, 1972). This technique is commonly used in agriculture and game management.

The dynamic models can be based either on solely time-dependent variables (often with constant time-steps as in the models generally presented in this section) or time- and space-dependent variables. When there is only one independent variable, the model can be based on ordinary differential equations–continuous in time, for example. Complex systems with more than one independent variable can be modeled by partial differential equations which, however, may be difficult to solve analytically. Both linear and non-linear differential equations are used in dynamic models, and so are difference or algebraic equations–discontinuous in time and/or space. The coefficients of the equations are constant in some models, but more often are variable (WIELGOLASKI et al., 1972). Most models in ecology are deterministic and state-oriented, i.e. "prior states of the state-variables determine succeeding ones with unit probability" (PATTEN, 1971); in the present section e.g. the models by SANDHAUG et al. (1975) and by STENSETH (1975). The behavior of the variables may also be probabilistic (even under the same conditions) and the system is then termed stochastic. An example of a partly stochastic model is given in the present section by GYLLENBERG (1975).

The equations in ecological models may be derived from information reported from investigations (inductive method) or by logical deduction from a set of testable hypotheses (deductive method) requiring knowledge of the causal relationship (STENSETH, 1975). This knowledge of biological processes must be obtained from physiological and biochemical experiments, which are of course most easily carried out with details of an ecosystem. Mechanistic models based on such knowledge are therefore mostly developed for small parts of the system (examples e.g. in WIELGOLASKI, 1975b, GATES and SCHMERL, 1975). However, a partly mechanistic approach has also been tried for total ecosystem models (ANWAY et al., 1972; WIELGOLASKI et al., 1974).

The deductive method is particularly important for the "experimental components approach" in modeling (HOLLING, 1966; WATT, 1968) where sets of processes determined by experiments are "transformed" to often elaborate difference equations. Another type of mathematical model is shown in compartment models (VAN DYNE, 1969) described by box and arrow diagram as shown in Fig. 1 and presented, for example, by SANDHAUG et al. (1975) in the present section. The emphasis in compartment modeling has been on description and summary of data collected over time without disturbing the system, but GORE and OLSON (1967) derived moorland model data from clipping and regrowth experiments. While the experimental component approach has tended to give realistic and precise models compartment models have tended to be general, but less realistic (WALTERS, 1971).

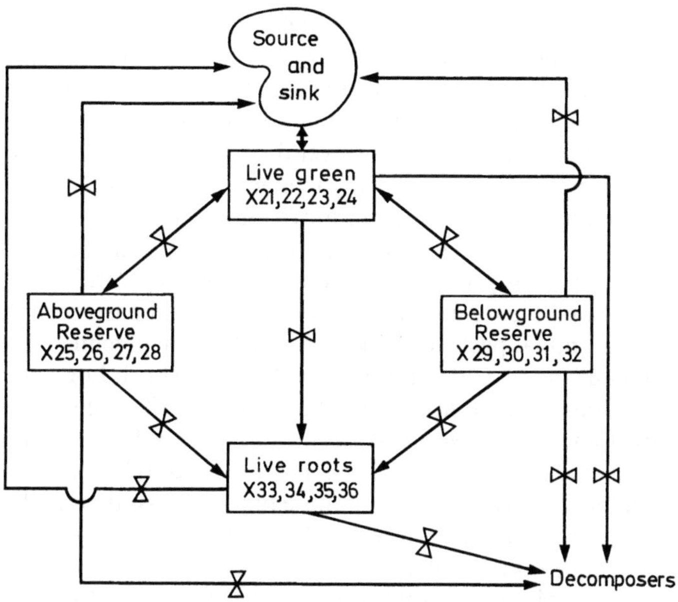


Fig. 1. Elementary diagram of the producer sub-system. X21,25,29,33 are state variables for shrubs. X22,26,30,34 are state variables for perennial herbs. X24,28,32,36 are state variables for cryptogams

## Wide-Scale Models Used on Tundra Data

Typical compartment models have, however, been found to give reasonable output over a wide range of environmental conditions. WIELGOLASKI (1975b) describes a general grazing-land model designed to be useful from relatively humid to dry ecosystems and from warm to cold regions of the globe. Both abiotic factors, plants and decomposers are state variables included in this compartment energy (carbon)–nutrient model, but animals are only introduced in the model as grazers, although they are important components of the ecosystems. To make this model useful over a wide range of conditions, it was found necessary at an interbiome IBP grazing-land symposium (WIELGOLASKI, 1972) to subdivide the plants into shrubs, perennial herbs, annual herbs and cryptogams. In some grazing-land regions, annuals are most important as forage for the animals, in others perennial herbs, while e.g. lichens are of great importance to reindeer in tundra regions (GAARE and SKOGLAND, 1975). The plant groups might preferably have been subdivided still more, e.g. into grasses and forbs, $C_3$–$C_4$ plants, deciduous and evergreen plants, legumes and nonlegumes, lichens and bryophytes etc. to get a higher resolution model with higher precision and more realistic output. The plants were compartmentalized in live green and roots and in above-ground, as well as below-ground reserves (Fig. 1). This division was found important, as the major components of vegetation in some regions have the reserve organs above ground, in other regions below ground.

The decomposer section of the model was subdivided in decomposition of herbaceous and woody above- and below-ground vascular plant material and of dead cryptogams, all compartments divided into simple (e.g. sugars), medium (e.g. cellulose) and difficult (e.g. lignin) decomposed material. The cycling of various nutrients was studied separately in several compartments of both live and dead plants and through various soil compartments, but competition between plants was not taken into account, either for nutrients or space.

To generalize the model, only coefficients of the different equations were changed under various grazing conditions around the world. The producer submodel worked well e.g. for an arid land's winter annual pasture in Israel, a humid, warm summer annual pasture in India, as well as for perennial cool- and warm-season grasslands in Canada and the U.S.A., heath shrub bogs in England and a Norwegian alpine lichen heath (WIELGOLASKI, 1972). The model, used on a Norwegian alpine lichen heath and wet meadow site is discussed in more detail by WIELGOLASKI et al., (1972). Even if this is a model of relatively low resolution, it is important for the understanding of these tundra ecosystems and to some degree predicts changes in the plant biomass by changes in abiotic factors or grazing pressure (WIELGOLASKI, 1975b).

Another example of the application of wide-scale mathematical models on tundra data is given in WIELGOLASKI et al. (1974). The model was originally designed for desert conditions. Even if both temperature and normal precipitation are quite different, desert and tundra do have some similarities. Both types usually have no trees and the ground vegetation may also be sparse in both areas. Very often the amount of soil organic matter is very low both in desert and in tundra. When organic matter is accumulated in tundra, the nutrients in the peat are often not available to living organisms, due to slow decomposition, which may also cause a low diversity of tundra ecosystems. The idea that a desert and a tundra may after all have many characteristics in common was one of the reasons for using Norwegian tundra data in the Desert model.

This partly mechanistic ecosystem model is built up of several versions of plant, animal and climate-soil submodels within a main program. The flexibility is very high by combination of different versions ot the various submodels. Within the compartments, amounts of the building stones of plant and animal material may be studied simultaneously (e.g. ash elements, protein carbon, reserve carbon and structural carbon). However, the problems posed by including such biochemical variables are that the functions describing their flows are often incompletely known.

When the model was first run with Norwegian tundra data, more reasonable output was obtained from the dry lichen heath than from the organic wet meadow site. The desert model was of course originally not designed to handle wet organic conditions, but with some slight alterations these problems were relatively well solved. The differences in dry matter found throughout the season by the model seemed to be of the same order as those evaluated from observed values, as did the amounts of the various plant constituents. Reduction during summer was stronger for protein and reserve carbon in the soil than for structural carbon. Because of input from precipitation (see KALLIO and VEUM, 1975) there was some increase in nitrogen, phosphorus, and cations to the system during winter, while the amount of total carbon was of course reduced during the same period. A detailed output on a daily basis was given for apparent photosynthesis as a function of radiation, air temperature during the day, and relative humidity, all in good agreement with observed values.

Similar tests of wide-scale models by data from quite different regions will probably be of great value in improving the models. But such general models have of course several limitations in their ecological significance, and require several kinds of improvement to help answer particular questions like those raised in the sections that follow.

## References


ANWAY, J. C., BRITTAIN, E. G., HUNT, H. W., INNIS, G. S., PARTON, W. J., RODELL, C. F., SAUER, R. H.: ELM Version 1.0. Tech. Rep. **156**, Fort Collins: US IBP Grassland Biome 1972.

GAARE, E., SKOGLAND, T.: Wild reindeer food habitats and range use at Hardangervidda. In: WIELGOLASKI, F. E. (Ed.): Fennoscandian tundra ecosystems. Part 2: Animals and systems analysis, pp. 195–205. Berlin-Heidelberg-New York: Springer 1975.

GATES, D. M., SCHMERL, R. B.: Perspectives of biophysical ecology. Berlin-Heidelberg-New York: Springer 1975.

GORE, A. J. P., OLSON, J. S.: Preliminary models for accumulation of organic matter in an *Eriophorum/Calluna* ecosystem. Aquilo. Ser. Botanica **6**, 297–313 (1967).

GYLLENBERG, G.: Systems analysis on invertebrates. In: WIELGOLASKI, F. E. (Ed.): Fennoscandian tundra ecosystems, Part 2: Animals and systems analysis, pp. 267–272. Berlin-Heidelberg-New York: Springer 1975.

HOLLING, C. S.: The functional response of invertebrate predators to prey density. Mem. Entomol. Soc. Canada **48**, 1–86 (1966).

KALLIO, P., VEUM, A. K.: Analyses of precipitation in Fennoscandian tundra sites. In: WIELGOLASKI, F. E. (Ed.): Fennoscandian tundra ecosystems. Part 1: Plants and microorganisms, pp. 333–338. Berlin-Heidelberg-New York: Springer 1975.

KJELVIK, S., WIELGOLASKI, F. E., JAHREN, A.: Photosynthesis and respiration of plants studied by field technique at Hardangervidda, Norway. In: WIELGOLASKI, F. E. (Ed.): Fennoscandian tundra ecosystems. Part 1: Plants and microorganisms, pp. 184–193. Berlin-Heidelberg-New York: Springer 1975.

LIESTØL, K., ØSTBYE, E., SKAR, H. J., SWARTZMAN, G.: A simulation model of a small rodent population. In: WIELGOLASKI, F. E. (Ed.): Fennoscandian tundra ecosystems, Part 2: Animals and systems analysis, pp. 273–282. Berlin-Heidelberg-New York: Springer 1975.

PATTEN, B. C. (Ed.): Systems analysis and simulation in ecology. New York-London: Academic Press 1971.

RAU, J. G.: Optimization and probability in systems engineering. New York: Van Norstrand Reinhold 1970.

SANDHAUG, A., KJELVIK, S., WIELGOLASKI, F. E.: A mathematical simulation model for terrestrial tundra ecosystems. In: WIELGOLASKI, F. E. (Ed.): Fennoscandian tundra ecosystems. Part 2: Animals and systems analysis, pp. 251–266. Berlin-Heidelberg-New York: Springer 1975.

SONESSON, M., WIELGOLASKI, F. E., KALLIO, P.: Description of Fennoscandian tundra ecosystems. In: WIELGOLASKI, F. E. (Ed.): Fennoscandian tundra ecosystems. Part 1: Plants and microorganisms, pp. 3–28. Berlin-Heidelberg-New York: Springer 1975.

STENSETH, N. C.: Energy model for individual small rodents and its significance to general population theory. In: WIELGOLASKI, F. E. (Ed.): Fennoscandian tundra ecosystems. Part 2: Animals and systems analysis, pp. 283–299. Berlin-Heidelberg-New York: Springer 1975.

SWARTZMAN, G. L. (Ed.): Optimization techniques in ecosystem and land use planning. Techn. Rep. **143**, Fort Collins: US IBP Grassland Biome 1972.

VAN DYNE, G. M.: Grassland management, research and teaching viewed in a systems context. Range Sci. Dept. Sci. Ser. **3**, Fort Collins: Colorado State Univ. 1969.

WALTERS, C. J.: The systems approach and mathematical models in ecology. In: ODUM, E. P. (Ed.): Fundamentals of ecology, pp. 277–292. Philadelphia-London-Toronto: Saunders 1971.

WATT, K. E. F.: Ecology and resource management: A quantitative approach. New York: McGraw Hill 1968.

WIELGOLASKI, F. E.: IBP Grassland/tundra international modeling-synthesis workshop: Summary of output workbook. Tech. Rep. **197**. Fort Collins: US IBP Grassland Biome 1972.

WIELGOLASKI, F. E.: Functioning of the Fennoscandian tundra ecosystems. In: WIELGOLASKI, F.E. (Ed.): Fennoscandian tundra ecosystems. Part 2: Animals and systems analysis, pp. 300–326. Berlin-Heidelberg-New York: Springer 1975a.

WIELGOLASKI, F. E.: A grazing land's simulation model. In: CHARNES, A., LYNN, W. R. (Eds.): Mathematical analysis of decision problems in ecology, pp. 45–83. Lecture Notes in Biomathematics, Vol. 5. Berlin-Heidelberg-New York: Springer 1975b.

WIELGOLASKI, F. E., HAYDOCK, K. P., CONNOR, D. J.: A grazing land's plant-decomposition, carbon-mineral simulation model. Tech. Rep. **203**. Fort Collins: US IBP Grassland Biome 1972.

WIELGOLASKI, F. E., RADFORD, J., VALENTINE, W. G.: Study on generality of a Desert Simulation model by use of tundra data. In: 1973 Progress Report, Vol. I. Genual-purpose model section, pp. 187–216. US/JBP Desert Biome Res. Memo 74–54. Logan: Utah State Univ. 1974.

# A Mathematical Simulation Model for Terrestrial Tundra Ecosystems

A. SANDHAUG, S. KJELVIK, and F. E. WIELGOLASKI

Mathematical simulation in ecology is a very powerful tool in testing the "long-term" effects of different hypothesis and/or parameter values. Several models simulate processes such as photosynthesis (DUNCAN et al., 1967; DE WIT et al., 1970; MILLER and TIESZEN, 1972) or greater or smaller parts of an ecosystem as in population dynamic models (TAYLOR, 1967; PIELOU, 1969; BRENNAN et al., 1970). Relatively few deal with whole ecosystems. The main reason being one of complexity and the necessity of data from all levels of the system, a fact which demands close and expensive teamwork between systems analysists and biologists from different fields. Examples of greater ecosystem models are those presented by BLEDSOE et al. (1972), SWARTZMAN and VAN DYNE (1972) and WIELGOLASKI et al. (1972).

The model presented here is intended to cover the structure of terrestrial alpine tundra ecosystems at Hardangervidda, Norway (studied by Norwegian IBP) and their production of carbon under different simulated environmental conditions or biological variations (e.g. reindeer and sheep grazing); further, by simulation and sensitivity analyses, to clarify effects of the functional relationships within the system, and to show where essential biological information for the model is missing, and thus the areas where more research is needed. Problems of age and population dynamics have only been touched on briefly. The model is a dynamic (time-dependent), deterministic, differential equation model with variable coefficients.

The present version of the model has been run at 24-hrs intervals for the wet meadow site described by SONESSON et al. (1975) and WIELGOLASKI (1975). The model has been made as general and flexible as possible. However, quite a lot of details pertain to the wet meadow, e.g. abiotic data, starting values of state variables and assumptions. Uncertainties are involved concerning both mode of function of some processes and many of the parameter values used. The model thus gives only a crude picture of the ecosystem. Parts of the model are close to "black box" simulation due to little or no field information. This is especially true for the process "death of plant material". Further studies and experiments are needed to build a more realistic model.

The model can easily be extended to include nutrient uptake and transport through the ecosystem. No nutrient stress is assumed in the present version except for forbs, which are more eutrophic than the dominating species within other plant groups at the site; also no growth limitations are expected or provided due to lack of water. It is possible that soil water may be overoptimal for some of the processes at the wet meadow due to lack of oxygen, but this is not taken into account in the present model. Abiotic variables are, as a whole, never expected to be overoptimal in the version described, and the system is said to be in a steady-state situation. A detailed description of the model is given by SANDHAUG (1974).

## Model Description

The variables in the model are divided into four groups according to mode of function and content:

(1) Driving variables represent environmental factors greatly influencing the ecosystem, but changing independently of the system, e.g. solar radiation, temperature and precipitation.

(2) State variables are elements comprising the basic structure of the system, e.g. standing crop of the different taxonomic groups and fractions (the compartments in the model diagram Fig. 1).

(3) Flow variables represent transfer of energy or compounds between state variables per unit time. Most flow variables represent biological processes, e.g. photosynthesis and respiration of different plant groups and fractions per $m^2$ per day.

(4) Intermediate variables may be important in controlling flow of compounds or energy (e.g. the effect of chlorophyll concentration or shading on apparent photosynthesis). Other intermediate variables are included as a result of programming and mathematical causes (e.g. the effect of temperature on decomposition (DE) at different soil temperatures).

The driving variables used in the present model are:

(a) Net solar radiation in langleys (ly) $(2\ \text{hrs})^{-1}$ is calculated by empirical formulae developed by SANDHAUG (1974), based on the observed sum of incoming radiation per day at 2 m (height) above ground, taking into account the variation in solar inclination in the course of a day and throughout the growing season.

(b) Air temperature at 10 cm height at any time of a 24-hrs period is also calculated by formulas developed by SANDHAUG (1974) based on the variation in day length throughout the season, and measured daily maximum and minimum temperatures at this height. The 24-hourly mean temperature is approximated by the integrated mean.

The following driving variables are based directly on observed values:

(c) Mean 24-hourly soil temperature at $-2$ cm.

(d) Precipitation in mm per 24-hrs period.

(e) Mean 24-hourly relative air humidity at 10 cm height.

(f) Mean 24-hourly soil water in 0–10 cm depth as percentage of fresh weight.

(g) Mean 24-hourly wind speed at 10 cm height in m $s.^{-1}$.

Primary producers included in the present model (Fig. 1) are the four groups, bryophytes, forbs, monocotyledons (grasses and sedges) and woody plants

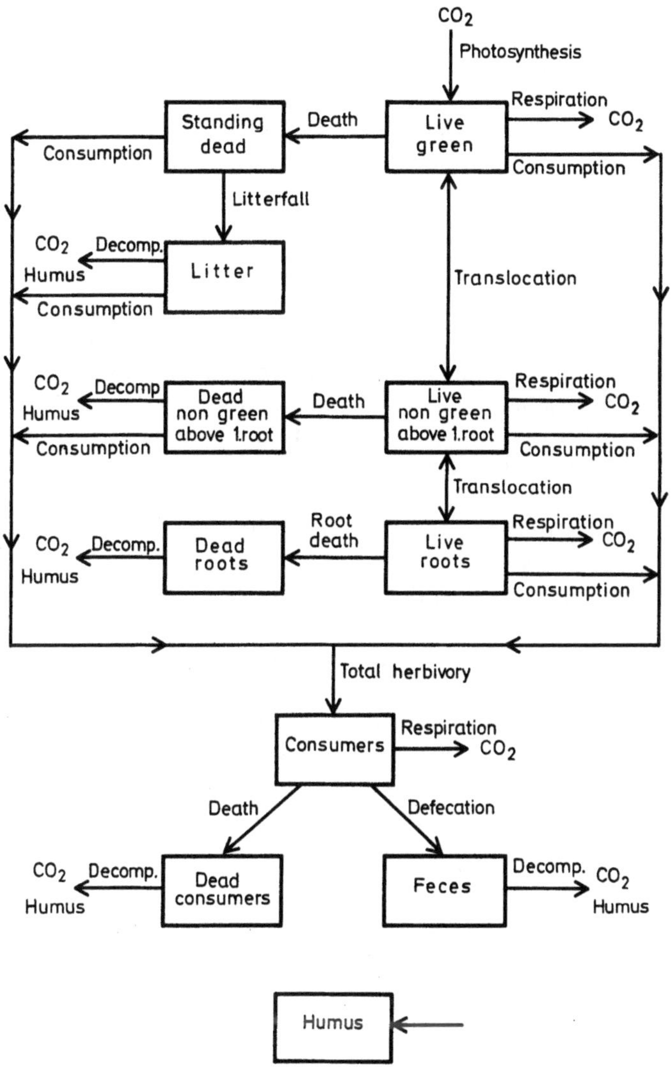


Fig. 1. Flow diagram representing state and flow variables in the present model

(shrubs). Lichens are neglected on the wet meadow. In animals, only invertebrates are included, although birds and small rodents are also found at the site.

The wet meadow may also be used for sheep grazing, but neither this nor reindeer grazing is included in the present version. Bryophytes are divided into green, non-green live, and dead material. The subdivision of vascular plants and consumers is shown in Fig. 1. All weights are given in g oven dry weight (80° C). Values for apparent photosynthesis and all respiration rates of plants (expressed according to Šesták et al. (1971) as g dry matter per g total dry material) are calculated every 2 h and summed up over a 24-hrs period.

All other flow variables in the model are calculated as 24-hourly means (e.g. translocation, death, decomposition and all rates for invertebrates). The flows are mainly controlled by external driving variables in the model, but internal factors, such as, for example, amount of biomass in the different compartments are also important.

The mathematical interpretations of processes are, when possible, based on the principle of optimal rates. The light dependency for apparent photosynthesis ($PH$) of vascular plants at optimal conditions can be expressed by the equation:

$$PH = X1\left(1.0 - e^{-\frac{S - S_0}{k}}\right) \tag{1}$$

$X1$ is maximum photosynthesis (g g$^{-1}$ (2 hrs)$^{-1}$), $S$ net solar radiation (ly), $S_0$ light at compensation point (ly) and $k$ a factor depending on the variation in apparent photosynthesis with radiation; all are given per time unit. This inverse exponential expression for photosynthesis, as a function of radiation, is based on the assumption that apparent photosynthesis starts at a minimum light intensity ($S_0$) and approaches a maximum level with increased radiation. (In the model $S_0 = 1.6$ ly, $k = 12.2$ ly and values for maximum photosynthesis are $X1 = 0.035$ g g$^{-1}$ (2 hrs)$^{-1}$ for forbs and woody plants and $X1 = 0.0389$ g g$^{-1}$ (2 hrs)$^{-1}$ for grasses and sedges.) The values used in the model are somewhat higher than data from phytotron experiments with vegetation mats from the site (SKRE, 1975).

The optimal rate of apparent photosynthesis, as found by Eq.(1), is reduced multiplicatively by factors depending on temperature, relative air humidity, shading and chlorophyll concentration. Each of these factors varies from one, at optimal conditions, to zero (when no photosynthesis can take place).

The air-temperature factor for vascular tundra plants is assumed to be zero at 0° C and 1.0 at 15° C based on observations by TIESZEN (1971) and SKRE (1975). Between these temperatures the modifying temperature factor ($TE$) is calculated by:

$$TE = 1.1\left(1.0 - e^{-\frac{TA}{4.55}}\right) \tag{2}$$

$TA$ is actual air temperature (°C). High temperature at the site never limits photosynthesis in the model.

Mean relative air humidity ($RH$) below 70% per 24-hrs period reduces the apparent photosynthesis of vascular plants according to the formula:

$$HE = 0.552 + 0.0106\,(RH - 35.0)\,. \tag{3}$$

This is suggested from field measurements (KJELVIK et al., 1975). $RH$ below 35% never occurs at the wet meadow site. Shading effects are assumed when the total green biomass exceeds 100 g m$^{-2}$. Above this value no further increase in net production takes place. The chlorophyll content per g dw of leaves is nearly constant from early spring to mid-July (BERG, 1975). It then decreases until the end of the growing season according to the equation:

$$CF = \frac{280 - D}{85} \tag{4}$$

$D$ is day number and $CF$ the chlorophyll factor. Apparent photosynthesis is assumed reduced in proportion to reduction in chlorophyll concentration.

Apparent photosynthesis in bryophytes is calculated by the expression:

$$PH = P\,(0.23 + TF)\,HF \cdot LGS \tag{5}$$

$P = 0.0086\ \mathrm{g\,g^{-1}\,(2\,hrs)^{-1}}$ which corresponds to a maximal photosynthetic rate of $0.0688\ \mathrm{g\,g^{-1}\,(2\,hrs)^{-1}}$ at 15° C, $TF$ is a temperature factor for bryophytes, $HF$ humidity factor and $LGS$ shading factor.

$$TF = 0.006\,(S - 4.6)\,(TA + 5.0) \tag{6}$$

$S$ is solar radiation (ly) (2 hrs) and $TA$ air temperature (° C). This function is used until light saturation at $S = 16.0\ \mathrm{ly\,(2\,hrs)^{-1}}$. At higher radiation $S$ is set equal to 16.0 ly. This dependency on light and temperature for apparent photosynthesis of bryophytes is an approximation to the curves given for the lichen species *Nephroma articum* by Kallio and Heinonen (1971). The humidity factor ($HF$) for bryophytes is assumed to be expressed by:

$$HF = 1.1\,\left(1.0 - e^{-0.028\,(RH - 25)}\right) \tag{7}$$

$RH$ is mean 24-hourly relative air humidity. The shading factor ($LGS$) is the same as for vascular plants.

Respiration rates ($RE$) of green material at the site are based on data by Skre (1975). It is a function only of air temperature ($TA$), and is given by:

$$RE = R\,e^{kTA}\,, \tag{8}$$

The basal respiration rate at 0° C is $R = 0.00036\ \mathrm{g\,g^{-1}\,(2\,hrs)^{-1}}$, and $k = 0.128°\ \mathrm{C^{-1}}$ for forbs and monocotyledons. Respiration rates of green of woody plants and of bryophytes are respectively 90% and 10% of the rate given above. Similar exponential expressions are used for respiration of non-green above 1. root and roots of vascular plants where $R$ and $k$ vary with plant groups and fractions. For respiration of below-ground parts, mean 24-hourly soil temperature at −2 cm depth is used instead of air temperature. As an example, root respiration of monocotyledons is calculated by Eq. (8), using the basal respiration at 0° C $R = 0.00001\ \mathrm{g\,g^{-1}\,(2\,hrs)^{-1}}$ and $k = 0.1322°\ \mathrm{C^{-1}}$. Root respiration was studied by water culture (Skre, 1975) for some species from the site. The values found by this method seem to be too high, probably caused by higher activities in the non-limiting nutrient solution used in the experiments or to some degree a wounding effect. No measurements have been carried out at the site for respiration of non-green above 1. root, for any species. The rates used in the model for these fractions are therefore mainly based on the assumption of a maintenance respiration rate and that a portion of the material translocated to these fractions is lost by respiration.

Growth in spring is assumed to start when soil temperature ($TS$) at −2 cm over a 24-h period is above 2° C. This also initiates upward translocation of photosynthates in vascular plants (although the function for translocation starts at 0° C) and the rate increases with increase in temperature. Variations in translocation rates with temperature for the different vascular plant groups are given by SANDHAUG (1974). The total amount translocated upwards is estimated to be about 20 g $m^{-2}$ dry material in a period of about 10–12 days. This amount is divided between the various vascular plant groups at the site proportionally to their live non-green above 1. root and live root biomass.

When the 20 g $m^{-2}$ has been translocated upwards, downward translocation starts from green material. Translocation downwards is taken as percent of apparent photosynthesis. This is 25, 15, and 10% to live non-green above 1. root for forbs, monocotyledons and woody plants, respectively, and to live roots 50, 75, and 75%, respectively. These rates are somewhat higher than reported from translocation studies at the site (BERG et al., 1975), but had to be used in order to obtain the high non-green production and support the great living root biomass at the site (WIELGOLASKI, 1975). It is assumed that about 25% of the total green biomass is translocated downwards in the fall within a period of about 30 days. The rest, mainly cellulose and lignin, is resistant to translocation (SKRE et al., 1975).

In bryophytes, there is only a weak translocation from green to non-green parts at a rate of 20% of apparent photosynthesis. Further, green parts of bryophytes are transferred to non-green live parts by senescence at a rate varying from 0.003 g $g^{-1}$ (24 hrs)$^{-1}$ in summer to 0.0675 g $g^{-1}$ (24 hrs)$^{-1}$ in late fall, estimated from production studies (WIELGOLASKI, 1975), but the values are very doubtful.

Some information on death of plant material is obtained from harvesting data (WIELGOLASKI, 1975; WIELGOLASKI and KJELVIK, 1975), but there is a need for more biological knowledge and data for this important process. Death in the model has no functional relationship with biotic or abiotic variables. Rates are taken as constants changing with plant fraction, species, group, and time of year. Death of live green of e.g. monocotyledons changes from 0.05 g $g^{-1}$ (24 hrs)$^{-1}$ in spring to 0.015 g $g^{-1}$ (24 hrs)$^{-1}$ in summer, and to 0.08 g $g^{-1}$ (24 hrs)$^{-1}$ in late fall.

Standing dead is transferred to litter at a rate depending on plant group, wind speed and time of year. The highest transfer occurs in late fall.

Decompensation is strongly dependent on site temperature as shown by soil respiration studies (SVENSSON et al., 1975). The general effect of soil temperatures ($TS$) on decomposition of plant material is expressed by a temperature factor ($DF$):

$$DF = 0.02347\,(TS + 2.0)\,e^{0.1486\,(TS+2.0)} \tag{9}$$

if $TS$ is below 4° C. If $TS$ is above 4° C, the equation changes to:

$$DF = 1.0 - 0.672\,e^{-0.1612\,(TS-2.0)} \tag{10}$$

$DF$ is then multiplied by a factor depending on plant group and fraction under decomposition. Litter bag studies (ROSSWALL et al., 1975) were the basis for decomposition rates used in the model, adjusted to avoid too high accumulation of peat (SANDHAUG, 1974).

Invertebrate biomass at the site was determined several times during the growing season (SOLHØY, 1972). Invertebrates are separated into one group (90% of total biomass) which lives in litter and soil, and a second group (10%) living in vegetation above the soil surface. Live biomass of the two groups follows a sine curve from June 1st to February 1st with a maximum in late September of about 2.0 g dw $m^{-2}$ and a minimum on February 1st of about 0.2 g. Between February 1st and June 1st, live biomass remains constant. At 15° C, the invertebrate consumption of dry plant biomass per 24 hrs is estimated to be 0.025 g (g dw)$^{-1}$. Consumption of plant material by both invertebrate groups is temperature-controlled and assumed to follow the function:

$$0.025\, e^{0.25(TA-15.0)}\ \mathrm{g\,g^{-1}\,(24\,hrs)^{-1}} \tag{11}$$

$TA$ is mean daily air temperature. Of consumption, 20% is assumed used in respiration, 20% in growth and 60% goes to feces. Between October 1st and February 1st, the death rate of both groups is 0.03 g (g dw)$^{-1}$ (24 hrs)$^{-1}$, and 1/10 of this value during the rest of the year. Decomposition of dead invertebrate biomass and feces are assumed to be the same as for litter from forbs (SANDHAUG, 1974).

From the daily process rates of the different plant groups, plant parts and invertebrates, calculated as exemplified above, new values are calculated each day for state variables of plants (weights) by a system of several first order and first degree differential equations (SANDHAUG, 1974).

As an example from this system of Equations, the change in live green ($LG'$) per 24-hrs period is given by:

$$LG' = (PH - DG - TRZ - TR)\, LG - HO \cdot C\,. \tag{12}$$

In spring (when there is a net upward translocation) Eq. (12) is changed to

$$LG' = (PH - DG)\, LG + RTM(TRZ \cdot SR + TR \cdot LR) - HO \cdot C\,. \tag{13}$$

In Eqs. (12) and (13) all symbols are vectors having different components, identified by indices for the different plant and animal groups. $PH$ = rate of apparent photosynthesis, $DG$ = rate of death of green, $TRZ$ = rate of translocation between live green and live non-green above 1. root, $TR$ = rate of translocation between live green and live roots, $HO$ = rate of consumption of live green from various plant groups by herbivores, $C$ = consumer biomass, $RTM$ = amount of translocated matter converted to green biomass (the rest $1-RTM$ being respired), $SR$ = live non-green biomass above 1. root and $LR$ = live root biomass.

Several numerical methods of solving systems of linear differential equations have been developed. In this model, however, the equations are solved successively analytically. The reasons for this are: (1) there are relatively small changes in the state variables over the chosen time step of 24 hrs; (2) numerical instabilities are avoided; (3) computer time is saved.

An equivalent form of Eq. (12) or Eq. (13) is:

$$y' = -Q(t)\,y + P(t) \tag{14}$$

$y$ is the state variable in question, $Q(t)$ the coefficient of $y$ and $P(t)$ the sum of the other terms in the equation (e.g. those terms not containing the particular state variable $y$). Since $P(t)$ and $Q(t)$ are given constant values each day (computed either as a summed or ordinary mean over a 24-hrs period), say $P$ and $Q$, the general solution of Eq. (14) is given (e.g. APOSTOL, 1961) by

$$y = \frac{P}{Q} + \left(C - \frac{P}{Q}\right) e^{-Qt}. \tag{15}$$

$C$ is the starting value of $y$ for the time period over which Eq. (15) is used. Since the Equations [e.g. Eqs. (12) or (13)] are solved once every 24-h period ($t = 1$), $C$ is the preceding day's value of $y$. By choosing shorter time steps, this method gives values closer and closer to an exact solution of the equation system.

## Results and Discussion

Table 1 gives the simulated flow of matter (in g dry weight $m^{-2}$ $year^{-1}$) through the different compartments of primary producers at the wet meadow as simulated by the model under "normal" conditions, i.e. with observed values (or estimated values if observations are lacking) of parameters, driving variables and starting values of state variables. High primary production is simulated for 1969 and 1971, lowest for 1970 (Table 1). Generally, temperatures in the first part of summer seem to be most important for annual primary production. 1969 and 1971 had higher July temperatures than 1970 (SKARTVEIT et al., 1975), and this is one reason for the higher production (Table 3). Earlier snow melt at the wet meadow site in 1969 and 1971 than in 1970 and 1972 (about one week difference) also contributed. High June temperatures in 1970 do not fully compensate for the late snow melt and low July temperatures in 1970, nor do high July temperatures 1972 fully compensate for the late snow melt and relatively low temperatures just afterwards that year. A comparison of production of green in e.g. monocotyledons in the year of highest production (1971) with the year of lowest production (1970) (Fig. 2) shows e.g. 37.5 g $m^{-2}$ green biomass on day 182 (July 1st) in 1971 and 30.5 g $m^{-2}$ on the same day in 1970. The same pattern is seen for green of other plant groups.

Figure 3 shows examples of the simulated annual course of biomass of four other plant compartments. Live roots of monocotyledons and woody plants have the greatest live non-green biomass at the site. In the simulation model output, both compartments clearly reflect the fact that 1969 and 1971 were the years of highest, and 1970 of lowest production. Dead roots and litter of monocotyledons are the most important dead compartments at the site and are therefore chosen to illustrate the annual course of dead material (Fig. 3) in the four research years.

Table 1. Model output under "normal" conditions. Plant group 1 = mosses, 2 = forbs, 3 = monocotyledons and 4 = woody plants. All results in g dw $m^{-2}$ $year^{-1}$. For further explanation see text

| Year | Plant group | Apparent photo-synthesis | Translocation from green to | | Respiration | | | Death | | | Transfer from standing-dead to litter | Decomposition | | |
|---|---|---|---|---|---|---|---|---|---|---|---|---|---|---|
| | | | non-green ab. l. root | roots | green | non-green ab. l. root | roots | green | non-green ab. l. root | roots | | litter | non-green ab. l. root | roots |
| 1969 | 1 | 128.2 | 116.6 | | 4.7 | 10.5 | | 13.0 | 102.7 | | 13.0 | 6.1 | 15.9 | |
| | 2 | 63.5 | 14.0 | 28.1 | 17.9 | 4.0 | 6.6 | 20.4 | 10.4 | 15.5 | 20.4 | 16.9 | 1.1 | 7.8 |
| | 3 | 591.8 | 56.2 | 407.3 | 79.6 | 25.6 | 74.9 | 105.1 | 26.3 | 271.8 | 111.1 | 109.1 | 25.5 | 252.4 |
| | 4 | 144.1 | 14.4 | 90.7 | 19.7 | 3.7 | 51.8 | 32.1 | 6.4 | 33.5 | 31.5 | 23.6 | 4.9 | 29.6 |
| 1970 | 1 | 88.4 | 79.8 | | 2.7 | 9.6 | | 9.3 | 90.1 | | 9.3 | 7.1 | 66.3 | |
| | 2 | 48.6 | 11.0 | 21.8 | 9.5 | 2.6 | 6.4 | 14.4 | 8.6 | 15.8 | 14.6 | 15.1 | 3.2 | 9.7 |
| | 3 | 452.2 | 42.3 | 303.3 | 49.6 | 22.4 | 70.8 | 83.4 | 23.4 | 271.1 | 91.0 | 90.3 | 23.4 | 238.4 |
| | 4 | 77.7 | 7.5 | 46.2 | 8.3 | 3.6 | 47.3 | 17.4 | 5.8 | 31.6 | 18.0 | 24.8 | 4.8 | 27.4 |
| 1971 | 1 | 140.9 | 128.6 | | 4.0 | 9.0 | | 14.2 | 100.0 | 15.7 | 14.2 | 9.0 | 83.3 | |
| | 2 | 52.8 | 10.9 | 21.8 | 12.5 | 2.7 | 6.4 | 19.1 | 8.3 | 276.4 | 19.2 | 17.8 | 4.9 | 12.1 |
| | 3 | 596.5 | 56.0 | 405.6 | 64.2 | 24.0 | 71.7 | 113.2 | 26.8 | 31.9 | 106.1 | 102.2 | 24.5 | 256.0 |
| | 4 | 159.1 | 15.8 | 99.1 | 17.1 | 4.0 | 46.5 | 37.2 | 7.4 | | 37.0 | 24.4 | 5.3 | 28.7 |
| 1972 | 1 | 112.2 | 103.7 | | 3.2 | 10.1 | | 9.6 | 99.0 | 15.5 | 9.6 | 9.5 | 93.9 | |
| | 2 | 50.4 | 10.9 | 21.9 | 12.1 | 2.5 | 6.4 | 16.1 | 8.5 | 274.5 | 16.3 | 16.7 | 5.7 | 13.0 |
| | 3 | 464.2 | 43.2 | 311.0 | 58.1 | 23.5 | 72.9 | 88.5 | 23.9 | 32.0 | 94.4 | 104.3 | 24.0 | 255.5 |
| | 4 | 119.9 | 11.7 | 72.0 | 14.5 | 4.5 | 47.8 | 27.8 | 7.5 | | 28.3 | 31.9 | 5.6 | 28.0 |

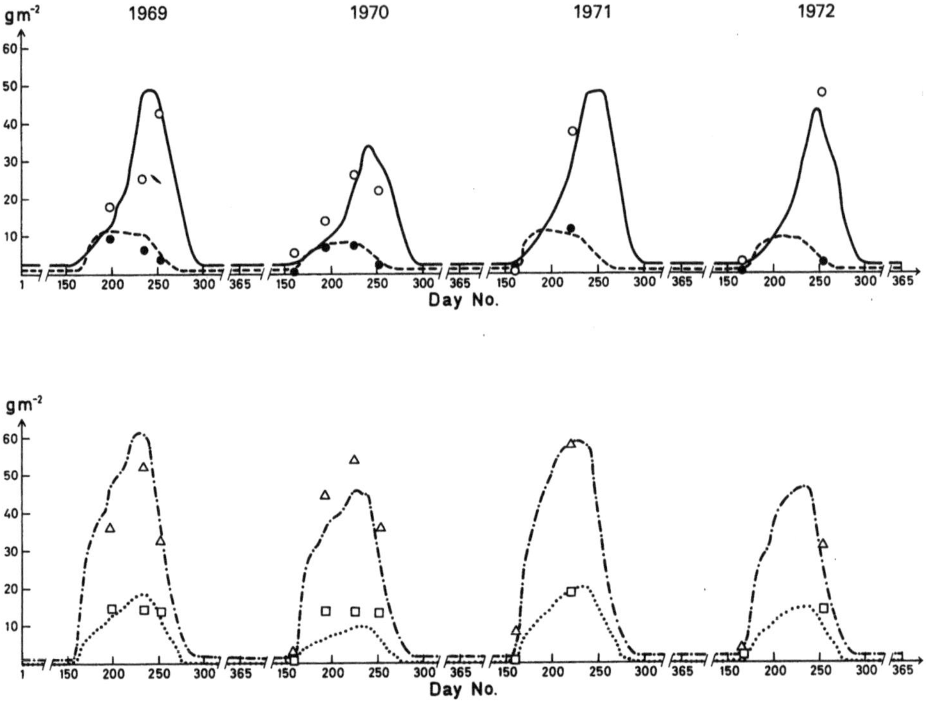


Fig. 2. Simulated and observed green biomass of different plant groups at the Hardangervidda wet meadow in the growing seasons 1969–1972. —— Bryophytes simulated, ○ Bryophytes observed values, ---- Forbs simulated, ● Forbs observed values, —·— Monocotyledons simulated, △ Monocotyledons observed values, ···· Woody plants simulated, □ Woody plants observed values

Simulated total annual invertebrate consumption in the period 1969–1972 are 8.7, 6.0, 5.9, and 7.9 g m$^{-2}$, respectively. The effect of invertebrates on the wet meadow community is therefore small if considered only by their consumption. Less than 10% of the consumption is green material and more than 90% litter and below-ground material.

Table 2 shows some examples of sensitivity tests on the model, i.e. how apparent photosynthesis (expressed in g dw photosynthates produced) changes under different hypothetical conditions (a–g) deviating from the "normal" conditions (Table 1).

(a) Doubling the rate of upward translocation in the spring for all vascular plants (while the amount translocated upward before downward translocation begins is the same as under "normal" conditions 20 g m$^{-2}$ as dw) causes the total production per m$^{-2}$ to be reduced for the main plant groups (Table 2). The reason for this is an earlier start of downward translocation. Less new green material is produced by photosynthesis in the now shorter period of upward translocation, and due to the positive feedback loop in the green compartment of the model, annual production is reduced. A more biologically true picture will probably be found if production of green material in the spring is simulated with no down-

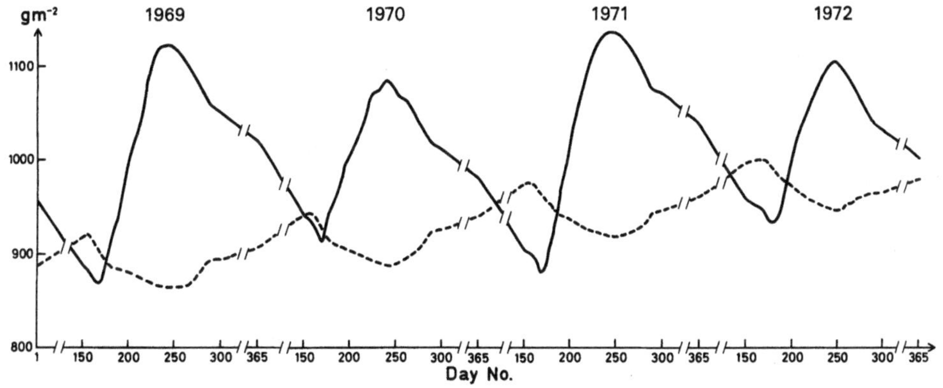


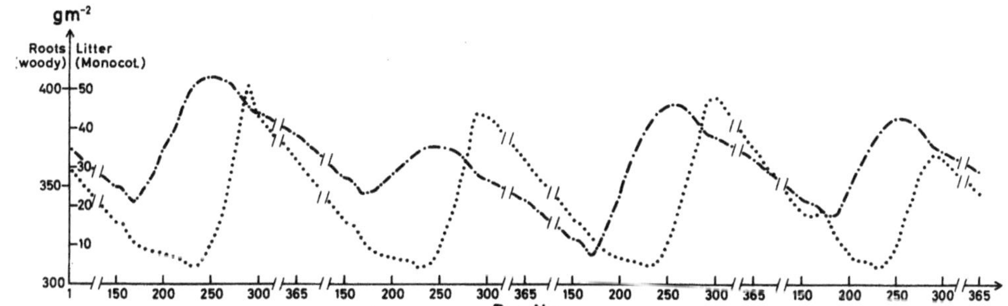


Fig. 3. Simulated live and dead root and litter biomass of monocotyledons and live root biomass of woody plants at the Hardangervidda wet meadow in the growing seasons 1969–1972. —— Live roots monocotyledons simulated, ---- Dead roots monocotyledons simulated, —·— Live roots woody plants simulated, ···· Litter monocotyledons simulated

Table 2. Model output under different conditions (sensitivity analysis) in g dw $m^{-2}$ $year^{-1}$ of apparent photosynthesis. Plant groups as Table 1. For further explanation see text

| Year | Plant group | a | b | c | d | e | f | g |
|---|---|---|---|---|---|---|---|---|
| 1969 | 1 | 139.3 | 94.7 | 136.4 | 127.6 | 134.4 | 129.2 | 38.6 |
| | 2 | 72.0 | 31.0 | 63.7 | 62.9 | 63.7 | 55.3 | 32.6 |
| | 3 | 422.5 | 860.5 | 614.6 | 634.6 | 639.5 | 721.9 | 405.7 |
| | 4 | 115.5 | 245.0 | 149.2 | 148.9 | 155.6 | 153.6 | 58.6 |
| 1970 | 1 | 88.1 | 85.8 | 94.3 | 91.1 | 95.3 | 116.5 | 43.7 |
| | 2 | 65.1 | 16.6 | 45.2 | 45.6 | 46.7 | 34.9 | 23.1 |
| | 3 | 327.1 | 822.5 | 548.6 | 586.2 | 525.0 | 706.5 | 381.2 |
| | 4 | 64.6 | 169.8 | 91.8 | 86.7 | 88.1 | 95.3 | 40.6 |
| 1971 | 1 | 145.9 | 120.2 | 149.6 | 122.8 | 144.3 | 142.1 | 55.7 |
| | 2 | 74.2 | 14.9 | 43.2 | 41.2 | 48.9 | 29.8 | 22.8 |
| | 3 | 414.7 | 857.7 | 740.0 | 731.7 | 656.6 | 853.6 | 504.0 |
| | 4 | 130.0 | 247.7 | 162.7 | 158.3 | 160.1 | 173.3 | 55.3 |
| 1972 | 1 | 104.7 | 108.3 | 113.3 | 112.4 | 117.5 | 135.9 | 51.4 |
| | 2 | 89.1 | 17.1 | 38.6 | 39.6 | 48.2 | 32.9 | 23.4 |
| | 3 | 284.0 | 669.9 | 627.4 | 615.6 | 559.9 | 708.9 | 392.4 |
| | 4 | 109.2 | 236.3 | 111.4 | 131.8 | 135.8 | 127.3 | 42.3 |

ward translocation until a certain level of new green biomass is reached. The annual production will then probably be somewhat higher when translocation rate upward is increased. Due to less shading by the lower biomass of vascular plants, a slightly higher bryophyte production (plant group 1) is found in the two most favorable years (1969 and 1971). Forbs (plant group 2) have increased production due mainly to reduced nutrient stress as a result of the lower production of other vascular plant groups and to less shading.

(b) A doubling of translocation rate and of the amount translocated upward in spring to 40 g $m^{-2}$ causes a great increase in production of monocotyledons and woody plants in the model (Table 2). Bryophyte photosynthesis, on the other hand, decreases somewhat due to increased shading. The relatively strong biomass reduction in forbs for all the research years is caused both by severe nutrient stress and shading from the high biomass of the dominating plant group, monocotyledons. A similar, but weaker effect, is seen on forbs under conditions c, d, and f.

The great increase in apparent photosynthesis for monocotyledons and woody plants refers to their extensive live root biomass. Since the proportion of upward translocated matter in spring is relative to the live non-green biomass of the different plant species, monocotyledons and woody plants are the greatest beneficiaries of the spring translocation. This results in monocotyledons and woody plants having a very rapid development of green tissue in spring, more than the other plant groups at the site.

(c) Air and soil temperatures, increased by 2° C between day 180–220 (June 29th to August 8th), give a relatively strong general increase in production of monocotyledons (plant group 3) and woody plants (plant group 4) as seen in Table 2, particularly in 1970 which had the lowest July temperature (Table 3). As optimal apparent photosynthesis is relatively high for all vascular plant groups, small changes in one of the abiotic factors most strongly affecting this process causes relatively great changes in production. Forbs show, in the model output, a slight decrease in apparent photosynthesis in 1970, 1971, and 1972 compared to "normal" conditions (for reasons given in b).

(d) Air and soil temperatures are increased by 2° C in the spring period (when translocation is upward). The average length of the period with upward translocation in the four years is 10–12 days, i.e. about one fourth of the time of increased temperature used in c. A comparison with the values under c in Table 2 shows, however, that the simulated values and changes under the present conditions are about the same as found in c, i.e. the same increase in total production per $m^2$ relative to normal conditions. This indicates that increased temperatures in spring have great influence on annual production in an otherwise normal year.

Comparison between results from the conditions used in d and c shows that the effect of an increase in temperature is dependent on the level from which the temperature is increased. An increase in temperature in a low temperature period has, as expected, a greater effect on production than in a relatively high temperature period. In the model, this is found to be true, particularly for woody plants (Table 2) in the spring of 1972 which was both late and cold (Table 3), indicating plant species of that group on the wet meadow to be more temperature-dependent in spring than species of the other groups.

Table 3. Snow melt dates at the wet meadow, Hardangervidda 1969–1972 and 10 days maximum temperature sums at 10 cm height for the period just after snow melt and in July from day 181 (June 30) to day 220 (August 8)

| | Date of snow melt | Maximum temperature sums at 10 cm height | | | | |
|---|---|---|---|---|---|---|
| | | 10 days after melting | June 30– July 9 | July 10– July 19 | July 20– July 29 | July 30– Aug. 8 |
| 1969 | June 3[a] | 116.8[a] | 91.0 | 131.3 | 145.9 | 208.2 |
| 1970 | June 10 | 193.4 | 105.6 | 103.7 | 87.1 | 148.6 |
| 1971 | June 4 | 159.9 | 164.8 | 90.4 | 124.7 | 129.7 |
| 1972 | June 12 | 135.3 | 161.6 | 179.3 | 195.5 | 130.7 |

[a] Estimated values.

(e) Light intensity, increased by 10%, causes only a moderate increase in production in the model output, the influence being greatest in the years of lowest production (1970 and 1972). Light intensities measured at Hardangervidda are close to or above light saturation most of the summer (SKRE, 1975; KJELVIK et al., 1975). The greatest part of the increase in production most probably occurs on overcast days, when light intensities are low. This is supported by the fact that the number of sunshine hours was considerably lower in the summer of 1970 than in 1969 and 1971 (234, 339, and 407-hrs, respectively, when totalled over July, August, and the first half of September).

(f) Decreased death of green material by 50% for all plant groups in spring and summer causes increased production for all plant groups, except forbs (Table 2). The increase is greatest for monocotyledons, less for bryophytes and woody plants. The combined effect of nutrient stress and shading causes a reduction in photosynthesis for forbs.

(g) Another "steady state".

In order to obtain the high values of production observed in the field and simulated by the "normal" conditions in Table 1, high maximum values of apparent photosynthesis had to be used. To keep green biomass down to values close to those observed, high death rates in spring and summer were applied. The only alternative way to restrict the amount of green biomass in the model is to increase the already high values of downward translocation.

When harvesting roots in the peat at the wet meadow, it is very difficult to separate living from dead, and visual separation is corrected for by using $^{14}C$ technique (WIELGOLASKI, 1975). The high values of live roots, and their relatively rapid turnover time (3–5 years for forbs and monocotyledons, and 10–20 years for woody plants) are the main reasons for the high production found at the wet meadow (WIELGOLASKI, 1975).

The main questions to be answered by the present run of the model are: if the amount of live-root biomass is lowered, death of green during the growing season reduced, and root respiration raised close to values found by SKRE (1975) in laboratory experiments, what changes must be made in the model, especially on rates of apparent photosynthesis of the different plant groups, in order to obtain the same turnover times of non-green material and the same biomass of green as under "normal" conditions?

Starting values for live roots were changed in the following way: forbs from 28.3 to 3.0 g dw $m^{-2}$; monocotyledons from 955.3 to 500 g $m^{-2}$, and woody from 369.9 to 230.0 g $m^{-2}$. Under "normal" conditions, death rates of green in spring are 0.015, 0.04, 0.05, and 0.018 g $g^{-1}$ $day^{-1}$ (expressed as dw) for bryophytes, forbs, monocotyledons and woody plants, respectively. In summer, the rates are 0.018, 0.015, 0.015, and 0.015, when given in the same order as above. In the present run, values for death of green are 0.00002 g $g^{-1}$ $day^{-1}$ in spring, and 0.0009 g $g^{-1}$ $day^{-1}$ in summer for all plant groups. To obtain values for root respiration rates close to the laboratory values found by SKRE (1975), the basal respiration rates had to be roughly increased by 300%. This gives an annual loss by root respiration close to annual root death for the different plant groups.

The rates of apparent photosynthesis are changed in several runs of the model until green biomass is of the same order as under "normal" conditions, and the live non-green compartments have a reasonable annual variation and the same turnover time as under "normal" conditions. The following changes in maximal rates of apparent photosynthesis at 15° C are obtained: Bryophytes change from 0.00344 to 0.00172 g $g^{-1}$ $h^{-1}$, forbs and woody plants from 0.0175 to 0.0100 g $g^{-1}$ $h^{-1}$, monocotyledons from 0.0195 to 0.0130 g $g^{-1}$ $h^{-1}$. The annual production given as apparent photosynthesis for the different plant groups in this version of the model are as given in the last column (g) in Table 2. It is easily observed that the variations from year to year are very similar to those under "normal" conditions, i.e. highest production in 1969 and 1971 and lowest in 1970 and 1972, but the total production is strongly reduced. In 1969 for example, the reduction in total production (all plant groups) is about 370 g $m^{-2}$.

A general feature, evident from Table 2, is that under all the different conditions, 1969 and 1971 are the years of highest production, and 1970 and 1972 the lowest.

## Conclusion and Some General Considerations on Ecosimulation

All the assumptions for the present version of the model are probably neither fulfilled for the wet meadow nor will be, generally, for any other site or ecosystem. For instance, most probably, primary production of all plant groups in an ecosystem will be nutrient-limited. Based upon chemical analysis of the available nutrients in the soil (HINNERI et al., 1975) and the content in different species and plant parts (WIELGOLASKI et al., 1975), nutrient pools and cycling (through plant uptake and release by decomposition) can be built into the model.

A general control of all ecosystems is exerted through what in this model has been called driving variables. In order to obtain realistic models, it is of primary importance to know how these variables affect the living organisms of an ecosystem. At present, there are great gaps in our knowledge in this respect. Stochastic models for generating climate and the abiotic regime in the soil have to be constructed in order to develop general models applicable to a great variety of ecosystems. The greatest value of ecosimulation today lies in testing system hypotheses over a long period of time, and in deciding where more detailed research is particularly needed in order better to understand the functioning of ecosystems. A fact often overlooked is that a simulation model for an ecosystem is not finished when it seems to produce satisfactory results relative to the real system. This should only be the first step in an iterative process going on into the future, closely linked to biological research, applications and literature studies.

*Acknowledgements.* The authors wish to thank all the scientists and technicians in the team working together on the Norwegian IBP alpine ecosystem project at Hardangervidda for cooperation in building the model and for permission to use often unpublished data. Thanks are particularly addressed to G. SWARTZMAN who helped in the initial model structure phase, and to A. BERG, A. SKARTVEIT, T. SOLHØY, J. GOKSØYR, O. SKRE and A. K. VEUM, although several other workers have also given valuable advice, built on their biological knowledge on the Hardangervidda system.

## References

APOSTOL, T. M.: Calculus, Vol. 1. New York: Blaisdell Publishing Company 1961.

BERG, A.: Pigment structure of vascular plants, mosses and lichens at Hardangervidda, Norway. In: WIELGOLASKI, F. E. (Ed.): Fennoscandian tundra ecosystems. Part 1: Plants and microorganisms, pp. 216–224. Berlin-Heidelberg-New York: Springer 1975.

BERG, A., KJELVIK, S., WIELGOLASKI, F. E.: Distribution of $^{14}C$ photosynthates in Norwegian alpine plants. In: WIELGOLASKI, F. E. (Ed.): Fennoscandian tundra ecosystems. Part 1: Plants and microorganisms, pp. 208–215. Berlin-Heidelberg-New York: Springer 1975.

BLEDSOE, L. J., FRANCIS, R. C., SWARTZMAN, G. L., GUSTAFSON, J. D.: PAWNEE. A grassland ecosystem model. U.S. IBP Grassland Biome Tech. Rep. **64**, 1–179 (1971).

BRENNAN, R. D., DE WIT, C. T., WILLIAMS, W. A., QUATTRIN, E. V.: The utility of a digital simulation language for ecological modeling. Oecologia **4**, 113–132 (1970).

DE WIT, C. T., BOUWER, R., PENNING DE VRIES, F. W. T.: The simulation of photosynthetic systems. In: Prediction and Measurement of photosynthetic productivity. Proc. IBP/PP Tech. Meeting, Trebon, Czechoslovakia, 14–21 September, 1969, pp. 47–70. Wageningen, The Netherlands: Center for Agr. Publ. Doc. 1970.

DUNCAN, W. G., LOOMIS, R. S., WILLIAMS, W. A., HANOU, R.: A model for simulating photosynthesis in plant communities. Hilgardia **38**, 181–205 (1967).

HINNERI, S., SONESSON, M., VEUM, A. K.: Soils of Fennoscandian IBP tundra ecosystems. In: WIELGOLASKI, F. E. (Ed.): Fennoscandian tundra ecosystems. Part 1: Plants and microorganisms, pp. 31–40. Berlin-Heidelberg-New York: Springer 1975.

KALLIO, P., HEINONEN, S.: Influence of short-term low temperature on net photosynthesis in some subarctic lichens. Rep. Kevo Subarctic Res. Stat. **8**, 63–72 (1971).

KJELVIK, S., WIELGOLASKI, F. E., JAHREN, A.: Photosynthesis and respiration of plants studied by field technique at Hardangervidda, Norway. In: WIELGOLASKI, F. E. (Ed.): Fennoscandian tundra ecosystems. Part 1: Plants and microorganisms, pp. 184–193. Berlin-Heidelberg-New York: Springer 1975.

MILLER, P. C., TIESZEN, L.: A preliminary model of processes affecting primary production in the arctic tundra. Arctic and Alpine Res. **4**, 1–18 (1972).

PIELOU, E. C.: An Introduction to Mathematical Ecology. New York: Wiley (Interscience) 1969.

ROSSWALL, T., VEUM, A. K., KÄRENLAMPI, L.: Plant litter decomposition at Fennoscandian tundra sites. In: WIELGOLASKI, F. E. (Ed.): Fennoscandian tundra ecosystems. Part 1: Plants and microorganisms, pp. 268–278. Berlin-Heidelberg-New York: Springer 1975.

SANDHAUG, A.: En matematisk simuleringsmodell for terrestre økosystemer på Hardangervidda. Univ. of Oslo: Unpubl. thesis 1974.

ŠESTÁK, Z., ČATSKÝ, J., JARVIS, P. G.: Plant photosynthetic production. The Hague: Junk Publ. 1971.

SKARTVEIT, A., RYDÉN, B. E., KÄRENLAMPI, L.: Climate and hydrology of some Fennoscandian tundra ecosystems. In: WIELGOLASKI, F. E. (Ed.): Fennoscandian tundra ecosystems. Part 1: Plants and microorganisms, pp. 41–53. Berlin-Heidelberg-New York: Springer 1975.

SKRE, O.: $CO_2$ exchange in Norwegian tundra plants studied by infrared gas analyzer technique. In: WIELGOLASKI, F. E. (Ed.): Fennoscandian tundra ecosystems. Part 1: Plants and microorganisms, pp. 168–183. Berlin-Heidelberg-New York: Springer 1975.

SKRE, O., BERG, A., WIELGOLASKI, F. E.: Organic compounds in alpine plants. In: WIELGOLASKI, F. E. (Ed.): Fennoscandian tundra ecosystems. Part 1: Plants and microorganisms, pp. 339–350. Berlin-Heidelberg-New York: Springer 1975.

SOLHØY, T.: Quantitative invertebrate studies in mountain communities at Hardangervidda, South Norway. Norsk Ent. Tidsskr. **19**, 99–108 (1972).

SONESSON, M., WIELGOLASKI, F. E., KALLIO, P.: Description of Fennoscandian tundra ecosystems. In: WIELGOLASKI, F. E. (Ed.): Fennoscandian tundra ecosystems. Part 1: Plants and microorganisms, pp. 3–28. Berlin-Heidwlberg-New York: Springer 1975.

SVENSSON, B., VEUM, A. K., KJELVIK, S.: Carbon losses from tundra soils. In: WIELGOLASKI, F. E. (Ed.): Fennoscandian tundra ecosystems. Part 1: Plants and microorganisms, pp. 279–286. Berlin-Heidelberg-New York: Springer 1975.

SWARTZMAN, G. L., VAN DYNE, G. M.: An ecologically based simulation-optimization approach to natural resource planning. Ann. Rev. Ecol. Syst. **3**, 347–398 (1972).

TAYLOR, N. W.: A mathematical model for *Tribolium confusum* populations. Ecology **48**, 290–294 (1967).

TIESZEN, L. L.: Comparative photosynthesis and stand structure and photosynthesis in Arctic tundra. In: BROWN, T., BOWEN, S. (Eds.): The structure and function of the tundra ecosystem, Vol. 1. Progress Report and Proposal Abstracts, pp. 38–43. Hanover, N.H.: U.S. Tundra Biome Program, USA CRREL 1971.

WIELGOLASKI, F. E.: Primary productivity of alpine meadow communities. In: WIELGOLASKI, F. E. (Ed.): Fennoscandian tundra ecosystems. Part 1: Plants and microorganisms, pp. 121–128. Berlin-Heidelberg-New York: Springer 1975.

WIELGOLASKI, F. E., HAYDOCK, K. P., CONNOR, D. J.: A grazing land's plant-decomposition, carbon-mineral simulation model. U.S. IBP Grassland Biome Tech. Rep. **203**, 1–75 (1972).

WIELGOLASKI, F. E., KJELVIK, S.: Plant biomass at the Norwegian IBP sites at Hardangervidda 1969–1972. In: VIK, R. (Ed.): IBP in Norway. Methods and results. Sections PT-UM Grazing project, Hardangervidda. Botanical investigations, pp. 1–88. Oslo: Norwegian Nat. IBP Comm. 1975.

WIELGOLASKI, F. E., KJELVIK, S., KALLIO, P.: Mineral content of tundra and forest tundra plants in Fennoscandian. In: WIELGOLASKI, F. E. (Ed.): Fennoscandian tundra ecosystems. Part 1: Plants and microorganisms, pp. 316–332. Berlin-Heidelberg-New York: Springer 1975.

# Systems Analysis on Invertebrates

G. GYLLENBERG

This paper presents the basic ideas used in constructing a model based on the ODUM energy circuit language (ODUM, H. T., 1971). A flow diagram of the energy flow through the main invertebrate consumer groups is given in Fig. 1. This language is especially convenient for energy-flow specifications, and has the advantage that it clearly defines the function of the system. This is not the case with the frequently used compartment flow diagrams.

However, the mathematical approach taken directly from the ODUM model has some disadvantages since:

1) The ODUM language does not permit the inclusion of stochastic parameters with the exception of environmental abiotic factors. Probability functions have

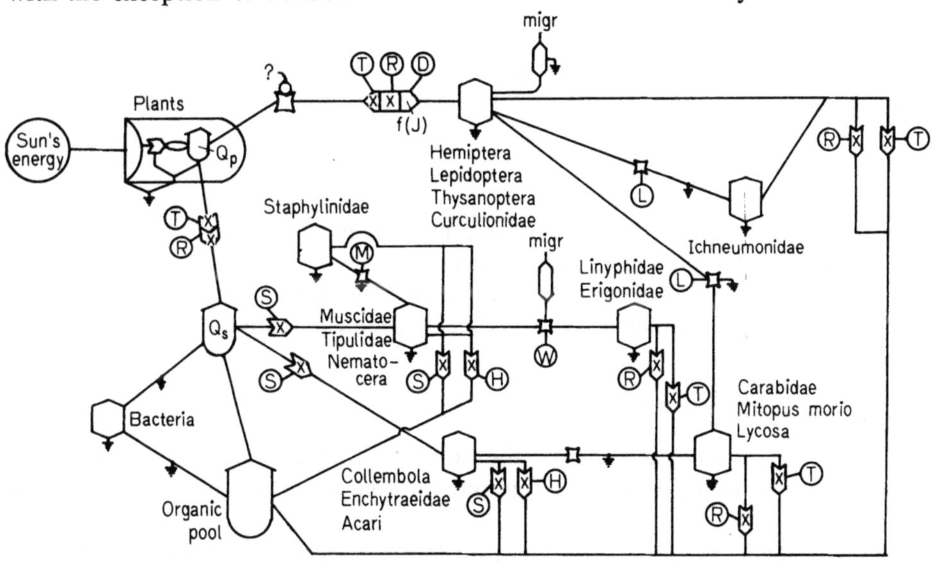


Fig. 1. Flow diagram for the invertebrates at Stigstuv, Hardangervidda. Interpolation is performed between the three community types (wet meadow, dry meadow and lichen heath). Explanation of characters: ? = a logical not function, is the plant not of species a?, $T$ atmospheric temperature, $R$ atmospheric relative humidity, $S$ soil temperature, $H$ soil relative humidity, $D$ probability of finding preferred food, $W$ probability of falling into a spider net, $L$ probability of finding Lepidoptera larvae, $M$ probability of finding Muscidae larvae, $Q_s$ pool of standing dead vegetation and litter, $Qp$ standing crop of vegetation, $X$ multiplication mark, $f(J)$ indicates functional relationship of the energy flow

been indicated in the flow diagram, but the solution of the equations cannot be interpreted from this diagram.

2) Density-dependent functions, such as the LOTKA-VOLTERRA models, are difficult to build into the ODUM language, since the basic unit is energy, and not the individual.

For these reasons, the model referred to above has been developed to test the following ideas:

1) Maintaining the advantage of the ODUM language: specification of the energy flow on individual basis.

2) Aggregation indices ($k$) according to a negative binomial distribution gives the chance of an animal finding its preferred food species ($p_x$):

$$p_x = \frac{(k+x-1)!}{x!(k-1)!} \cdot \frac{p^x}{(1+p)^{k+x}}$$

where: $p = \bar{x}/k$, and $\bar{x}$ is the mean distance between plant specimens or tufts.
$x =$ the distance the herbivores have to move before finding their preferred food species.

3) Owing to a pronounced territorial competition between the individuals of spiders in particular, the frequency distribution of spider nets $f(x)$ with the distance ($d$) is computed according to a POISSON distribution:

$$f(x) = e^{-\bar{x}} \cdot \frac{\bar{x}}{d!}$$

where: $\bar{x} =$ mean distance between nets.
The probability of a prey falling into a spider net is then calculated as:

$$p_x = e^{-\bar{x}} \cdot \frac{\bar{x}}{d!} \cdot A$$

where: $A =$ area occupied by net outside movement area of the prey. This area is different at different heights in the vegetation.

4) The life cycle of populations, including natality and mortality, is computed from density-dependent models through maximum likelihood fitting (see READ and ASHFORD, 1968):

$$X_i = \min(Y_i, Z_i)$$

where: $Y_i =$ a random variable for the time spent in the stage before transition into next stage ($i+1$), and $Z_i =$ a random variable for death in stage $i$ (negative exponential).

5) Stochastic variability of the transfers (processes) in order to test the variation of the empirical values.

## Results

The field data for the model were obtained from Norwegian IBP Annual Reports. Some data for the life cycles of representative species of the different groups were obtained from personal communication with investigators involved

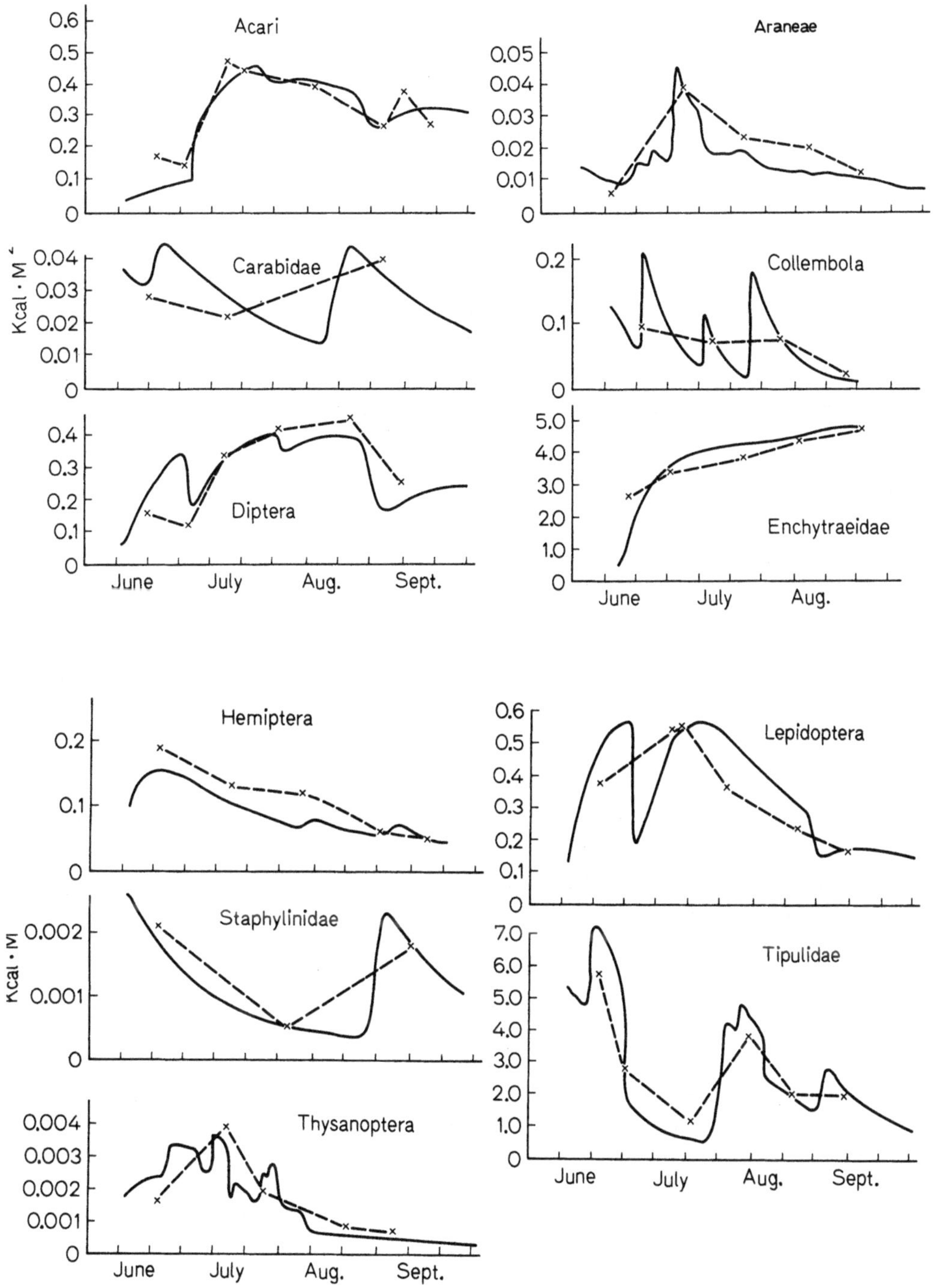


Fig. 2. A comparison between the simulated outputs of a model outlay, and field data. ------ = field data of 1970, from the dry meadow type of community, ——— = corresponding simulated curve. *All* values are given as kcal $m^{-2}$. For details, see text

in the project (FJELLBERG, KJELSEN, NÆSS, and SOLHØY). The main basis for testing the model is provided by data for:

1) The number of individuals $m^{-2}$ for all groups.

2) Biomass $m^{-2}$, and accordingly biomass $ind.^{-1}$ through the season. Thus an estimate is obtained for the individual increase in biomass throughout the season.

3) Life cycle and individual biomass increase of different instars. These data were available for Collembola (FJELLBERG, 1972, 1973, 1975), Hemiptera (NÆSS, 1972, 1973), Thysanoptera (KJELSEN, 1975) and Tipulidae (HOFSVANG, 1972, 1973). Lepidoptera, Acari (SOLHØY, 1975a) and Diptera (incl. Muscidae and Nematocera) were divided into larvae and adults. Enchytraeidae (SOLHØY, 1975b), Araneae, Carabidae and Staphylinidae have been treated collectively.

4) Respiration data were only available for Collembola, Tipulidae and *Mitopus morio*. Consumption, egestion and respiration were therefore based on best estimates from the literature (e.g. EDWARDS et al., 1970). Normally, production of biomass was estimated as 15% of the consumption, respiration as 15% and egestion as 70%. For the respiration data, allowance was made for respiratory losses involved in searching for food and during flight.

5) Food preference and food selection data have been based on field observations when available, and completed with information from the literature.

The first outlay of the model still omits much information. It was constructed for the dry meadow type of community at Hardangervidda. Other programs were then constructed for wet meadow and lichen heath, and an interpolation can be made between the sites. It was not possible to include the abiotic factors at this first step of simulation. Furthermore, the length of season must be included as a time-dependent variable that has an effect on the development of the animals.

The program was run for 127 days, from June 11 to October 16, as this period comprises roughly the growing season at Hardangervidda. A comparison between the simulation outputs of this outlay and the field data is given in Fig. 2. The calculation of biomass for the different groups is based on a set of ordinary difference equations.

The introduction of abiotic factors in the program is maintained in two ways:

1) The outputs from the compartments are non-linear, the abiotic factors being included as simple cross-products, according to the laws of irreversible thermodynamics (see ODUM, H.T., 1971). When the abiotic factors are left out, this procedure assumes a mean value for each factor during the growing season.

2) The energy flows are functions of the abiotic factors. This second alternative provides a possibility to test the stochastic variability in model outputs. This type of model predicts energy flow only within optimal environmental conditions of the animals (GYLLENBERG, 1974). Including the environmental factors according to this procedure would only flatten out the biomass fluctuations in Fig. 2, but not change the fit to the biomass field data.

## Discussion

The task of the systems analyst nowadays is to build ecological models on predictable bases, i.e. including the experimental components approach. Most models dealing with energy flow on the ecosystem level are deterministic. This is also the case with the ODUM energy-circuit model, although some testing of stochastic behavior can be maintained trough noise generators.

The question is now whether the ecosystem can be regarded as a unit, with its own energy dynamics, as suggested by ODUM, E. P. (1971) and ODUM, H. T. (1971). During optimal environmental conditions, most niches in the ecosystem are filled, and the system behaves like a unit. However, the most difficult task for the systems analyst is to predict how the system behaves when catastrophic factors occur. Catastrophes often increase the stochastic pattern in animal behavior.

Evolution tends to decrease the stochastic element in animal behavior, and therefore minimize the entropy production of the species. But if an environmental factor exceeds the tolerance limits of a species, the stochastic part in its behavior increases. The pattern of behavior depends on how often the same type of catastrophe occurs, and if there has been enough time for the mapping from genotypes into a multidimensional space of phenotypes (WADDINGTON, 1968).

Therefore it is thought that a truly realistic simulation of the animal world is obtained by including the stochastic variability of different energy-flow parameters. The stochastic variability could be made a function of catastrophic factors, each of them acting in a different way on different groups of animals. Furthermore, the amount of stochasticity for a specific animal species is inversely proportional to the development of sense organs.

*Acknowledgements.* I wish to express my thanks to prof. H. KAURI, Director of the Zoological Museum in Bergen, who has provided me with the material for simulation, given economic support and valuable criticism of the ecological basis for simulation.

## References

EDWARDS, C. A., REICHLE, D. E., CROSSLEY, D. A., JR.: The role of soil invertebrates in turnover of organic matter and nutrients. In: REICHLE, D. E. (Ed.): Analysis of temperate forest ecosystems, pp. 147–172. Berlin-Heidelberg-New York: Springer 1970.

FJELLBERG, A.: Populasjonsutvikling hos Collembola. IBP i Norge: Årsrapport 1971, 134–141 (1972).

FJELLBERG, A.: Collembol-faunaen på forsommeren 1970. IBP i Norge: Årsrapport 1972, 194–201 (1973).

FJELLBERG, A.: Organization and dynamics of Collembola populations on Hardangervidda. In: WIELGOLASKI, F. E. (Ed.): Fennoscandian tundra ecosystems, Part 2: Animals and systems analysis, pp. 73–79. Berlin-Heidelberg-New York: Springer 1975.

GYLLENBERG, G.: A note on the use of irreversible thermodynamics in population ecology. Entomologiske Meddelelser **42**, 78–79 (1974)

HOFSVANG, T.: *Tipula excisa Schum.* (Diptera, Tipulidae), life cycle and population dynamics. Norsk Ent. Tidsskr. **19**, 43–48 (1972).

HOFSVANG, T.: Energy flow in *Tipula excisa Schum.* (Diptera, Tipulidae) in a high mountain area, Finse, South Norway. Norw. J. Zool. **21**, 7–16 (1973).

KJELLSEN, E. K.: Dynamics of Thysanoptera populations on Hardangervidda. In: WIELGOLASKI, F. E. (Ed.): Fennoscandian tundra ecosystems, Part 2: Animals and systems analysis, pp. 80–83. Berlin-Heidelberg-New York: Springer 1975.

NÆSS, S. J.: Kalorimetring og tørrvektbestemmelser. IBP i Norge: Årsrapport 1971, 141–156 (1972).

NÆSS, S. J.: Kalori- og tørrvektbestemmelser. IBP i Norge: Årsrapport 1972, 212–219 (1973).

ODUM, E. P.: Fundamentals of ecology. 3rd ed. Philadelphia-London-Toronto: W. B. Saunders 1971.
ODUM, H. T.: Environment, power and society. New York: Wiley and Sons 1971.
READ, K. L. Q., ASHFORD, J. R.: A system of models for the life cycle of a biological organism. Biometrika **55**, 211–221 (1968).
SOLHØY, T.: Dynamics of Oribatei populations on Hardangervidda. In: WIELGOLASKI, F. E. (Ed.): Fennoscandian tundra ecosystems, Part 2: Animals and systems analysis, pp. 60–65. Berlin-Heidelberg-New York: Springer 1975a.
SOLHØY, T.: Dynamics of Enchytraeidae populations on Hardangervidda. In: WIELGOLASKI, F. E. (Ed.): Fennoscandian tundra ecosystems, Part 2: Animals and systems analysis, pp. 55–59. Berlin-Heidelberg-New York: Springer 1975b.
WADDINGTON, C. H.: Does evolution depend on random search? In: WADDINGTON, C. H. (Ed.): Towards a theoretical biology 1. Prolegomena, 111–119. Edinburgh: University Press 1968.

# A Simulation Model of a Small Rodent Population

K. LIESTØL*, E. ØSTBYE, H.-J. SKAR, and G. SWARTZMAN

Small rodent populations show distinct cyclic fluctuations with three to four years cycle length in many different ecosystems. One of the classic problems in population ecology has been to explain these cycles.

One reason for this interest in small rodent dynamics is that population cycles present an interesting situation in which to study population regulation. Cyclic populations, which show a pattern of increase, peak, decline and low, allow examination of the population during a continously repeated sequence of contrasting phases. This contrast is quite conspicuous because the peak population may be several hundred times greater than the minimum population and because the animals may display a quite different behavior in the different phases.

Another interesting fact is that cyclic fluctuation has been described for a variety of ecological communities in arctic, boreal and temperate areas. Despite this variation in environmental conditions, a remarkable similarity in cycle pattern can be registered. This uniformity has led to the idea that one or a few basic regulating mechanisms are involved, and that the same kind of control might act in other populations.

The first systematic investigation of population dynamics in small rodents was presented by ELTON (1924). Since then a great amount of work has been carried out to describe and explain these fluctuations. Nevertheless there is at present no agreement about the causes of the fluctuations. This combination of great effort and few reliable results clearly reflects the complexity of the systems.

One possible way to investigate systems involving complex interactions is to use mathematical models and simulation on the digital computer. The aim of the present study was therefore to investigate whether a simple mathematical model could give any new information about the cyclic fluctuations of the small rodent populations on the mountain plateau, Hardangervidda, Norway.

This paper is focused on what we believe can be achieved by this model. The description of the basic model is therefore only summarized. As an example, a more detailed description is given of the reproduction in order to give an idea of the simplifications and uncertainties. A detailed description of the total model, together with a listing of one version of the computer program has been published earlier (HAGEN et al., 1975)[1].

* Responsible for the manuscript.

[1] Extracts are included here with the permission of the Reports from the High Mountain Ecology Research Station, Finse.

Some possible causes of population fluctuation are included in the model. We have chosen to simulate the action of predation, rodent-vegetation interaction and density-dependent behavioral changes because these factors include the most commonly suggested causes of population fluctuations and because the causality of other possible factors is even less well known. We will focus especially on how we believe the results from these simulations should be interpreted.

## The Model

During the planning and building of the model some serious problems arose, caused by the very limited knowledge of the system. We have tried to summarize some of these difficulties:

(a) Despite the extensive work that has been carried out on small rodent subsystems, there is a general lack of reliable quantitative data.

(b) There are considerable dissimilarities between different communities in which cycling rodent populations are found, making it risky to use data from different areas in the same model. Even within tundra ecosystems the differences may be significant.

(c) Within the same area, different species may dominate at different times. This forces the model builder either to make the model extremely complex by including all important rodent species, or to operate with a representative but nonexistent „average" rodent.

(d) The small rodent subsystem is an integrated part of the whole ecosystem, and the structure and function of the rodent population are closely connected with other parts of the system. To get a good model of the rodent population, it is therefore probably necessary to have a relatively good model of the whole ecosystem.

(e) The long cycle-time (3–4 years) makes it difficult to base the work on interaction between model and field work.

(f) The rodent population and the connected parts of the ecosystem on Hardangervidda have been under close observation for only the last five years.

Such problems are probably of a general nature and apply to most ecosystem models.

Comparing the complexity of this system with our knowledge of its structure and function, it is obvious that the number of variables included in the model must be strongly limited and that the interaction between these variables must be greatly simplified. We have therefore omitted all factors that we believe have a minor effect. In addition, some factors that we know may have a significant influence on the rodent population are omitted, for example migration and epidemics. The reason for this omission is that our knowledge of these elements is so limited that the only effect of including them would be to make the model more complex without making it more reliable.

A flow chart of the model is presented in Fig. 1.

Most of the interactions in the model are supposed to be of a linear type, the rest being either exponential or of an on/off-type. Since the interactions in the real

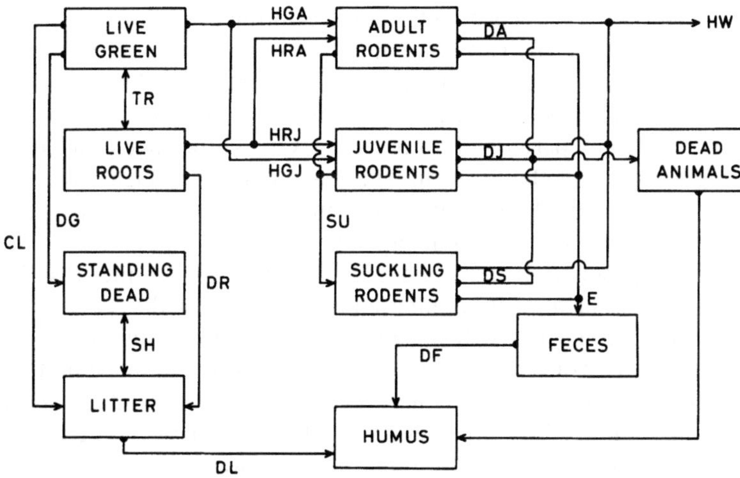


Fig. 1. Flow chart of the model. *Hw* harvest function of predators

system are so poorly understood, the choice of interactions in the model is in many cases partly based on guesswork.

The rodents on Hardangervidda are the root vole *(Microtus oeconomus)*, the field vole *(Microtus agrestis)*, the bank vole *(Clethrionomys glareolus)*, the grey-sided vole *(Clethrionomys rufocanus)* and the lemming *(Lemmus lemmus)*.

The available data precluded the possibility of representing each of these species as a separate unit. On the other hand, there are significant differences between these species. The lemming especially differs from the others with regard to diet, size and probably also reproduction. Since the lemming was less numerous during most of the IBP period, it was decided to operate with what may be regarded as an „average" of the other species. Because cycling is observed both in communities with one and with several species, the interaction between species seems to be neither an obstacle nor a requirement for cycling.

The input to the model consists of parameter values, initial values and climatic conditions.

The vegetation subsystem constitutes a relatively great part of the model. It is, however, only modeled to serve as background for the small rodent subsystem. The vegetation consists of live green, live roots, standing dead and litter. No distinction has been made between different plant species. The production of the vegetation is mainly controlled by the climatic conditions. The parameters of the vegetation subsystem are adjusted so that the plant biomass dynamics of the model under normal climatic conditions resembles the observed values from Hardangervidda.

The small rodents treated in this model are active throughout the year. During winter they inhabit the subnivean space, i.e. the space between soil and snow. The subnivean space has a fairly stable microclimate with temperatures of about 0° C.

Data from different areas show marked dissimilarities in reproduction. It is therefore necessary to use data mostly from Hardangervidda, which reduces the model data base.

First a description of the reproduction under optimal conditions will be given: The females become sexually mature at an age of about two weeks. Pregnancy lasts three weeks, and a new pregnancy may begin shortly after breeding.

In our model we suppose that there are two reproductive seasons, summer and winter, separated by periods of no breeding. Our knowledge of what initiates the beginning and end of the reproductive seasons is insufficient. Weather and nutritive conditions probably play an important part. It was decided, however, to regulate these events solely by the time of the year.

Most of the reproduction takes place during the summer months. The first breeding seems to occur almost simultaneously in most of the females. For the rest of the summer no such synchronism is noticed. The following values were chosen for summer reproduction: 60% of the females breed in the first fourteen days of June. In the next two weeks the rest of the females reproduce and for the remainder of the summer, 30% breed every fourteen-day period. These values are based on quite uncertain estimates from Hardangervidda.

In fall, breeding seems to be rare, probably as a result of unfavorable weather and food conditions. In the model, the breeding ceases completely for two months, beginning with mid-September.

On Hardangervidda there are rodents which, judging from eye lens weight, were born during the winter (A. HAGEN et al., in prep.). They are only a small minority. In most of the runs of the model it is supposed that only 1–5% of the females reproduce per fourteen-day period in this season.

It is assumed that in spring, breeding ceases for fourteen days during snow melt.

Based on counts of fetuses, litter size has been estimated to 6.4 animals per litter.

Although data from Hardangervidda have shown that the ratio of males to females may vary greatly, we have not been able to connect this either to the growth or the collapse of populations. The ratio of males to females is therefore kept equal to 1.0 throughout the cycle.

Under unfavorable environmental conditions, reproduction may be reduced. However, our knowledge of what should be called unfavorable conditions and how these factors act, is very sparse. When we introduced such factors in the model, it was because we thought that by giving these factors a strongly varying effect, it might be possible to investigate whether such kind of factors were possible population-density regulators.

SCHWARZ (1963) has stated that low temperatures stop reproduction in certain small mammals. The rodents treated in this model are probably adapted to temperatures somewhat below 0° C. In their microhabitat, average temperatures much below 0° C are rare, but may take place during the winter. In this model we have somewhat arbitrarily supposed that this happens if the average temperature over the snow cover drops below −12° C. In this case, no reproduction takes place.

After the snow has settled, periods of mild weather, accompanied by rain, may occur. This rain filters through the snow and fills the subnivean space, later freezing to ice. This considerably reduces the activity area of small mammals and secondarily leads to starvation. Reproduction is supposed to cease completely in this case.

In the model, reproduction is also influenced by limited food uptake and social factors. These effects will be discussed later.

Information about the different causes of death in natural small rodent populations is very sparse. This forced us to use "curve fitting" in the determination of mortality rates.

The model distinguishes between increase in mortality due to unfavorable environmental conditions and what is called "general mortality". The "general mortality" is due to disease, age and unfavorable factors not included in the model. The "curve fitting" produced death rates of about 8% every two weeks, somewhat higher for young animals than for mature animals. Seasonal changes are included, since data from Hardangervidda indicate that mortality is reduced during the winter. In the model this reduction is 50%.

The model also computes food uptake, metabolic rate and growth of the small rodents. Metabolic rate is based on average daily metabolic rate and is temperature-dependent. The model is implemented on a CDC 3300 computer. It is written in Fortran IV and treated as a non-linear difference equation system. Time step is fourteen days.

## Results and Discussion

In the following we will consider three current hypotheses for rodent cycling: predation, rodent-vegetation interactions and density-dependent behavioral changes.

When the model is run with no effect from any of the abovementioned factors, the population will, dependent on parameter values, either become extinct or increase to infinity. This can be prevented by introducing a nonage-specific density-dependent control-mechanism (increase in mortality and decrease in natality when population density increases). The model then shows seasonal variations but still no annual cycles.

Interactions between species on different trophic levels have long been known as a possible cause of population fluctuations. One hypothesis based on such interactions is the predation hypothesis, summarized as follows: When the small rodent population is low, it can only support a low predator population. An increase in the small rodent population is followed by a delayed increase in predator density. This continues until the predator population is large enough to begin reducing the prey population. The predator population continues to grow until starvation brings the predator population back to a level below the carrying capacity determined by the prey populations.

We do not think that the predators, with their limited reproductive potential, are able alone to limit prey populations to the degree reported in prey-populations fluctuations. Therefore the population has to be reduced either by a seasonal stop in reproduction or by some other factors before the effect of increased predation may occur.

To reveal if a combination of predation and seasonal variations in reproduction can be responsible for the cyclic fluctuations, the predatory effect was introduced as follows: During the increase of the small rodent populations the predators harvest a constant fraction of the population. When the small rodent population is reduced after a seasonal peak, predation is increased for six weeks. Predation then remains constant until the next predator-breeding season when the predation again increases. High predation pressure is then maintained for one year, after which it returns to normal levels. Predation is reduced during winter due to snow cover and emigrating predatory birds.

This predation effect is roughly in agreement with the findings of PEARSON (1964) and with limited observations from Hardangervidda.

By gradually increasing this predation pressure on small rodents, a cyclic population pattern resulted. The period of the cycles varied between two and four years, depending upon the parameters.

This indicates that if predation, together with the seasonal variations in rodent reproduction acts as presumed in this model, this could be sufficient to cause the cyclical patterns observed for small rodents. The following points must, however, be noted:

(a) Evidence that the predation effect works in the way we have supposed is not satisfactory.

(b) Collapses of population in the virtual absence of predators have been reported (KREBS, 1970).

(c) The predatory pressure necessary to induce cycles resembling the empirical data is higher than we expected to be likely on Hardangervidda. It has been shown, however, that predation in some communities can be fairly severe (PEARSON, 1964).

(d) The model is very sensitive to parameter changes.

Even small changes in predatory pressure may produce great changes in model output.

Considering these points and the great dissimilarities between communities in which small rodent cycles are found, it seems unlikely that predation can account for all observed small-rodent fluctuations. It might, however, be reasonable to suppose that predation can alter the length and pattern of the cycles.

Another ecosystem-level hypothesis to explain the cycles is based on the interaction between small rodents and the primary producers. This theory is based on the fact that if no other factors stop the population's growth, food limitations eventually will. A hypothesis for small rodent fluctuations must, howewer, explain both what causes the peak population to collapse and also what prevents the population from rebuilding its numbers in one single year. One hypothesis trying to explain both these phenomena is SCHULTZ's (1964) "nutrient recovery hypothesis". According to this, heavy grazing and clipping during the peak year ties up the nutrients in organic components and reduces the availability of nutrient in the following years. The food therefore becomes sparse and has a low nutrient content, which leads to reduced reproduction and low survival rates (especially among juveniles). After two or three years when the nutrients are recycled, the vegetation recovers and the ecosystem is again able to support a high density of small rodents.

To test such a hypothesis using simulation requires a good description both of the needs of the small rodents and of the structure and function of vegetation. Our only possibility therefore was to investigate whether rather simple interaction between rodents and vegetation could produce results approximating the empirical data. This interaction was implemented as follows: If the vegetation biomass was reduced below a certain threshold and the small rodent population consumed more than a certain fraction of the biomass, the mortality increased and the natality decreased. To simulate the prolonged deficiency in nutrients, the low natality and high mortality were maintained for one year.

The results may be summarized as follows:

(a) By using suitable parameter values, the small rodent population showed three to four years fluctuation.

(b) When winter breeding was increased, the effect of the rodent-vegetation interaction increased significantly. As stated earlier, winter breeding is probably low for the small rodents incorporated into this model. It may, however, be extensive for other species and even for the same species in other areas (LARSSON et al., 1973; MACLEAN et al., 1974).

(c) Addition of predation to this model causes better parameter stability and makes it easier to get good agreement with empirical data.

Another group of hypotheses for small rodent population cycles connects the fluctuations to intrinsic factors in the population.

There is mounting evidence that behavioral, physiological and genetic changes accompany the changes in population density. However, there is at present no agreement as to whether these changes cause or are merely adaptions to the fluctuations.

Based on SELYE'S (1946) theory of stress (i.e. the general adaptive syndrome), CHRISTIAN (1950) proposed that cycles might be caused by stress and that declines could be associated with changes in adrenal-pituitary functions. There is, however, some evidence against this hypothesis. It has been pointed out that since there is low reproduction during and after the decline, the effects of stress produced among the adults by mutual antagonism must be transfered to the offspring. There is no evidence for any such transfer (NEWSON and CHITTY, 1962).

To account for the prolonged quality changes observed in a declining population, CHITTY (1958, 1960) proposed a "polymorphic behavioral hypothesis". This hypothesis has been supported and further developed by KREBS et al. (1973). According to this, the population is comprised of animals of different genotypes (two broad genotypes according to KREBS). Some animals have greater reproductive success but are intolerant to high-population densities, while other animals are adapted to survival under crowded conditions but have a much lower rate of reproduction. When the population increases, the density-intolerant individuals die or emigrate. The peak population therefore consists of animals with low reproductive capacity. The sharp decline is then explained as a result of this low reproduction and high susceptibility to ordinary causes of mortality. While the density is minimal, the population of animals with high reproductive capacity recovers and a new cycle may start.

In an attempt to simulate this hypothesis, we introduced the following effects into the model: When the population density increased beyond a certain threshold, mortality was increased and natality reduced. To simulate the prolonged genetic shifts we added an effect which started to work after the population had passed a peak and continued for one year. This effect also caused increased mortality and reduced reproduction.

The results of this simulation may be summarized as follows:

(a) Good agreement with the empirical data could be obtained. It should however, be noted that the effects are introduced in such a way that they almost certainly had to induce cycling. Even though most investigators agree that behavioral changes can at least greatly modify periodicity and amplitude of the fluctuations, there is no conclusive evidence that selective forces really work in a way similar to what is supposed in this model.

(b) Addition of predation resulted in a model somewhat more stable to parameter changes, and agreement with empirical results could then be obtained with less strong effects from "genetic" factors.

(c) When effects from rodent-vegetation interaction and "genetic" effects were included simultaneously, one of the effects "masked" the other. The model with both sets of effects is more stable to parameter changes, since if one of the effects "failed" to work, the other became operative.

Even though some investigators believe that cyclic fluctuations may be caused by climatic factors, we find this hard to believe until climatic phenomena with the suitable cycle time have been demonstrated. As a consequence of this view, we have regarded shifts in climatic conditions merely as "noise". This however, does not exclude the possibility that weather may play an essential part in synchronizing fluctuations over large areas.

The model was run with different sets of climatic conditions. If our estimates of the effects of unfavorable weather conditions are not too erroneous, the annual variations in climatic conditions may significantly influence both amplitude and period of the fluctuations. Even relatively strong changes in climatic conditions could, however, not destroy the cyclic pattern of the fluctuations.

Most of the hypotheses of the cyclic rodent fluctuations have concerned the decline of the population. Very often, however, hypotheses on the population decline are rejected because they are not able to give a plausible explanation of the relatively regular periodic length.

Run with optimal climatic conditions, the rodent population in the model showed a peak every third year. Under more realistic conditions, the peak occurred every third to fourth year, independently of what caused the breakdown. For this model the conclusion therefore must be that the cause of the population decline and the periodic length may be unrelated.

The validation of ecological models is problematic, and few well-developed methods are available. Because of this scarcity of validation methods, and also for other reasons we are at present not able to give reliable information about the validity of the model.

Some sensitivity analysis has been carried out to determine the stability towards parameter changes. When a model containing all the above-mentioned factors was used, the stability was relatively high, and quite large parameter variations were needed to destroy the cyclic pattern. The system was, however, quite sensitive to variations in the regulating factors, and with less strong action from these factors, the system became more sensitive to parameter changes.

The system was as expected, most sensitive to changes in essential parameters like birth rate, and mortality rate. It may be noted that when both birth and mortality rates were reduced (corresponding to a longer life span of the small rodents) the system fluctuated less violently.

## Conclusion

(1) The model was not able to give any conclusive answer to the question of which factors caused the fluctuations.

(2) The sensitivity to changes in the regulating factors seems to indicate that the system has a relatively low stability. This is probably partly due to the short life span of the small rodents and to their dominating position in the system.

(3) Due to this low stability, it seems possible that under suitable conditions several factors may be able to induce fluctuations.

(4) When several regulating factors act simultaneously, the fluctuations seem to adopt a more uniform pattern. (It is tempting to believe that if still more regulating factors had been introduced, a more "stable" system would have been established with less violent fluctuations and greater stability against environmental stress.)

(5) Points (3) and (4) in no way exclude the possibility that one single factor may cause all the fluctuation pattern in all areas. These points may, however, be interpreted as a support to the view that the cyclic pattern is caused by combinations of factors and that the relative importance between these factors may vary.

(6) The complexity of the system and the paucity of the data forced us to use "curve fitting" and to try to test several hypotheses simultaneously. The same problems would probably be encountered in any modeling effort of this type. We therefore presume that at present, more valuable information will be obtained from mathematical modeling if such models are used as integrated parts of investigations of subsystems, for example, on ecophysiological problems or on problems concerning the effect of selection and/or behavior.

*Acknowledgements:* We wish to thank all those who made valuable contributions during the construction of the model; especially A. HAGEN and S. KJELVIK, and L. WALLØE for critical reading of the manuscript.

## References

CHITTY, D.: Self-regulation of numbers through changes in viability. Cold Spring Harbor Symp. Quant. Biol. **22**, 277–280 (1958).

CHITTY, D.: Population processes in the vole and their relevance to general theory. Canad. J. Zool. **38**, 99–113 (1960).

CHRISTIAN, J. J.: The adrenal-pituitary system and population cycles in mammals. J. Mammal. **31**, 247–259 (1950).

ELTON, C.: Periodic fluctuations in the number of animals: their causes and effects. Brit. J. Biol. Exp. **2**, 119–163 (1924).

HAGEN, A., KJELVIK, S., LIESTØL, K., ØSTBYE, E., SKAR, H.-J., SWARTZMAN, G.: A simple ecosystem-model with emphasis on small rodents. Rapp. Høyfjellsøkol. Forsk. Stn., Finse, Norge 1975 **1**, 1–104 (1975).

KREBS, C. J.: The lemming cycle at Baker Lake, Northwest Territories, during 1959–62: Discussion. In: ADAMS, L. (Ed.): Population ecology, pp. 8–31. Blemont, California: Dickenson Pub. Co. 1970.

KREBS, C. J., GAINES, M. S., KELLER, B. L., MYERS, J. H., TAMARIN, R. H.: Population cycles in small rodents. Science **179**, 35–41 (1973).

LARSSON, T.-B., HANSSON, L., NYHOLM, E.: Winter reproduction in small rodents in Sweden. Oikos **24**, 475–476 (1973).
MACLEAN, S. F. JR., FITZGERALD, B. M., PITELKA, F. A.: Population cycles in Arctic lemmings: Winter reproduction and predation by weasels. Arctic Alpine Res. **6**, 1–12 (1974).
NEWSON, J., CHITTY, D.: Haemoglobin levels, growth and survival in two *Microtus* populations. Ecology **43**, 733–738 (1962).
PEARSON, O. P.: Carnivore-mouse predation: an example of its intensity and bioenergetics. J. Mammal. **45**, 177–188 (1964).
SCHULTZ, A. M.: The nutrient-recovery hypothesis for arctic microtine cycles: Ecosystem variables in relation to arctic microtine cycles. In: CRISP, D. J. (Ed.): Grazing in terrestrial and marine environments, pp. 57–68. Oxford: Blackwell 1964.
SCHWARZ, S. S.: Ways of adaptation of terrestrial vertebrates to the condition of existence in the subarctic. Vol. 1, Mammals. Trudy Inst. Biol. Ural Fil. Acad. Sci. U.S.S.R. **33**, 1–132 (1963). English translation by ISSAKOFF, E., FULLER, W. A. (Eds.).
SELYE, H.: The general adaptive syndrome and the diseases of adaption. J. Clin. Endocrin. **6**, 117–230 (1946).

# Energy Model for Individual Small Rodents and Its Significance in General Population Theory

N. C. STENSETH

Voles and lemmings show cyclic fluctuations in most of the populations studied. Currently, however, there is no agreement on which factors cause these regular fluctuations (KREBS et al., 1973). Factors proposed range from food quantity and quality (SCHULTZ, 1964, 1969; BATZLI and PITELKA, 1971), predators (PEARSON, 1964, 1966, 1971), physiological stress (CHRISTIAN and DAVIS, 1964), to aggressive behavior (CHITTY, 1960, 1967; KREBS, 1970, 1971; MYERS and KREBS, 1971; KING, 1973). The small rodent systems are of such complexity that it seems natural to employ systems analysis and modeling in studies of cyclic fluctuations. The overall objective of this study is therefore an analysis of these fluctuations, combining a review of the relevant literature describing the system with modeling. This theoretical study is then combined with field and laboratory studies on the population dynamics of small rodents. The root vole *Microtus oeconomus* (Pallas), serves as an example for testing the mathematical model(s) under development.

This paper describes a simulation model (METVOLE) for energy acquisition and utilization by individual root voles. The specific purpose of this model is to predict the change in weight and fat percentage of an animal of specified reproductive category, age, weight, and fat percentage under a variety of environmental conditions. Analysis has shown that the population dynamics of a species are difficult to comprehend without an understanding of the processes related to energetics (COLLIER et al., 1975; see also GOLLEY et al., 1975). The present model is being used as a submodel for a population dynamics model of the same species (Fig. 1). This latter model calculates for each day a series of average rates of death, transfer from one reproductive category to another etc., employing the results calculated in METVOLE. A random number is then drawn for each individual to ascertain whether the possible change will occur.

This technique of modeling population dynamics is called the Monte Carlo technique (GARFINKEL et al, 1964) and was used by COLLIER et al. (1974) for a brown lemming *(Lemmus trimucronatus)* population. This technique is especially suited to model fluctuating populations where random events during population lows may be of great interest.

I feel that an analysis of METVOLE as a separate model is more fruitful than an analysis of the whole model. The types of analysis needed are: (i) Analysis of factors to be included in the model, e.g. the least number of variables and

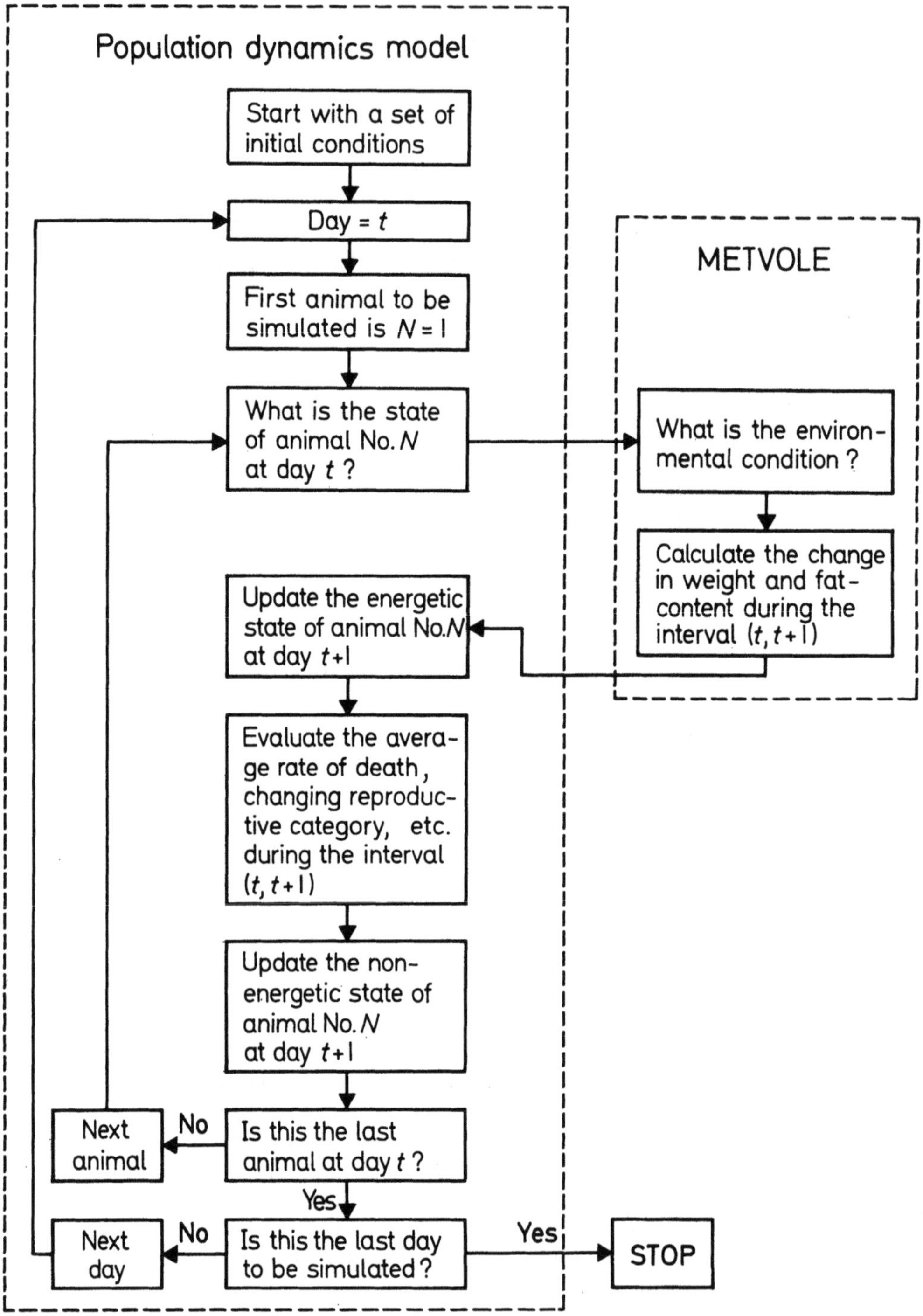


Fig. 1. Basic structure of the models used for studying the population dynamics of small rodents. Dependence on the environment external to the system is not shown

parameters required for estimation of the growth rate; and (ii) analysis of the specific functional relationship to be used in the model. Implied in these first two items is the analysis of the consistency and sufficiency of our present knowledge in a given area, By such analysis, areas to be studied may be revealed. Other types of analysis are (iii) sensitivity analysis of the parameters used; (iv) sensitivity analysis of the components included in the model, showing which components have little effect and thus may be excluded; and (v) evaluation of the model by testing its predictions against experimental data.

Such procedure for model development is based on the hypothetico-deductive method (e.g. FRETWELL, 1972). The steps in this research spiral are: (1) summarize experience (published, personal etc.) in a set of hypotheses [corresponds to item (i)]; (2) formalize the hypotheses into a mathematical model [corresponds to items (ii) to (iv)]; (3) deduce a testable prediction from the model (corresponds to making one or more computer runs with the model); and (4) test the prediction with data not used for building the model [corresponds to item (v)]; if data matches prediction, return to step (3), otherwise to step (1). As this method is closely tied to data, "guesses" of parameter values should be avoided. Otherwise, such quantification may produce faulty hypotheses which only *seem* plausible.

Building and analyzing a model as described above should inspire confidence that the essentials have been properly extracted from our present knowledge of energetics. It is important to note that such an analysis is an iterative process emphasizing the fact that simulation modeling is not an end in itself, but rather a continual process aiding critical biological thinking.

The main objective of this paper is to give a graphical representation of the relations used in METVOLE. Implicit in this is the indication of where more research is needed. At this point it should be remembered that one of the first steps in developing a model is to draw graphs showing the relationships between different factors to be included in the model. Therefore, showing the graphs of the mathematical relations used in the model should give a meaningful way of judging the model's feasibility. Finally an example will show whether METVOLE is able to imitate the growth pattern of individuals whose development was followed by live trapping in the field (Finse, Hardangervidda, South Norway). Reports on more general aspects of the root vole popualtion at Hardangervidda have been given by COLLETT (1911–12), STEVEN (1952), WILDHAGEN (1952, 1953), LINN (1954), HAGEN (1973), HAGEN et al. (1975a). A model for an "average" small rodent population (composed of *Microtus* spp. and *Clethrionomys* spp.) at Hardangervidda is reported by LIESTØL et al. (1975).

## „Total" Models versus Submodels

One of the approaches taken in this study of cyclic small rodent populations is the development of realistic and precise, but not general, models (LEVINS, 1966). With this approach, it was decided to build up a population dynamics model from a series of submodels, one of which is METVOLE. The first analysis of a "total" population dynamics model can probably best be performed for each submodel, or even components of submodels. For realistic and precise models, the larger the system represented, the more parameter values and functional relations have to be guessed at in order to obtain the best correspondence with observations. There is then a high prior-probability of the whole being satisfactory, even though the submodels are not valid, a fact often overlooked by some ecological modelers, as discussed by SMITH (1952).

Having performed such an analysis, it may be possible to draw conclusions on the population level of organization with a model operating at the individual level of organization. Thus by simulating the future fate of individual animals we can at least reach a conditional conclusion on whether they can in fact obtain the amount of food required under certain environmental conditions. An analysis of this type was carried out by COLLIER et al. (1975) with a model similar to METVOLE for the brown lemming. There are several phenomena at the population level, however, (e.g. dispersal, impact of predators etc.) which can only be studied at that level. One population model for small rodents at Hardangervidda is developed by LIESTØL et al. (1975) and serves as a preliminary analysis of the dynamics of this population. These authors simulate an "average rodent" species, i.e. the "average" of the root vole *(M. oeconomus)*, the field vole *(M. agrestis)*, the bank vole *(Clethrionomys glareolus)*, and the grey-sided vole *(C. rufocanus)*, but not the sympatric lemming *(Lemmus lemmus)*. As indicated by LIESTØL et al. (1975), this assumption disregards the growing evidence of the importance of species interactions when studying each species' population dynamics. More detailed and comprehensive models are therefore needed for a closer analysis of the demographic and energetic changes commonly observed in cyclic rodents. The model depicted in Fig. 1 is an attempt to implement a more detailed population model for the root vole. As the field vole is ecologically quite similar, the model may also serve as a model for this species.

## Description and Discussion of the Model

**General.** The system modeled with METVOLE is defined as being one animal. For reproducing females we consider the fetuses and/or sucklings as external to the system, thus emphasizing the open character of the system. Other external factors influencing the system are temperature and plant density in the microhabitat of the animal. The density of animals in the area also influences the *inductive* al's energy requirements through huddling.

METVOLE was developed partly from empirically-derived equations taken from literature, or from equations derived from information reported or personally communicated, by investigators working within IBP (i.e. *inductive* method only requiring hypotheses about the correlation between phenomena), and partly by logical deduction from a set of testable hypotheses (i.e. *deductive* method requiring knowledge of the causal relationship). The model may then be classified as an inductive-deductive model (WATT, 1961). In using empirical equations, one should avoid using them outside their range of definition, and should this happen, then logical deduction must be employed.

The general structure of METVOLE is the same as that of the model presented by COLLIER et al. (1975). It is a state-orientated, deterministic, constant time-step model in its mathematical structure. That is, each simulation begins at an arbitrary point in time with the assigning of values to each variable in the model (these being the "initial conditions"). By following a set of rules a calculation of how much each variable changes in one time-step is made, thus arriving at the new state of the system. A time-step of one day is used. As the system develops through time, it is only the present state which determines the next, that is, METVOLE has the Markov property. The computer coding for METVOLE is structured in such a way that any new information can easily be incorporated into the model. A manual for METVOLE (STENSETH, 1975) may be obtained from the author. In the following the weight of animals refers to fresh weight, whereas weight of plants refers to dry weight.

**Growth and Body Composition.** Following ENZMANN'S (1935) work with *Mus musculus*, the growth curve of individual fetuses during the gestation period is described by a sigmoid curve (Fig. 2A). The growth curve of the sucklings (Fig. 2B) up to an age of 20 days is assumed to be linear (with respect to time) with

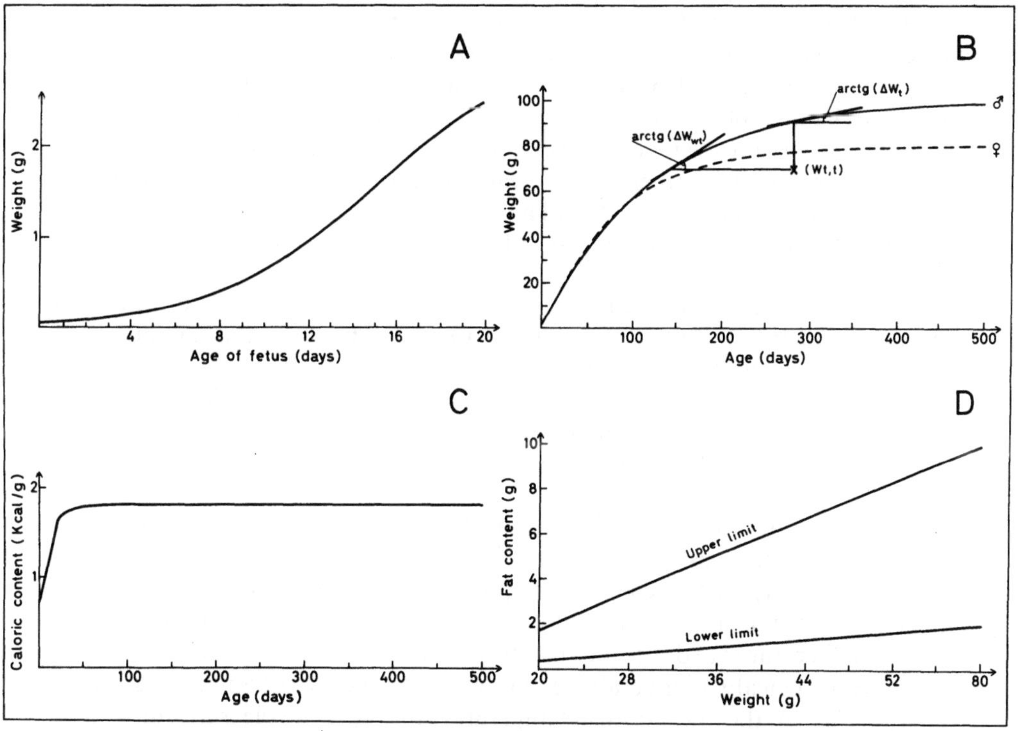


Fig. 2A-D. Graphs showing the realized growth-curve for an individual fetus (A), and for an individual suckling up to 20 days old (B). For animals older than 20 days, the maximal weight as a function of age is shown (B). Here the point (*Wt*, *t*) indicates the position of a hypothesized animal whose predicted growth rate as a function of its weight is $\Delta W_{wt}$, and as a function of its age $\Delta W_t$. The caloric content as a function of age is shown in (C). The maximal and minimal fat content as a function of weight is is shown in (D)

a constant growth rate equal to 0.7 g/day (MORRISON et al., 1954; FRANK and ZIMMERMANN, 1956; HAGEN et al., 1975b).

Following the definition of the system, the weight of a pregnant female does not include the weight of her fetuses. Fig. 2B shows the maximum weight for a given age of animal kept under optimal conditions up to that age. However, several authors have shown, that weaned animals may not grow as rapidly as shown in Fig. 2B (e.g. SCHWARZ et al., 1964; TAST, 1966; FULLER, 1969; HAGEN, ØSTBYE, SKAR, unpubl. data). This may be due to suboptimal environmental conditions such as low temperature and insufficient food (both in quality and quantity). Then, in order to predict the actual maximum growth rate of weaned animals, the geometric mean of (i) the maximum growth rate predicted by its age, $\Delta W_t$, and (ii) the maximum growth rate predicted by its weight, $\Delta W_{wt}$ is taken (COLLIER et al., 1975). For illustrations see Fig. 2B.

Having found the maximum possible weight increment, it is necessary to predict the body composition of the animal, i.e. percentage of water, fat, and non-fat tissue. There are two obvious reasons for being interested in this: first, the caloric content of the animal varies with its body composition; and secondly, it is believed that body composition reflects the physiological status of the animal (KREBS, 1964; BATZLI, G. O., pers. comm.). Although several studies on seasonal changes in body composition have been reported (e.g. SAWICKA-KAPUSTA, 1968, 1970; EVANS, 1969, 1973; HANSSON, 1971; FLEHARTY, et al., 1973), none was found on the mechanism of fat storage and fat usage. Since most of the studies cited above follow only the average fat level in the population, they are difficult to use when building a model; rather they are valuable for testing a proposed model for fat metabolism. COLLIER et al. (1975) have proposed four different models for fat allocation for the brown lemming. In METVOLE I have adopted their model No. 2 for fat allocation and usage.

The components of this fat model are shown in Fig. 2: the caloric content is shown as a function of the animal's age in Fig. 2C. The low caloric content for new-born root voles is due to their high water percentage (GORECKI, 1965). For any given weight Fig. 2D shows the maximum and minimum fat content. Thus an animal's fat content may be anywhere between these two curves. A starving animal will lose weight so that 90% of the weight loss is fat.

**Metabolism.** In order to estimate the average daily metabolic rate (Fig. 3A) of the root vole, I have used the information given by GEBCZYNSKA (1970) and GRODZINSKI (1971) on two root vole populations, and by HANSSON and GRODZINSKI (1970) on a field vole *(Microtus agrestis)* population. Data on metabolism reported in literature are difficult to use because metabolic rates are often given in cal $g^{-1}$. A biologically more realistic unit would be energy per animal. As can be seen from Fig. 3A, the lower critical temperature for the root vole is assumed to be 15°C (GEBCZYNSKA, 1970). Because of the snow-cover and the development of a subnivean space, the environmental temperature seldom falls below 0°C (COULIANOS and JOHNELS, 1962). With a temperature of −15°C outside the nest, COADY (1975) found that the nest temperature of the brown lemming was approximately 0°C. Assuming that the root voles are able to raise the temperature in their nests by a similar amount, these animals will, when they are inside the nest always, be

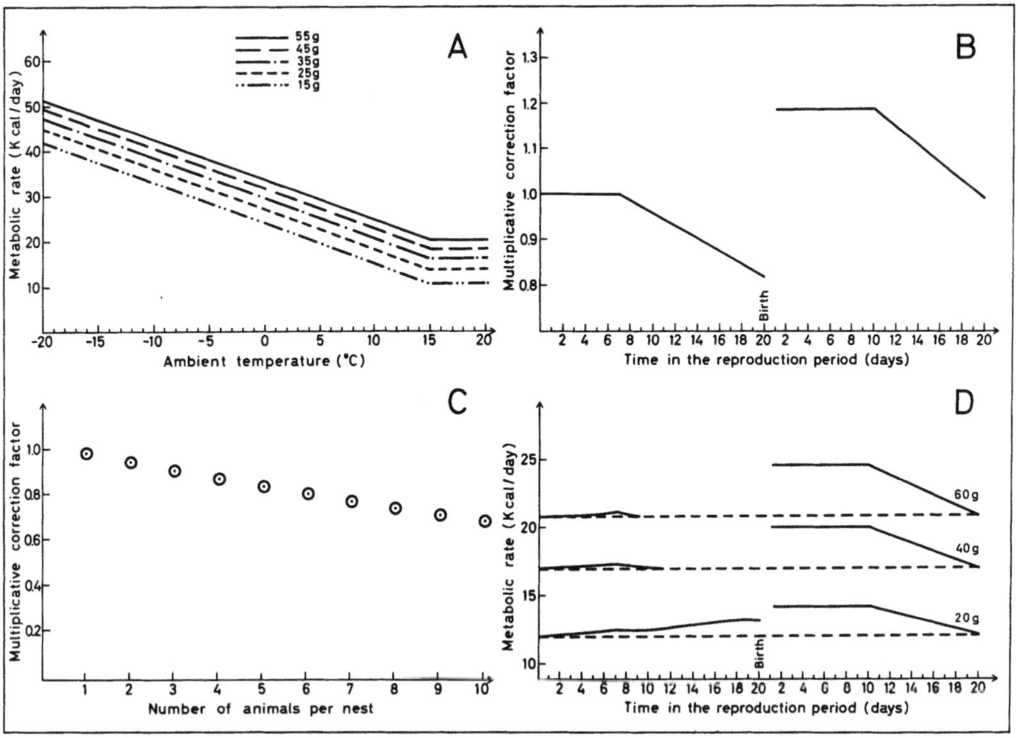


Fig. 3A-D. The average daily metabolic rate is shown as a function of the ambient temperature and the animal's weight (A). Multiplicative correction factors due to reproduction (B) and huddling (C) are shown together with the results of employing the factor due to reproduction (D). In this last case 6 fetuses and ambient temperature equal to 15° C were assumed

within their thermoneutral zone. Thus I am using the horizontal part of the curves in Fig. 3 A as the nest metabolism.

The metabolic rate found in this way may be changed by two multiplicative correction factors, i. e. (i) a correction factor due to reproduction, and (ii) a correction factor due to huddling.

Analyzing the data given by TROJAN and WOJCIECHOWSKA (1967) in their study of the common vole (*M. arvalis*), COLLIER et al. (1975) found the first factor to be as shown in Fig. 3 B. The linear decrease of this factor in the last ten days of the lactation period is due to the weaning process (TAST, 1966: FRANK and ZIMMERMANN, 1956). Not shown in this figure is the restriction that the metabolism for a reproducing female of a given weight is not allowed to fall below that of a non-reproducing female of similar weight.

Using information on the common vole as reported by TROJAN (1969) it is proposed that the function given in Fig. 3C represent the huddling factor. TAST (1966) reported that root voles may stay together in colonies of 10 or more individuals. He did not state whether so many individuals actually occupy the same nest. However, in Norwegian experiments up to 9 individuals caught in the same area (less than 250 m$^2$) and brought to the laboratory did actually construct one nest in a cage and stay together without fighting.

The resulting metabolic rate, after employing the correction factor due to reproduction, is shwon in Fig. 3D. Although the model for metabolism during the reproductive period follows the results given by TROJAN and WOJCIECHOWSKA (1967, Fig. 2), these, when used to find the total metabolic rate, do not seem correct. The chance in total metabolism during lactation should not increase with the increasing size of the mother. It would be more realistic to assume that the metabolic load is reduced with increasing size. One way of implementing this would be to make the factor shown in Fig. 3B a decreasing function of weight. However, no information on this has been found in the literature, so that our present knowledge in this context is shown to be insufficient. Furthermore, the peak on the 7th day of the gestation period for the heavier animals does not seem consistent with general biological knowledge. Analysis of the model has shown that this "error" is more pronounced for lower temperatures, a phenomenon related to the increase in metabolism in a colder environment.

**Reproduction.** The length of both gestation and lactation periods is assumed to be 20 days (MORRISON et al., 1954; FRANK and ZIMMERMANN, 1956; FRANK, 1957). Here only the energy needed directly by the new individual will be considered.

A pregnant female needs to assimilate energy for growing her fetuses and for building up the placenta. Based on information from ENZMANN (1935), the function illustrated in Fig. 4A, is proposed for the additional energy needed by a pregnant female.

The energy a lactating female must give to her sucklings can be divided into the following components: (i) energy needed for the sucklings' respiration, and (ii) energy needed for the sucklings' growth. No study on *Microtus* spp. concerning the first component has been found. The measurements reported by HISSA (1968) for the Norwegian lemming *(L. lemmus)* are therefore adopted here. As Norwegian lemmings attain thermoregulation before root voles (HISSA, 1964, 1968; MORRISON et al., 1954; ØSTBYE, 1965) this assumption may result in an overestimation of the suckling's energy requirements. Based on this is the amount of energy which must be supplied to the sucklings by the mother as given in Fig. 4B.

With the functions illustrated in Figs. 3D and 4A and B the total energy needed by a reproducing female (maintaining 20 g and having 6 fetuses, then 6 sucklings) can be calculated. The result is shown in Fig. 4C. Comparing this with results from the study by MIGULA (1969) who worked with the common vole, there is a high degree of correspondence, both in the general form of the curve and in the numerical values.

As experimental data is lacking, it is assumed that the effects of reproduction illustrated in Fig. 4A and B are additive, whereas those shown in Fig. 3D are multiplicative. These hypotheses need to be tested.

**Food Consumption.** The summer food of the root vole consists of various live, above-ground parts of graminoids (OGNEV, 1950; TAST, 1966; TAST and KALELA, 1971). Towards winter there seems to be a partial change to storage organs (TAST and KALELA, 1971) and roots of the same plants, dandelion (*Taraxacum* spp.)

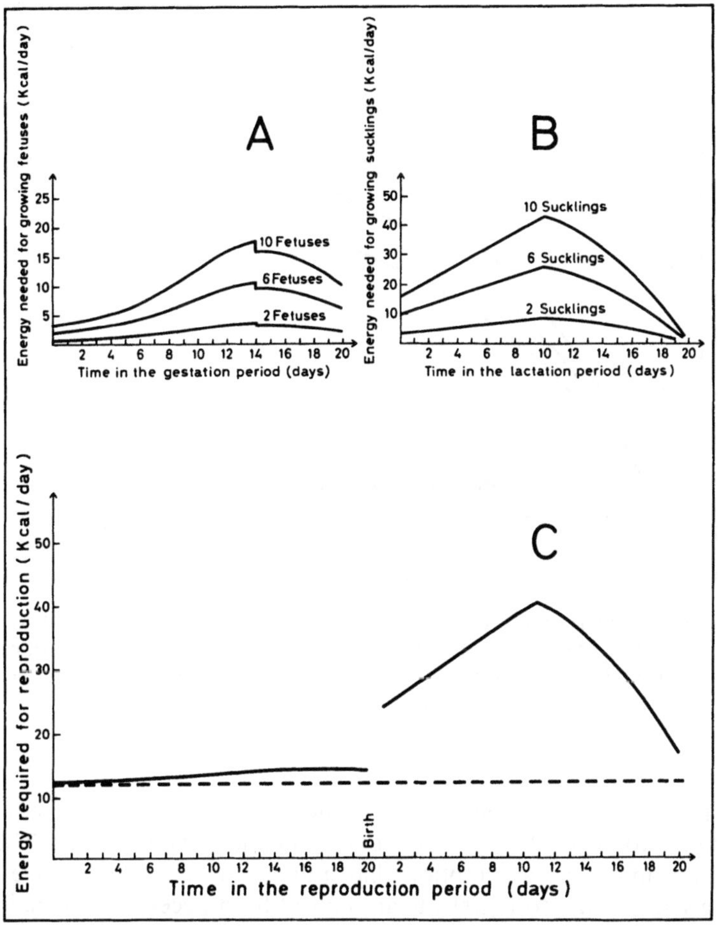


Fig. 4A-C. The energy needed (excluding what shows up in the metabolism) by a pregnant female in order to grow her fetuses and placenta is shown for different numbers of fetuses (A). The equivalent relation for a lactating female is shown in (B). The total energy needed during reproduction is shown in (C) for a female weighting 20 g and having 6 fetuses, then 6 sucklings

(GEBCZYNSKA, 1970), and bark (TAST, 1966; HANSSON, 1971), especially in a peak occurrence of small rodents (HANSSON, 1972).

The actual rate of food assimilation may depend on food quantity and quality, ambient temperature, and the status of the animal (given by its age, weight, fat, and reproductive status). It has been assumed in this model that the rate of food intake is described by a functional response curve (Fig. 5) similar to the one proposed by HOLLING (1959). Even though this relationship has not been studies for any herbivorous small rodent, the evidence for its validity in other groups is ample. Thus this relation has been assumed to be valid for various herbivorous mammals (e.g. BUNNELL, 1973; BUNNELL et al., 1973; COLLIER et al., 1975). To plot the assimilated energy in Fig. 5, an assimilation efficiency of 0.5 (HANSSON,

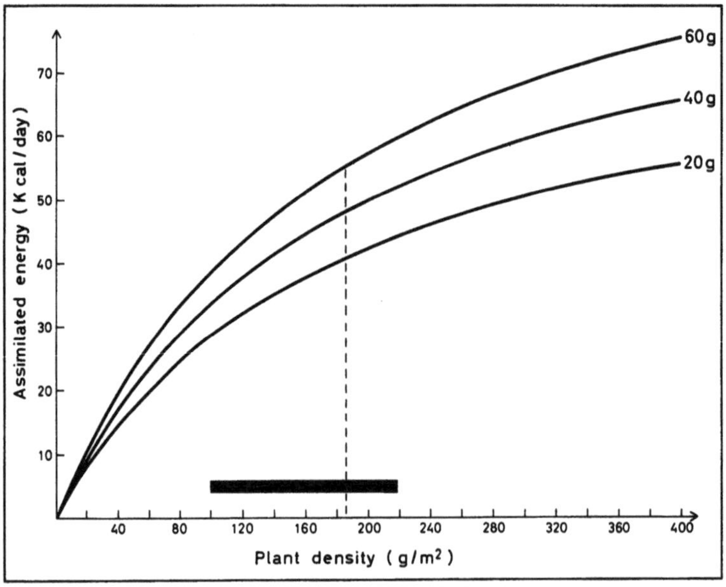


Fig. 5. The functional response curve with a half-saturation constant equal to 185 g m$^{-2}$ (indicated by broken line). The horizontal bar gives the range of plant densities in g m$^{-2}$ (above first root) at Hardangervidda (WIELGOLASKI and KJELVIK, 1973). See text for further explanations

1971; BATZLI, 1975) has been assumed, and the caloric value of edible plant material taken to be 4500 cal g$^{-1}$ (WIELGOLASKI and KJELVIK, 1975). The asymptote is found under the assumption that animals of all weights have approximately the same ability to ingest the amount of food necessary for maintenance.

COLLIER et al. (1975) argued that the animals will not spend more time outside the nest than is necessary to obtain the food needed for growth and development of the young and maintenance metabolism. Furthermore they assumed the maximum time fraction out of the nest to be 0.8 of each day. In order to predict the time budget, the same model as reported by COLLIER et al. (1975) has been apdopted in METVOLE. Experience from several simulations both with METVOLE and the model presented by COLLIER et al. (1975) shows that generally the animals will spend around 0.2 of each day outside the nest, and only in periods of very unfavorable environmental conditions will the time-fraction outside the nest be close, or equal to its upper limit. These results are in good agreement with GEBCZYNSKA (1970) and HANSSON and GRODZINSKI (1970).

The half-saturation constant is defined as that plant density which yields a consumption rate that is half the maximum possible, i.e. the asymptote of the functional response curve. This parameter is estimated to be 185 g m$^{-2}$ (see below). This is within the range of densities of edible food items at Hardangervidda. Based on information presented by COLLIER et al. (1975), the estimated half-saturation constant seems, however, too high. This high value may be supposed to be an artifact of wrong parameter values in other parts of METVOLE, and at present the absolute maximum weights of animals is believed to be too high. The values given by TAST (1966), have been adopted as no other reliable data seem to be available.

**Temperature in the Microhabitat.** As emphasized above, the temperature in the root vole's microhabitat is a factor external to the system being studied. A preliminary model for this microenvironmental factor has been developed: Analysis of the information presented in JOHNSON (1954), WOLD (1957), and COULIANOS and JOHNELS (1962) shows that when the snow depth is at least 10 to 20 cm, the temperature in the subnivean space is between $-3$ and $+2°$ C, this being almost independent of the supranivean temperature. Thus the temperature in the microhabitat is assumed to be 0° C whenever more than 25% of the area is covered with snow of an average depth of at least 15 cm. It is also assumed that temperatures measured 2 m above ground level are a reasonable approximation of the temperature in the microhabitat during the snow-free season, here defined as either a coverage of less than 25% or an average snow depth of less than 15 cm.

## A Simulation Example

Employing METVOLE as described above, with the same parameter values as in Figs. 2 to 5, the meteorological data for the period August 1969 to September 1970 (Meteorological stations at Slirå and Finse), and the plant density (above first root) at a dry meadow community at Hardangervidda in the same period (WIELGOLASKI and KJELVIK, 1973), the changes in weight of two females were simulated over several months (Fig. 6). Animal No. 1 is assumed born on August 1, 1969. Animal No. 2 is assumed born on September 20, 1969. Even though animal No. 2 may be assumed to have given birth to a litter on August 21, 1970, no breeding for the animals simulated during the time period given will be assumed. As the main point is to investigate whether the model is able to imitate the weight changes during a winter period as found for wild animals, it is believed that the error introduced by such an assumption is negligible. The root vole is generally assumed not to have any winter reproduction (e.g. TAST, 1966). These simulated animals are supposed to represent two animals followed by live trapping in the field at Finse, Hardangervidda, over just this period of time (HAGEN, ØSTBYE and SKAR, unpublished data).

All parameter values except the half-saturation constant in the functional response curve are based on experimental data from Hardangervidda or on data reported in the literature. The half-saturation constant is, however, estimated so that the deviation between simulated and observed values was minimized (Fig. 7). The estimated half-saturation constant based on animal No. 1 is greater than 188 g $m^{-2}$, whereas that based on animal No. 2 is 186 g $m^{-2}$. Animal No. 2 did not survive 188 g $m^{-2}$. The overall minimizing value is 185 g $m^{-2}$ (or 188 g $m^{-2}$).

Fig. 6 shows that METVOLE is not able to imitate the decrease in weight during fall, a commonly-observed phenomenon in small rodents (SCHWARZ et al., 1964; FULLER, 1969; BROWN, 1973). In METVOLE it is mainly the total quantity

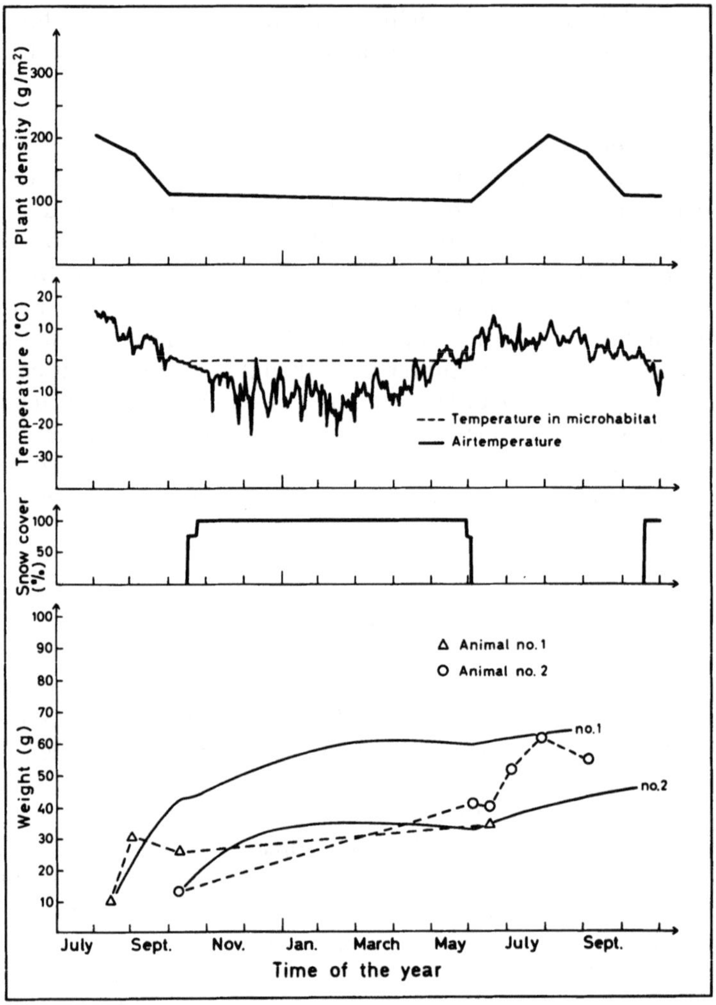


Fig. 6. The simulated (solid lines) and observed (broken lines) weight change for two animals. See text for further discussion. Superimposed on this graph are the plant density in $g\,m^{-2}$ (abova first root), air temperature, ground temperature and snow cover

of food and temperature that control the growth and survival of the animals. Several other factors may however be assumed to control the growth pattern of small rodents, e.g. (i) the quantity of some relatively rare food items, or (ii) the quality of the food. Very little research has been done on these two factors.

## Conclusions and Recommendations for Further Work

I have shown that our knowledge concerning the mother's metabolism during the reproductive phase is insufficient and inconsistent with general biological knowledge. More important, however, for the overall models is that experimen-

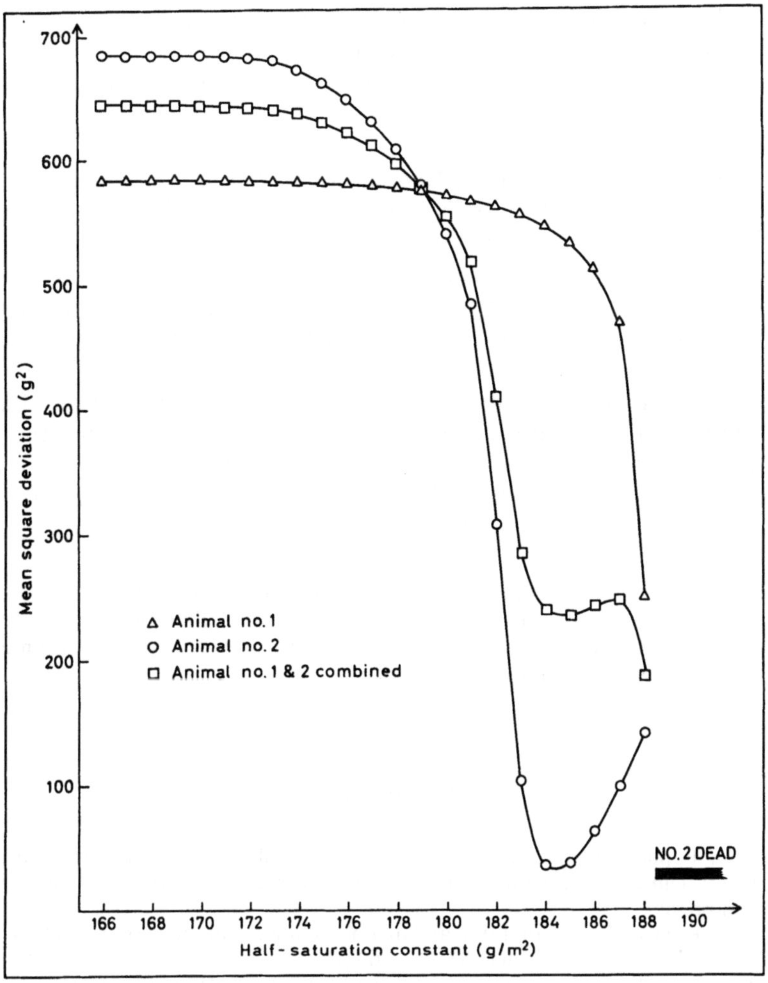


Fig. 7. The mean square deviation $\left(\frac{1}{n}\sum_{i=1}^{n}(W_{\text{pred}}-W_{\text{obs}})^2\right)$ as a function of the half-saturation constant. See text for further discussion

tal data related to the functional response curve seem non-existent. In conclusion I can say that before METVOLE can be used as a submodel for a population dynamics model, more research is needed on (i) the food uptake (kind of food, quantity and quality), (ii) the energy requirement during reproduction, and (iii) the animal's growth pattern. More specifically the following questions need to be asked:

**(i) Food Uptake.** How much energy needs to be spent on acquiring a given amount of energy for metabolism, reproduction etc. at different food densities? Is the functional response curve as used in METVOLE valid for herbivorous small rodents? How does the temperature affect food uptake? How does reproduction affect food uptake? What is the quantity and composition (quality) of newly-added tissue of animals feeding on foods of different density and quality? What are the mechanisms regulating food selection? Is food selection only a function of what is available or are there some physiological mechanisms?

**(ii) Energy Requirement during Reproduction.** What is the total metabolic rate of a pregnant female as a function of its weight (mother's weight per se plus weight of the fetuses and the placenta), number of fetuses, age of the mother, age of the fetuses, and ambient temperature? What is the equivalent function for lactating females? Is the additional energy needed during pregnancy and lactation additive? What additional energy does a male spend during reproduction?

**(iii) Growth Pattern.** What are the growth curves (as a function of age) for animals born in the field at different times of the year, removed into the laboratory and kept under identical conditions? How does energy stress on the mother affect the growth pattern of her offspring? What form has the growth pattern when an animal kept on suboptimal diet is given plentiful high-quality food?

*Acknowledgements.* The author wishes to thank S. BARKLEY, G. O. BATZLI, B. D. COLLIER, A. HAGEN, K. LIESTØL, E. ØSTBYE, R. OSBORN, A. SANDHAUG, A. SEMB-JOHANSSON, J. SKAPPEL and H.-J. SKAR for the information and ideas they have provided. The EDB-center at the University of Oslo is thanked for good cooperation.

## References

BATZLI, G. O.: Populations and energetics of small mammals in tundra ecosystem. In: MOORE, J.J. (Ed.): Tundra: Comparative analysis of ecosystems. Cambridge: University Press 1976 (in press).

BATZLI, G. O., PITELKA, F. A.: Condition and diet of cycling populations of the California vole, *Microtus californicus*. J. Mammal. **52**, 141–163 (1971).

BROWN, III, E. B.: Changes in patterns of seasonal growth of *Microtus pennsylvanicus*. Ecology **54**, 1103–1110 (1973).

BUNNELL, F. L.: Computer simulation of nutrient and lemming cycles in an arctic tundra wet meadow ecosystem. U.S. Tundra Biome Report **73**, 1–312 (1973).

BUNNELL, F. L., KÄRENLAMPI, L., RUSSELL, D. E.: A simulation model of lichen-*Rangifer* interactions in Northern Finland. Rep. Kevo Subarctic Res. Stat. **10**, 1–8 (1973).

CHITTY, D.: Population processes in the vole and their relevance to general theory. Canad. J. Zool. **38**, 99–113 (1960).

CHITTY, D.: The natural selection of self-regulatory behaviour in animal populations. Proc. Ecol. Soc. Austr. **2**, 51–78 (1967).

CHRISTIAN, J. J., DAVIS, D. E.: Endrocrines, behaviour, and population. Science **146**, 1550–1560 (1964).

COADY, J. W.: Bioenergetics of the brown lemming. Univ. Alaska: Unpubl. Ph. D. Thesis, 1975.

COLLETT, R.: Norges pattedyr. Kristiania (Oslo): H. Aschehoug & Co. 1911–12.

COLLIER, B. D., STENSETH, N. C., BARKLEY, S., OSBORN, R.: A simulation model of energy acquisition and utilization by the brown lemming *(Lemmus trimucronatus)* at Barrow, Alaska. Oikos **26**, 276–294 (1975).

COLLIER, B. D., STENSETH, N. C., OSBORN, R., BARKLEY, S.: Models of the brown lemming at Barrow, Alaska. U.S. Tundra Biome Report **73-1**, 1–53, 1974.

COULIANOS, C.-C., JOHNELS, A. G.: Note on the subnivean environment of small mammals. Arkiv för Zoologi, Ser. 2, **15**, 363–370 (1962).

ENZMANN, E. V.: Intrauterine growth of albino mice in normal and in delayed pregnancy. Anat. Rec. **62**, 31–45 (1935).

EVANS, D. M.: The effects of changes in nutrition on populations of *Microtus agrestis* L. In: PETRUSEWICZ, K., RYSZKOWSKI, L. (Eds.) Energy flow through small mammal populations, pp. 157–158. Warszawa: Polish Scientific Publishers 1969.

EVANS, D. M.: Seasonal variations in the body composition and nutrition of the vole *Microtus agrestis*. J. Anim. Ecol. **42**, 1–18 (1973).

FLEHARTY, E. D., KRAUSE, M. E., SPINNETT, D. P.: Body composition, energy constant, and lipid cycles of four species of rodents. J. Mammal. **54**, 426–438 (1973).

FRANK, F.: The causality of microtine cycles in Germany. J. Wildl. Mgmt. **21**, 113–121 (1957).

FRANK, F., ZIMMERMANN, K.: Zur Biologie der Nordischen Wühlmaus (*Microtus oeconomus strimmingi* Nehring). Z. f. Säugetierkde. **21**, 58–83 (1956).

FRETWELL, S. D.: Populations in seasonal environments. Princeton, New Jersey: Princeton University Press 1972.

FULLER, W. A.: Changes in numbers of three species of small rodent near Great Slave Lake, N. W. T. Canada, 1964–1967, and their significance for general population theory. Ann. Zool. Fenn. **6**, 113–144 (1969).

GARFINKEL, D., MACARTHUR, R. H., SACK, R.: Computer simulation and analysis of simple ecological systems. Ann. N. Y. Acad. Sci. **115**, 943–951 (1964).

GEBCZYNSKA, Z.: Bioenergetics of a root vole population. Acta theriol. **15**, 33–66 (1970).

GOLLEY, F. B., PETRUSEWICZ, K., RYSZKOWSKI, L. (Eds): Small mammals: their productivity and populations dynamics. New York: Cambridge University Press 1975.

GORECKI, A.: Energy values of body in small mamals. Acta theriol. **10**, 333–352 (1965).

GRODZINSKI, W.: Energy flow through populations of small mammals in the Alaskan taiga forest. Acta theriol. **16**, 231–275 (1971).

HAGEN, A.: Energibudsjett for en fjellrottepopulasjon [*Microtus oeconomus* (Pall.)] i den lav/mellomalpine sone på Finse, Hardangervidda. Univ. Oslo: Unpubl. Thesis 1973.

HAGEN, A., ØSTBYE, E., SKAR, H.-J.: Energy budget of a population of the root vole [*Microtus oeconomus* (Pall.)] in a high mountain habitat, Hardangervidda. In: WIELGOLASKI, F. E. (Ed.): Fennoscandian tundra ecosystems. Part 2: Animals and systems analysis, pp. 170–173. Berlin-Heidelberg-New York: Springer 1975a.

HAGEN, A., ØSTBYE, E., SKAR, H.-J., STENSETH, N. C.: Age determination of *Microtus oeconomus* (1975b) (in prep.).

HANSSON, L.: Habitat, food and population dynamics of the field vole *Microtus agrestis* (L.) in South Sweden. Viltrevy **8**, 267–378 (1971).

HANSSON, L.: Seasonal changes in physiology and nutrition of herbivorous small rodents. Aquilo **13**, 53–55 (1972).

HANSSON, L., GRODZINSKI, W.: Bioenergetic parameters of the field vole *Microtus agrestis* L. Oikos **21**, 76–82 (1970).

HISSA, R.: The postnatal development of homoiothermy in the Norwegian lemming (*Lemmus lemmus*). Experimentia **20**, 326–330 (1964).

HISSA, R.: Postnatal development of thermoregulation in the Norwegian lemming and the golden hamster. Ann. Zool. Fennici **5**, 345–383 (1968).

HOLLING, C. S.: Some characteristics of a simple type of predation and parasitism. Canad. Ent. **91**, 385–398 (1959).

JOHNSON, H.: Winter microclimates of importance to Alaskan small mammals and birds. Cornell University. Ph. D. Thesis 1954.

KING, J. A.: The ecology of aggressive behaviour. Ann. Rev. Ecol. Syst. **4**, 117–138 (1973).

KREBS, C. J.: The lemming cycle at Baker Lake, Northwest Territories, during 1959–62. Arctic Inst. North Amer. Tech. Paper **15**, 1964.

KREBS, C. J.: *Microtus* population biology: behavioural changes associated with the population cycle in *M. ochrogaster* and *M. pennsylvanicus*. Ecology **51**, 34–52 (1970).

KREBS, C. J.: Genetic and behavioural studies an fluctuating vole populations. Proc. Adv. Study Inst. Dynamics Numbers Popul. (Oosterbeek) 243–256 (1971).

KREBS, C. J., GAINES, M. S., KELLER, B. L., MYERS, J. H., TAMARIN, R. H.: Population cycles in small rodents. Science **179**, 35–41 (1973).
LEVINS, R.: The strategy of model building in population biology. Amer. Scient. **54**, 421–431 (1966).
LIESTØL, K., ØSTBYE, E., SKAR, H.-J., SWARTZMAN, G.: A simulation model of a small rodent population. In: WIELGOLASKI, F. E. (Ed.): Fennoscandian tundra ecosystems. Part 2: Animals and systems analysis, pp. 273–282. Berlin-Heidelberg-New York: Springer 1975.
LINN, I.: Some Norwegian small mammal faunas; a study on trappings in West and North Norway. Oikos **5**, 1–24 (1954).
MIGULA, P.: Bioenergetics of pregnancy and lactation in European common vole. Acta theriol. **14**, 167–179 (1969).
MORRISON, P. R., RYSER, F. A., STRECKER, R. L.: Growth and development of temperature regulation in the tundra redback vole. J. Mammal. **35**, 376–386 (1954).
MYERS, J. H., KREBS, C. J.: Genetic, behavioural, and reproductive attributes of dispersing field voles *Microtus pennsylvanicus* and *Microtus ochrogaster*. Ecol. Monogr. **41**, 53–78 (1971).
OGNEV, S. I.: Mammals of the USSR and adjacent countries. Vol. 7 Rodents. Moscow-Leningrad: Idzabel'stvo Akademii Nauk SSR 1950. (Israel Program for Scientific Translations, Jerusalem 1964).
ØSTBYE, E.: Development of thermoregulation in relation to age and growth in the Norwegian lemming (*Lemmus lemmus* L.). Nytt Mag. Zool. (later Norw. J. Zool.) **12**, 65–75 (1965).
PEARSON, O. P.: Carnivore-mouse predation: An Example of its intensity and bioenergetics. J. Mammal. **45**, 177–188 (1964).
PEARSON, O. P.: The prey of carnivores during one cycle of mouse abundance. J. Anim. Ecol. **35**, 217–233 (1966).
PEARSON, O. P.: Additional measurements of the impact of carnivores on california vole (*Microtus californicus*). J. Mammal. **52**, 41–49 (1971).
SAWICKA-KAPUSTA, K.: Annual fat cycle of field mice, *Apodemus flavicollis* (Melchior, 1834). Acta theriol. **13**, 329–339 (1968).
SAWICKA-KAPUSTA, K.: Changes in the gross body composition and the caloric value of the common voles during their postnatal development. Acta theriol. **15**, 67–79 (1970).
SCHULTZ, A. M.: The nutrient-recovery hypothesis for arctic microtine cycles. II Ecosystem variables in relation to arctic microtine cycles. In: CRISP, D. J. (Ed.): Grazing in terrestrial and marine environments, pp. 57–68. Oxford: Blackwell Scientific Publications 1964.
SCHULTZ, A. M.: A study of an ecosystem: the arctic tundra. In: VAN DYNE, G. M. (Ed.): The ecosystem concept in natural resource management, pp. 77–93. New York: Academic Press 1969.
SCHWARZ, S. S., POKROVSKI, A. V., ISTCHENKO, V. G., OLENJEN, V. G., OVTSCHINNIKOVA, N. A. PJASTOLOVA, O. A.: Biological pecularities of seasonal generations of rodents, with special reference to the problem of senescence in mammals. Acta theriol. **8**, 11–43 (1964).
SMITH, F. E.: Experimental methods in populations dynamics: A critique. Ecology **33**, 441–450 (1952).
STENSETH, N. C.: User's manual: METVOLE-version 1. Rep. High Mountain Ecology Res. Stat., Finse, Norway 1976 (in press).
STEVEN, D. M.: Notes on a collection of small mammals from western Norway. Univ. Berg. Årbok 1952, Naturvitenskapelig rekke No. **6**, 1–13 (1952).
TAST, J.: The root vole, *Microtus oeconomus* (Pallas), as an inhabitant of seasonally flooded land. Ann. Zool. Fenn. **3**, 127–171 (1966).
TAST, J., KALELA, O.: Comparisons between rodent cycles and plant production in Finnish Lapland. Ann. Acad. Sci. Fenn. Series A, IV Biologica **168**, 1–14 (1971).
TROJAN, P.: An ecological model of the cost of maintenance of *Microtus arvalis* (Pall.). In: PETRUSEWICZ, K., RYSZKOWSKI, L. (Eds.): Energy Flow Through Small Mammal Populations, pp. 113–122. Warszawa: Polish Scientific Publishers 1969.
TROJAN, P., WOJCIECHOWSKA, B.: Resting metabolism rate during pregnancy and lactation in the European common vole — *Microtus arvalis* (Pall.). Ekol. Polska-Seria A. **15**, 811–817 (1967).
WATT, K. E. F.: Mathematical models for use in insect pest control. Canad. Entomol., Suppl. No. **19**, 1–62 (1961).

WIELGOLASKI, F. E., KJELVIK, S.: Production of plants (vascular plants and cryptogams) in alpine tundra, Hardangervidda. In: BLISS, L. C., WIELGOLASKI, F. E. (Eds.): Primary production and production processes. Tundra Biome, pp. 75–86, Edmonton and Oslo: IBP Tundra Biome Steering Committee 1973.

WIELGOLASKI, F. E., KJELVIK, S.: Energy content and use of solar radiation of Fennoscandian tundra plants. In: WIELGOLASKI, F. E. (Ed.): Fennoscandian tundra ecosystems. Part 1: Plants and microorganisms, pp. 201–207. Berlin-Heidelberg-New York: Springer 1975.

WILDHAGEN, AA.: Om vekslingene i bestanden av smågnagere i Norge 1871–1949. Drammen: Statens Viltundersøkelser 1952.

WILDHAGEN, AA.: On the reproduction of voles and lemmings in Norway. Drammen: Statens Viltundersøkelser 1953.

WOLD, K.: Temperaturforhold i snø og is og beregning av varmeledningsevnen i og varmetransporten gjennom isen. Univ. Oslo: Unpubl. Thesis 1957.

# Functioning of Fennoscandian Tundra Ecosystems

F. E. WIELGOLASKI

Fennoscandia is strongly influenced by the warm Gulf Stream along the coast of Norway causing higher temperatures than in other parts of the world at similar latitudes (about 60–70° N). Most of the tundra (defined as areas with annual mean air temperatures below 0° C) in Fennoscandia is alpine (about 50% of Norway as a whole). The annual mean temperature is −1 to −2° C at 1200 m in the south-western parts about 200 m above the tree line and is as low as about 0° C even in the upper birch forest zone. Tundra areas close to sea level are only found in the northernmost regions. Annual mean temperatures in the birch forest at Kevo, northern Finland, about 100 m above sea level, are about −2° C (KALLIO, 1975). The dominating westerly winds and the Fennoscandian mountain range north-south through most of Norway in western Fennoscandia cause high precipitation in the west, and leaside effects east of the mountains. At the mountain plateau, Hardangervidda, in southern Norway, annual precipitation decreases for instance

Fig. 1. Steep mountains of western Norway. (Photo: I. BU)

from above 1500 mm in the west to about 500 mm in the east (ØSTBYE et al., 1975), while only about 300 mm per year is registered at Abisko, northern Sweden, just east of the mountain range (ROSSWALL et al., 1975b). Permafrost is only found in peaty soil in northern parts of Fennoscandia with moderate winter precipitation. Polar days (midnight sun) last for more than two months in the north (e.g. the Kevo area) while day length never exceeds about 20 hrs in southern Fennoscandia. This means that the combination of thermal and radiation conditions are unique in northern Fennoscandia compared to lower latitudes around the world.

Hard and poor gneiss and granite bedrocks predominate in Fennoscandia (SONESSON et al., 1975), but younger Cambro-Silurian bedrocks, often schisty phyllite, are found in many places in connection with the Caledonian folding zone and are better parent materials to more nutrient-rich soils. Soils from the richer bedrocks may have a pH ($H_2O$) of about 6 and a C/N ratio in humus about 15, while pH is about 4 in poor soil (quartsitic mineral soil or ombrotrophic peat) and the C/N ratio then sometimes as high as 40–45 in humus. In the poorer areas, podzolization often occurs, except at the highest elevation, i.e. in the lowest temperatures.

## Vegetation

### General Structural-Functional Relationship

On a macroscale, temperature is the most important limiting factor to primary production and vegetation structure in Fennoscandian tundra, e.g. for the altitudinal belts described by DAHL (1975). The coniferous forest tree line is calculated to follow the 8.3° C isotherm for the mean diurnal temperature of the four warmest months (tetratherm) and the birch forest the tetratherm of 7.3° C (HELLAND, 1912), but temperature limits are decreased in continental areas with high day temperatures (in relation to the exponential function of metabolism at increasing temperature). At Hardangervidda, the upper limit for willows (*Salix lapponum* and related species) and *Vaccinium myrtillus* seems to be relatively close to the 5.5° C tetratherm. From the subarctic forest tundra in Finland, KALLIO (1975) reports about 500 vascular plant species, but the number is considerably lower at the low alpine heaths in the area. DAHL (1975) reports somewhat more than 100 vascular plants in the low alpine zone in southern Norway, but only about 30 in the high alpine zone above 2000 m. The relative number of chamaephytes, with perennating buds just above soil surface, strongly increases with elevation, indicating this life form to be favorable under extreme climatic conditions.

On a smaller scale, the causal relationship between the abiotic and biotic components of the Fennoscandian tundra ecosystems is much more complex. In addition to temperature are precipitation, soil moisture, snow cover and nutrient conditions of the soil also very important both to vegetation structure and production. Snow cover together with temperature are the main factors influencing length of the growing season. Much of the annual precipitation occurs as snow which blows off the macro and micro ridges by the often strong wind in the treeless areas and accumulates in the depressions. In western Norway the depth may be several meters thick in the depressions, but often less than 50 cm on the ridges. In alpine northwestern Sweden, just east of the mountain peaks, only 10–60 cm maximum snow cover is found in the depressions and 0–10 cm on the hummocks (ROSSWALL et al., 1975b). Also, in northern Finland, snow cover is very variable and often nearly non-existent at the mountain tops (KÄRENLAMPI, 1972).

## Structure and Functioning of Important Community Types

**Snow Beds.** The soil temperature even in the top layer seldom falls below 0° C when the snow cover is more than about 2 m thick. This protects plant species sensitive to low temperatures. On the other hand, the snow-free period and thus the growing season, is very short in extreme snow beds (sometimes with 10–15 m maximum snow cover), often less than two months. Only few vascular plant species (e.g. *Salix herbacea*) are able to develop in this short period. Several seasons are necessary in many species for differentiation of flower buds to flowering (RESVOLL, 1917). Total primary production (cryptogams and vascular plants, both tops and roots) in *Salix herbacea* snow beds at Hardangervidda is generally considerably below 100 g $m^{-2}$ $yr^{-1}$ which is significantly less than the calculated average primary dry weight production of 350 g $m^{-2}$ $yr^{-1}$ (Table 1) for Hardangervidda as a whole. The ratio production to biomass in snow beds is of the order 0.1–0.4, lowest when slow-growing lichens are dominating.

Table 1. Annual primary production as dry weight at Hardangervidda based on areas of various plant communities given (HESJEDAL, 1975) (slightly modified because of average higher unproductive area at Hardangervidda than within the study plots), primary production estimates (KJELVIK and KÄRENLAMPI, 1975; WIELGOLASKI, 1975a) and interpolation between values from observed community types. In all cases, death and decomposition are compensated during the growing season. Total area 10000 $km^2$ when the birch zone is included

| | Area % | $km^2$ | Prim. prod. g $m^{-2}$ or t $km^{-2}$ | Total tons[a] |
|---|---|---|---|---|
| Lichen heaths (oligotrophic) | 10 | 1000 | 100–300 | 150000 |
| Dwarf shrub heaths (oligotrophic) | 32 | 3200 | 250–400 | 1000000 |
| Moderate herbaceous snowbeds (oligotrophic and eutrophic) | 17 | 1700 | 350–600 | 720000 |
| Birch forests (oligotrophic and eutrophic) | 10 | 1000 | 600–900 | 700000 |
| Mires, bogs and fens (oligotrophic and eutrophic) | 17 | 1700 | 500–900 | 900000 |
| Extreme *Salix herbacea* snowbeds (oligotrophic) | 4 | 400 | <100 | 30000 |
| Unproductive (lakes and bare rocks) | 10 | 1000 | 0 | 0 |
| Total | | | | 3500000 |

[a] Generally oligotrophic communities dominate (>85%). Therefore, lower production values than average are used from vegetation types covering both eutrophic and oligotrophic communities.

**Mire Hummocks.** Similar low plant production is found on hummock tops nearly free of snow in ombrotrophic mires in northern Fennoscandia (production to biomass ratio 0.1–0.2). This is caused by the extreme weather conditions with low temperatures in winter and damage to above-ground plant parts by ice particles moving at considerable velocity across the surface, and aided by grazing and trampling of animals, principally reindeer and hare (ROSSWALL et al., 1975b). The relatively long growing season on the hummocks, about five months, does not compensate for the winter damage of the perennating buds; nor does the high summer soil temperature on the hummock tops (up to about 30° C, observed just below the surface). On the contrary, the high temperatures cause midday depressions in photosynthesis in the area (JOHANSSON and LINDER, 1975). The plant production is also limited by lack of nutrients under the ombrotrophic conditions on hummock tops. Nutrient input by precipitation is low, particularly in northern Fennoscandia (on an average 0.11% N and 0.14% K calculated at Abisko, while the values at Hardangervidda were 0.17% N and 0.80% K). The low annual precipitation on the leaside of the mountain range is of course also of importance.

**Lichen Heaths.** In many respects, similar functional limitations for plant growth are found on many nutrient-poor moraine and esker ridges and other windswept macro- and micro-uplift areas in Fennoscandia with 120–150 snow-free days as at ombrotrophic mire hummock tops. Because of poor snow cover, soil winter temperatures may be $< -15°$ C in upper layers. This, together with the often extreme dry conditions (2–5% soil moisture as wet weight is measured both in northern Finland and southern Norway), and high soil-surface temperatures in summer (40–50° C observed in both areas) strongly limits the number of vascular plant species, to *Festuca ovina* and a few other monocotyledons, and some dwarf shrubs, e.g. *Empetrum hermaphroditum*, *Vaccinium vitis-idaea* and *Arctostaphylos alpina*. The production of vascular plant species is also often strongly limited by nutrient deficiency in these areas, e.g. on coarse, oligotrophic, quartzitic moraine and esker ridges restricted to external nutrient input from precipitation.

Cryptogams, however, and particularly lichens, e.g. *Cladonia* spp. and *Cetraria* spp., often dominate under these environmental conditions, probably mostly because of good light conditions, highly important to lichens. Light saturation for lichen photosynthesis is about 35000 lux, which is higher than for vascular plants, 20000–30000 lux, and bryophytes, 10000–15000 lux (SKRE, 1975). Both lichens and bryophytes start photosynthesis early in spring, probably even before snow melt, and it continues until late fall. Cryptogams growing under tundra conditions are often found to have a very low optimum temperature, about 5° C for *Rhacomitrium lanuginosum* (KALLIO and HEINONEN, 1975) and 7–10° C for lichens, depending on the temperature pretreatment (SKRE, 1975). Apparent photosynthesis in cryptogams can continue down to $-5$–$-10°$ C. This means that cryptogams in many ways are well-adjusted to growth in tundra conditions. Low mid-summer moisture is probably the strongest growth-restricting factor for lichens and bryophytes (KÄRENLAMPI and KAUHANEN, 1972; KALLIO und KÄRENLAMPI, 1975) in most Fennoscandian tundra regions. Inactive periods may be found in summer, while spring and autumn may be more favorable for their growth. In some cases, however, it is observed that the everlasting summer days in northern Fennoscandia may also be a restricting factor for growth, particularly for some boreal moss and lichen species penetrating northwards into arctic climate (KALLIO and VALANNE, 1975).

In alpine southern Norway, maximum lichen biomass about 1200 g $m^{-2}$ is found on unshaded, but not too dry, ridges in the more continental areas (GAARE, pers. comm.). In the western districts lichens are restricted by shading from bryophytes which grow under the better moisture conditions, e.g. *Rhacomitrium lanuginosum;* here the maximum lichen biomass is probably not above 500 g $m^{-2}$.

The lichen heaths are very important to reindeer grazing, particularly during winter, in large parts of Fennoscandian tundra. At Hardangervidda these vegetation types (Loiseleurio-Arctostaphylion alliance in phytosociological terminology), covering about 10% of the area (HESJEDAL, 1975), were used by the reindeer for 80–100% of the total grazing time during winter period (Fig. 2). At the central Hardangervidda plateau lichen dry weight biomass is usually of the order 200–400 g $m^{-2}$, often dominated by *Cetraria nivalis.* Visually, much of the strongly grazed and trampled lichen mat seems to be dead, but radioautographs have shown that only about 2% were not assimilating $CO_2$ (WIELGOLASKI, 1975a) and were unable to recover.

The primary production of lichen heaths is very dependent on the amount of vascular plants. In some cases total production may be as high as 250–300 g dry weight $m^{-2}$ $yr^{-1}$ (KJELVIK and KÄRENLAMPI, 1975). When dominated totally by lichens, however, it is only about 100 g $m^{-2}$ $yr^{-1}$ even in a dense lichen mat. The production to biomass ratio in lichens at Hardangervidda is about 0.2, but is still lower in more continental Fennoscandian tundra regions, and the maximum apparent photosynthesis observed in lichens about 0.1 mg $CO_2$ $g^{-1}$ $h^{-1}$ (SKRE, 1975).

**Grass-Vaccinium Heaths.** If the snow cover is somewhat better and the soil-moisture content slightly higher than at typical lichen heaths, increasing amounts of vascular plant biomass and production are observed even if the nutrient conditions are poor. In these very important heath communities (sheltered by about 0.5–1.5 m maximum snow cover in relatively moist southwestern alpine Norway, and covering about 1/3 of the total area of Hardangervidda) monocotyledonic species such as *Carex bigelowii* and *Nardus stricta* and others dominate, together with *Vaccinium myrtillus* in the low alpine and subalpine-subarctic regions. Lichens are still important; at Hardangervidda *Cladonia alpestris (C. stellaris)* dominates in these communities. The moss cover is usually stronger than in typical lichen heaths. Total annual primary production may be 250–400 g $m^{-2}$ at Hardangervidda, but is lower in similar communities under drier conditions in northern Fennoscandia. Soil moisture is still probably the strongest limiting factor for growth.

In both typical lichen and other oligotrophic heaths nitrogen fixation by some lichen species may in many cases be important in Fennoscandian tundra areas. According to calculations in northern Finland, the lichen species *Stereocaulon paschale* in dry areas, and *Nephroma arcticum* and *Peltigera aphthosa* in moister communities, are able to fix up to 3–4 g N $m^{-2}$ $yr^{-1}$ if the cover is 100% of these

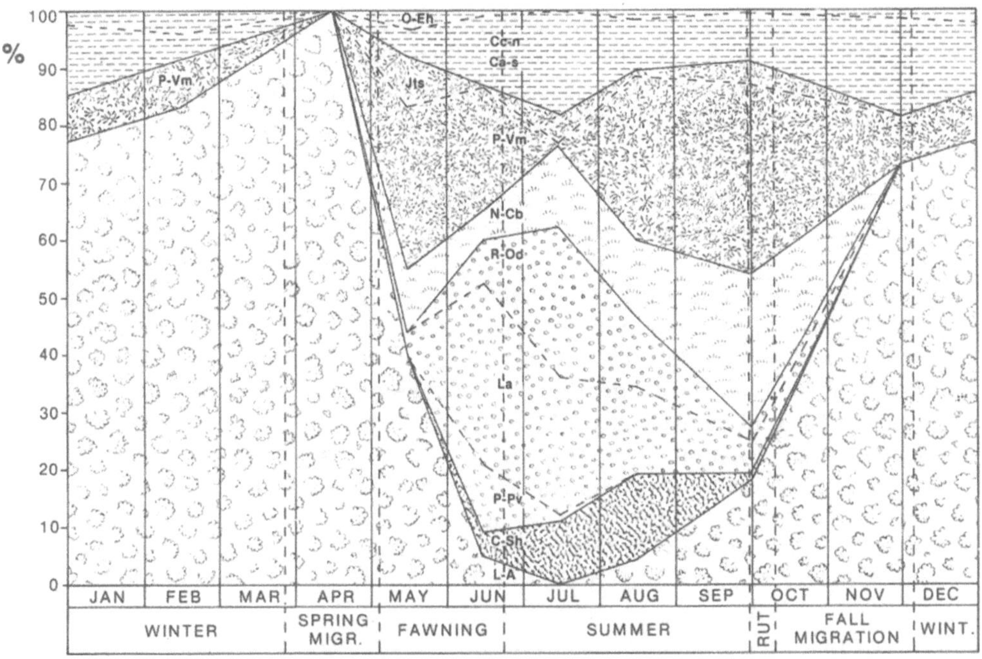


Fig. 2. Reindeer use of plant alliances, Hardangervidda 1971–73 in percent time at various alliances (ØSTBYE et al., 1975).

| | | |
|---|---|---|
| Mires | O – Eh | Oxycocco-Empetrion hermaphroditi, ombrotrophic mire or mire hummocks. |
| | Cc – n | Caricion canescentis-nigrae, minerotrophic fen often with *Salix* shrubs. |
| | Ca – s | Caricion atrofuscae-saxatilis, minerotrophic rich fen. |
| Heaths | Jts | Juncion trifidi scandinavicum, oligotrophic and moderately chionophilous grass heath. |
| | P – Vm | Phyllodoco-Vaccinion myrtilli, oligotrophic and moderately chionophilous dwarf shrub heath. |
| | L – A | Loiseleuro-Arctostaphylion, oligotrophic, chionofobous dwarf shrub heath, often with dominating lichens. |
| | N – Cb | Nardo-Caricion bigelowii, oligotrophic, chionophilous grass heath. |
| | C – Sh | Cassiopo-Salicion herbaceae, oligotrophic, strongly chionophilous dwarf shrub heath. |
| Meadows | R – Od | Ranunculo-Oxyrion digynae, oligotrophic to eutrophic, strongly to very strongly chionophilous meadow. |
| | La | Lactucion alpinae, eutrophic, chionophilous tall herb meadow. |
| | P – Pv | Potentillo-Polygonion vivipari, eutrophic, moderately chionophilous dry meadow |

species, and in field plots (with somewhat lower cover) up to about 0.4 g N $m^{-2}$ $yr^{-1}$ are found (KALLIO and KALLIO, 1975). N-fixation occurs only when temperature is above 0° C and maximum fixation is also dependent on high moisture in the thalli. The observed rate of fixation is much lower in southern Norway even under favorable climatic conditions than for the same species in northern Finland (KALLIO and WIELGOLASKI, unpubl.). One possible reason for this may be inhibition of N-fixation by the higher sulfur content of precipitation in southern Norway closer to the central-European pollution centers [about 600 mg $SO_4$-S $m^{-2}$ $yr^{-1}$, compared to less than 400 mg $m^{-2}$ $yr^{-1}$ in northern Fennoscandia (KALLIO and VEUM, 1975)]. In extreme cases, pH of precipitation at Hardangervidda was below 4, in other cases above 6.5, depending on the direction of the cyclones.

**Dry Meadows.** With increased snow cover and soil moisture (but still relatively dry), vegetation is more meadow-like, dominated by grasses, but also with several forbs. Maximum snow cover is usually 1–2 m in relatively oceanic regions and minimum soil temperature in the upper layer is seldom below $-1°$ C. This vegetation type is most frequent in western districts with relatively high precipitation but also at higher elevations with reduced evapotranspiration because of lower temperatures. At Hardangervidda, it occupies nearly 20% of the area (HESJEDAL, 1975). Most of these grass communities are relatively poor in nutrients and dominated by *Deschampsia flexuosa*, in some cases by *Nardus stricta;* somewhat richer types include more *Anthoxanthum alpinum* and in the richest, *Poa alpina* is common. Even the more eutrophic types of dry meadows are situated on leached mineral soil in the low alpine belt with a podzolic profile.

Primary production, in these dry meadow communities is dependent on both nutrients and soil moisture. In the richest types the total annual production is 500–600 g $m^{-2}$ at Hardangervidda (WIELGOLASKI, 1975a), but about 100–150 g $m^{-2}$ lower in the poorer types. The production-biomass ratio for vascular plants is relatively high in dry meadows, about 0.7, due to a high percentage of herbaceous plants.

These vegetation types are important for sheep grazing in Norway. Experiments have shown it possible to increase particularly the above-ground production of herbaceous plants by strong mixed fertilization, for important grasses 4–5 times (WIELGOLASKI, 1975a). Fertilization with only nitrogen and phosphorus in equal amounts shows less influence. One reason for the weak influence of N-fertilization may be the relatively strong N-fixation by anaerobic and particularly aerobic living bacteria in many of the dry meadows (above 100 mg N $m^{-2}$ $yr^{-1}$ observed at Hardangervidda), while symbiotic bacterial N-fixation is mostly weak (GRANHALL and LID-TORSVIK, 1975).

**Wet Meadows Mires.** In the moister regions humus-peat profiles develop, and the maximum snow cover will, in precipitation-rich areas, be above 2 m, giving good insulation in the winter, but may be considerably less in rain-shadow areas. The primary production in these vegetation types is not strongly limited by water shortage, but this may in short periods occur even there (by midday stomatal closing). Nutrient conditions are, however, very important to the production potential. Minerotrophic wet meadows and mires dominated by monocotyledons (particulary *Carex* spp., e.g. *C. nigra*) and also forbs and willows in the low alpine belt, have very high productivity. Maximum apparent photosynthesis is high for many of these monocotyledons and forbs under favorable water and nutrient conditions, sometimes above 20 mg $CO_2$ (g dw)$^{-1}$ $h^{-1}$ (SKRE and WIELGOLASKI, 1975). At Hardangervidda a total primary production of 800–900 g $m^{-2}$ $yr^{-1}$ is measured in relatively eutrophic wet meadows (WIELGOLASKI, 1975a) and willow thickets (KJELVIK and KÄRENLAMPI, 1975). About $^1/_2$ is root production and about 200 g $m^{-2}$ $yr^{-1}$ is bryophyte production. The production to biomass ratio is extremely high in many of the mosses at these wet areas, sometimes above 1.0, while the ratio is often relatively low in vascular plants, 0.3–0.5, due to high below-ground biomass compared to tops, sometimes also to a high percentage of woody species. The top/root ratio is 1:10 to 1:20 in the wet sedge communities, which is considerably lower than in drier types (WIELGOLASKI, 1972). High root biomass and production in the wet communities may in some way be related to the anaerobic conditions found just below the surface. Root hairs are poorly developed in the oxygen-limited atmosphere and this may be compensated for by a higher root production.

Mires and wet meadows are particularly common in the most oceanic parts of Fennoscandian tundra areas. Due to the high biomass, evapotranspiration is relatively high; actual evapotranspiration is about 200 mm during the growing season (SKARTVEIT et al., 1975), which is of the same order as precipitation in the same period. Mires cover nearly 20% of the Hardangervidda, but the majority are oliogotrophic (HESJEDAL, 1975) with an estimated total primary production sometimes as low as 500 g $m^{-2}yr^{-1}$. The soil is more peaty in poorer mires and the nutrient input by horizontal water transport may be low. In ombrotrophic mires or bogs *Rubus chamaemorus*, *Eriophorum vaginatum* and *Empetrum hermaphroditum* are the dominating vascular plants and *Sphagnum* spp. by far the dominating bryophyte. Particularly in wet depressions of ombrotrophic mires, relatively low production is found, estimated at about 250–300 g $m^{-2}yr^{-1}$ in northwestern Sweden (ROSSWALL et al., 1975b).

Nitrogen fixation, particularly by blue-green algae is often significant to the nitrogen input in mire communities and may influence the often high productivity. The annual N-fixation by blue-green algae living in close association with mosses is found to be within the range 0.2–12 g N $m^{-2}$ $yr^{-1}$ from ombrotrophic mires to minerotrophic wet depressions in northern Fennoscandia (GRANHALL and LID-TORSVIK, 1975).

**Birch Forests.** In most parts of Fennoscandia, birch (*Betula pubescens* spp. *tortuosa*) is found bordering both alpine and arctic tundra zones. The subalpine birch belt covers about 10% of the whole Hardangervidda area, and subalpine-subarctic birch forest is the absolutely dominant vegetation form in northern Fennoscandia (HÄMET-AHTI, 1963). Most of the Fennoscandian birch forests are relatively oligotrophic and dry with a ground vegetation often closely related to that of the grass-*Vaccinium* heaths in more extreme climatic conditions with no trees. In northern Fennoscandia reindeer lichens and *Empetrum* often dominate the understory of birch forests. The production is considerably higher in the birch forests than on the alpine heaths, mainly due to the higher temperatures and acclimatization to this temperature. This is clearly seen from maximum photosynthesis values for e.g. *Empetrum hermaphroditum* from alpine and subalpine areas, 2.2 and 6.3 mg $CO_2$ (g dw)$^{-1}$ $h^{-1}$, respectively (SKRE and WIELGOLASKI, 1975). The total primary production in oligotrophic birch forest is found to be 700–800 g $m^{-2}yr^{-1}$ in southern Norway (KJELVIK and KÄRENLAMPI, 1975), but somewhat lower in northern Fennoscandia, and probably higher in moist, eutrophic forests, estimated at 900–1200 g $m^{-2}yr^{-1}$. Total production of ground vegetation is higher than for the birch trees even in relatively dense stands (400–500 and 200–300 g $m^{-2}yr^{-1}$, respectively, in oligotrophic forest). The total live birch biomass in a birch forest with moderate tree cover (basal area 2.5 $m^2$ $ha^{-1}$) in northern Finland (Kevo) is found to be about 1100 g $m^{-2}$ of which about 60 g $m^{-2}$ green leaves (KJELVIK and KÄRENLAMPI, 1975). The basal areas of the trees, however, may be 3–5 times greater in the same area (KÄRENLAMPI and KAUHANEN, 1972), indicating that the maximum tree biomass may be at least as high as in a subalpine birch forest in southern Norway, which from investigations showed nearly 3500 g $m^{-2}$, of which nearly 100 g $m^{-2}$ was green leaves. The overall production to biomass ratio is low in birch forest, about 0.2, due to the high percentage of woody material. The ratio is higher in bryophytes, about 0.7 observed at Hardangervidda, because of relatively rapid growth of the dominating moss species, *Pleurozium schreberi.*

### Primary Production Summary

Temperature is obviously the overall limiting environmental factor for tundra primary production. The snow cover-soil moisture complex, however, strongly modifies the temperature influence, and snow cover also strongly limits the vegetation period together with temperature. With better moisture conditions, soil nutrients are the main growth-limiting factors.

Plant structure in dry conditions is dominated by slow-growing species; lichens, slow-growing bryophytes and woody vascular plants, particularly evergreens. These species can start photosynthesis early in spring and continue the process until late fall, but with low activity in periods with the most unfavorable water conditions during summer. To cryptogams, fall may be the best growth period due to the usually high moisture content. They are also normally adjusted to maximum photosynthesis at low temperatures. Evergreen vascular plants often store reserve food in their leaves in addition to roots and above ground woody parts (BLISS, 1971; JOHANSSON, 1974; A. BERG et al., 1975). The main storage period is in the fall; during summer only small amounts of assimilates are stored in evergreens, but are used for production. New shoot growth in spring generally starts later in evergreen shrubs than in other vascular plants.

On a daily basis, primary production in Fennoscandian tundra areas varies from less than 2 to about 9 g dw $m^{-2}$ when death and decomposition are compensated for. This shows that the photosynthetic efficiency may be relatively high in tundra, up to about 2.5% is calculated for Hardangervidda at 60° N (WIELGOLASKI and KJELVIK, 1975). The total primary production of a larger area is presented in Table 1. The relatively high average value, 350 g dw $m^{-2}yr^{-1}$, is caused by the relatively high percentage of moist, highly productive areas. In drier and oligotrophic parts of Fennoscandian tundra, lower average primary production figures are found.

## Consumers

### Vertebrates

Several vertebrate species are observed in many of the Fennoscandian tundra areas. In alpine regions of southern Norway, 148 species are registered, but only

84 of these can be regarded as breeders (SOLHØY et al., 1975). Quantitatively most important are the reindeer and in some years the small rodents. Lists of vertebrates in northern Fennoscandian tundra areas are given by e.g. ISO-IIVARI (1975) and JONSSON (1975). In the Kevo area 157 vertebrates are observed of which 79 breed within the area, while 240 species are registered in the whole Inari Lapland forest tundra, of which 152 breed (ISO-IIVARI, 1975). In the Torneträsk area in northern Sweden, 200 vertebrate species are observed of which 153 are relatively abundant and 147 breed within the area (JONSSON, 1975).

The 3–4 year cycles in productivity of certain vertebrates are well known in the arctic and alpine circumpolar zone, probably most pronounced in small rodents (KALELA, 1962). The high ecological diversity of Fennoscandia may be one reason why these cycles are not fully synchronized throughout the whole region. In Fennoscandia, the biomass peak in a small rodent cycle is usually first observed in alpine regions of southern Norway. The following year the maximum of the cycle has reached alpine regions in mid-Norway and northern Sweden and in the third year the northeastern parts of Norway and northern Finland (HAGEN, 1956). The reasons for the cyclicity are not fully understood, but may be related to the food availability or stress conditions. Many other vertebrates follow the same cycles as the small rodents, mostly because they prey on small rodents (both mammalian predators and raptorial birds), but similar cyclicity is also found in willow grouse and ptarmigan, probably also in passerines (LIEN et al., 1975).

It is often difficult to place vertebrates in distinct trophic levels. They may be carnivores for part of the year and omnivores the rest. For instance willow grouse and obviously also ptarmigans have a carnivore diet as juveniles and are herbivores as adults (ØSTBYE et al., 1975).

**Herbivores.** Reindeer is the most important herbivore for man in the Fennoscandian tundra areas (Fig. 3). Hunting and trapping of wild reindeer must have been of great interest already in pre-historic times as shown by the several pitfalls

Fig. 3. Reindeer are the most important herbivore mammal in Fennoscandian tundra ecosystems, domestic animals in the north and wild populations in the south

found along migration routes in various parts of the Fennoscandian tundra. The history of reindeer obviously goes back to the last ice age with several long-term fluctuations in population up to recent times (ØSTBYE et al., 1975). The wild reindeer population in southern Fennoscandia decreased so far by the end of the 18th century that domestic reindeer were introduced. They mixed, however, with the wild animals particularly during and after the second world war (GAARE and SKOGLAND, 1975). Today wild reindeer are only found in the southernmost alpine tundra regions of Fennoscandia, while domestic herds are still important in northern parts, and no wild reindeer live in that area.

The herbivory of reindeer, and often even more the trampling, have great influence on the vegetation in some Fennoscandian tundra areas. It is estimated that the daily plant dry weight intake per reindeer is 1500–2000 g (GAARE, pers. comm.), highest in summer. The annual birch consumption per domestic reindeer in northern Finland is of the order of 425 kg dry matter (HAUKIOJA and HEINO, 1974). At Hardangervidda, the total annual consumption of the herd within the home range is approximately 7000 tons dry weight when based on a population of 10000 animals (making a total body fresh weight of 550 t, or dry weight about 155 t), as calculated by GAARE and SKOGLAND (1975). They have estimated the trampling effect to be more than twice that of grazing at Hardangervidda, while a spillage factor of about 10 is estimated for reindeer in more continental areas of Norway with more reindeer lichens. When total primary production of Hardangervidda (about 10000 $km^2$) is estimated at 3.5 mill. t dw (Table 1), intake and trampling by reindeer make up about 0.6%. In an average year, however, the reindeer herd utilizes less than one third of the total Hardangervidda (ØSTBYE et al., 1975), and within the home-range area consumption and destruction of plants may be about 2%. The density for the gross area may vary from 1.0 to 1.7 ind. $km^{-2}$, but is about 4 ind. $km^{-2}$ for the net area. A still smaller area is utilized by the reindeer during winter as they concentrate for about 150 days in eastern parts of the plateau with a high percentage of lichen heaths. The density on the winter range is estimated at 6.4 ind. $km^{-2}$ (ØSTBYE et al., 1975). Food is available to the reindeer on only about 1/3 of the winter range, mostly on wind-blown lichen heaths. Primary production on the utilized winter range (about 500 $km^2$) is about 100000 tons. During winter, the reindeer herds consume and destroy by trampling 5–10% of the available annual primary production in this area. In northern Finland a reindeer lichen stand is renewed from bare soil after 5 years (KALLIO, 1975), the lichen thalli of *Cladonia alpestris*, however, being only 1 cm in diameter. There seems to be good agreement between spring herd size of 10000 reindeer at Hardangervidda and the critical winter food intake and destruction.

Summer food probably seldom limits the reindeer herd sizes in Fennoscandian tundra. The animals, however, then prefer plant species and developmental stages with a high sugar content and/or high mineral content (cf. WIELGOLASKI, 1975b; WIELGOLASKI et al., 1975), except nitrogen-fixing lichens which are avoided (KALLIO, 1975). The herds therefore move to south exposed slopes in spring and early summer, while northern exposures are favored in late summer and fall (ØSTBYE et al., 1975). The use of different plant communities at different times of the year is shown in Fig. 2. Grass heaths and tall forb meadows are obviously preferred in the summer but mires are also often visited. Graminoids, forbs and leaves of shrubs (*Betula nana* and *Salix* spp. particularly) are the main plant groups consumed in mires, also *Equisetum* with very high mineral content. In years with high mushroom production (warm and moist), their nutrient-rich fruiting bodies (WIELGOLASKI et al., 1975) are important to reindeer diet in August (KALLIO, 1975). Lichens are consumed by reindeer even in summer, but comprise a lower percentage of the total intake than in winter; they have a high digestibility, about 70% of dry matter, for reindeer, which therefore utilize the energy in lichens far better than, for instance, sheep (ØSTBYE et al., 1975). Although lichens (mainly *Cladonia* spp. of the Cladina group and yellow *Cetraria* spp.) are the main reindeer diet in winter, graminoids and woody plants of the wind-blown communities make up about 40% of the intake even in this period (GAARE and SKOGLAND, 1975), when based on rumen samples.

Average density of small rodents is estimated to 18000 ind. $km^{-2}$ at Hardangervidda in peak years (ØSTBYE et al., 1975) and the total fresh weight is about 5500 t for the whole area, about 10 times higher than for reindeer (or about 1550 t dw). The annual small rodent plant intake may be above 10 t $km^{-2}$ in these years. This means about 3% of the average annual plant production at Hardangervidda (Table 1), but more than 10% of preferred communities (WIELGOLASKI, unpubl.), and approximately the same percentage of the available plant material during winter. In a minimum year, densities of small rodents per $km^2$ are estimated to be within the range 0–300, and the plant intake is then negligible.

*Microtus* spp. usually inhabit relatively rich and wet places during summer, while in winter a migration to drier habitats is assumed (HAGEN et al., 1975). The main diet of above-ground material is sedges, grasses, forbs and dwarf *Salix* (particularly *S. herbacea*), as well as roots (ØSTBYE et al., 1975). *Clethrionomys* spp. are most abundant in drier communities rich in boulders and stones. They seem to prefer the above-ground material of forbs and ferns, as well as roots and seeds, and also to some extent invertebrates. The greatest number of *Lemmus* at Hardangervidda are found at oligotrophic dry heaths and pioneer communities (ØSTBYE et al., 1975), but they also live in relatively wet communities in all Fennoscandian tundra areas. Their main diet seems to be mosses, sedges and grasses. In addition to direct consumption all small rodents also clip a high amount of plant material e.g. for nest building, and disturb the ground with their burrowing activities. They cause local accumulation of nutrients both directly by their feces and by liberation of minerals from the clipped material. This may be important to the nutrient cycling of these relatively simple ecosystems (SKRE et al., 1975).

Other herbivorous mammals in Fennoscandian tundra systems are hare (*Lepidus timidus*), which browse on shrubs such as *Salix* spp., *Betula* spp., *Empetrum hermaphroditum* and *Juniperus communis*, and in subalpine and subarctic regions moose (*Alces alces*). Herbivorous birds of some importance are the willow grouse (*Lagopus lagopus*), which is mostly resident in the birch forests and in low-mid alpine regions (HAFTORN, 1971) and, in higher mountain regions particularly during summer, ptarmigans (*Lagopus mutus*), which migrate to valleys in the lower mountain regions and birch forests in winter.

**Carnivores.** The larger carnivore vertebrates play quantitatively a minor role today in Fennoscandian tundra areas. The wolf *(Canis lupus)* was earlier the major predator particularly on reindeer. Today only a few animals are left in the far north of Fennoscandia. The wolverine *(Gulo gulo)* is still found in the whole Fennoscandian region, but is rare, so is the lynx *(Lynx lynx)*, which is usually not observed above the tree line in Fennoscandia. Brown bears were earlier common in forest areas of many places in Fennoscandia. Sometimes they hunted in the tundra regions as well. Today only a few brown bears are left. Foxes *(Vulpes vulpes)* and arctic foxes *(Alopex lagopus)*, as well as weasels (*Mustela erminea* and other species) are found in the whole of Fennoscandia, particularly in years with small rodent maxima (ØSTBYE et al., 1975; KALLIO, 1975), and are obviously the most important mammalian predators.

Several bird species are carnivorous in Fennoscandian tundra. Invertebrates are the most important diet of most passerine birds, waders and plovers (ØSTBYE et al., 1975). Passerines make up the majority of bird biomass in summer in large areas of Fennoscandian tundra; densities range from 24–100 breeding pairs per $km^2$ (LIEN et al., 1975) or 4–15 kg fresh weight $km^{-2}$. Densities also for other birds in various parts of Fennoscandian birch forests are given by HOGSTAD (1975). In years with peaks of small rodents, maximum densities are also found in raptor populations (e.g. owls, falcons, buzzards and long-tailed jaegers), but they generally do not become very abundant, for each species usually less than one pair $km^{-2}$ ØSTBYE et al., 1975).

## Invertebrates

The most abundant groups vary with the plant communities and thus also with environmental factors, e.g. moisture, nutrient conditions and temperature. There seems also to be some variation between southern and northern Fennoscandia; Enchytraeidae are important in wet soils in south, but may be scanty in north both in wet and drier soil. In all tundra communities in the whole of Fennoscandia, Acari and Collembola dominate in the soil. In the vegetation layer, Hemiptera, Coleoptera, Diptera, Hymenoptera and Lepidoptera may be important invertebrate groups in Fennoscandian tundra areas, particularly in birch forest tundra. In the northern birch forests, the geometrid *Oporinia autumnata* has for instance periodic heavy outbreaks. All together several hundred species of invertebrates are observed in the Fennoscandian tundra areas.

A characteristic feature of the invertebrate composition in alpine southern Norway is the absence of Isopoda, Diplopoda, Chilopoda and Formicidae which are important in soil forming processes in the surrounding lowland (SOLHØY et al., 1975). There is also a scarcity of Lumbricidae and terrestrial Gastropoda in alpine Fennoscandian tundra compared to the adjacent lowlands. No species of these groups are found in late-melting snow beds or in dry lichen heaths, particularly because of lack of calcium (KAURI, pers. comm.).

Fennoscandian tundra invertebrates (like the plants mentioned earlier) need a longer time to complete their life cycle than the same groups in lowland areas and further south. In tundra, the number of generations per year is reduced in Aphididae (TAMBS-LYCHE, 1975), and a life cycle may last for more years than in

Table 2. Estimated life cycle of some invertebrate taxa from Hardangervidda

| Taxon | Duration of life cycle (years) | Ref. |
|---|---|---|
| *Pardosa palustris* (Araneae) | 3 | STEIGEN (1975) |
| *Mitopus morio* (Opiliones) | 1–2 | KAURI (unpublished) |
| *Oribatei* | 2–4 | SOLHØY (1975b) |
| *Collembola* | 1–2 | FJELLBERG (1975) |
| *Amara alpina* (Coleoptera) | 3 | SOLHØY (unpublished) |
| *A. praetermissa* (Coleoptera) | 3 | SOLHØY (unpublished) |
| *Melasoma collaris* (Coleoptera) | 1 | HÅGVAR and ØSTBYE(1975) |
| *Anapothrips secticornis* (Thysanoptera) | 1 | KJELLSEN (1975) |
| *Acyrthosiphon malvae* (Aphididae) | 1/3 | TAMBS-LYCHE (1975) |
| *Tipula excisa* (Tipulidae) | 2 | HOFSVANG (1975) |
| *Vitrina pellucida* (Gastropoda) | 3–5 | SOLHØY (unpublished) |

lowland and more southern areas for several groups, Table 2 and e.g. old observations on Lepidoptera (SANDBERG, 1883). Generally, the invertebrate abundance is lowest in spring and highest in fall in Fennoscandian tundra ecosystems, but great fluctuations occur between years. In dry communities, the populations are reduced in drought periods during summer, particularly in Enchytraeidae (SOLHØY, 1975a). The density of invertebrates may also change with the lemming cycle, both because of their herbivory, nesting, feces and decomposition. Similarly, there may be a reduction in soil invertebrate density in Fennoscandian tundra meadows after strong sheep grazing which has reduced the root biomass.

**Herbivores.** Invertebrate consumption of live plant biomass is low in most years in Fennoscandian tundra regions in field and bottom layers (ØSTBYE, et al., 1975). Less than 0.1 g $m^{-2}$ plant material (dw) is estimated to be consumed by invertebrates in alpine southern Norway (SOLHØY, pers. comm.), but the chrysomelid beetle *Melasoma collaris* may remove a substantial part of *Salix herbacea* leaves in snowbeds (HÅGVAR, 1975). Invertebrate consumption may, however, be considerably higher in birch forest tundra. In a "normal" year, about 4% of the birch foliage may be consumed (KALLIO, 1975) and often the percentage is much higher, 15% is mentioned by HAUKIOJA and KOPONEN (1975). These figures do not include losses caused by reactions of the plant to herbivory. Annual variations in invertebrate herbivory are considerable. In northern Fennoscandia, *Oporinia autumnata* may in some years, usually in 9–10 years cycles (TENOW, 1972), defoliate vast areas of birch (HAUKIOJA and KOPONEN, 1975; KALLIO and LEHTONEN, 1975) and is known to affect even the dwarf shrubs, principally *Vaccinium uliginosum* (ROSSWALL et al., 1975b). In other, more "normal" years, the most important leaf grazers, especially of birch, are Symphyta larvae, particularly those of the sawfly species *Dineura viridiorsata* (HAUKIOJA, 1973; KOPONEN, 1973b; HAUKIOJA and KOPONEN, 1975). Mining lepidopteran species, e.g. *Eriocrania* spp. are in some years more important to the vegetation than many of the vertebrates (KOPONEN, 1973a; HAUKIOJA and KOPONEN, 1975). In the northern Fennoscandian birch forest fall-inhabiting Diptera larvae of Cecidomyidae quickly build up large numbers and biomass and may cause a birch leaf material loss of nearly 5 g $m^{-2}$ during one season or about half the total herbivory in a normal year (HAUKIOJA and KOPONEN, 1975).

Generally, the invertebrate herbivory of live plant material is represented chiefly by Lepidoptera larvae (mining and often also free-living), sapsucking Hemiptera (including Homoptera), and Thysanoptera and Coleoptera adults and larvae; in many cases and particularly in birch forest tundra, Diptera larvae and Hymenoptera larvae are very important herbivores (HAUKIOJA and KOPONEN, 1975). The total annual secondary production sum of all these invertebrate groups is estimated to be about 1 g dw $m^{-2}$ in birch forests in northern Fennoscandia, but is considerably lower in treeless tundra areas. In the field layer, at least of northern Fennoscandian forest tundra and alpine heaths, sap-sucking Hemiptera (including Homoptera) domitate in abundance, while Symphyta larvae were sometimes the herbivore invertebrate group with the highest biomass (KOPONEN and OJALA, 1974).

**Carnivores.** Spiders, carabid- and staphylenid beetles and some chironomid larvae are the most important "macropredators" consuming a variety of other invertebrates from springtails and enchytraeids to large craneflies (ØSTBYE et al., 1975). Up to 20% of the Acari biomass may also be included in the predator group because of presence of Prostigmata and Mesostigmata (SOLHØY, unpubl.), but the amount is doubtful. In birch forests Araneae and parasitic Hymenoptera are the most important carnivores (HAUKIOJA and KOPONEN, 1975). Micropredators (the Acari and part of freeliving Nematoda) mostly consume Enchytraeidae, Collembola, Acari and Nematoda. Omnivorous Nematoda seem to make up 40% of the nematode biomass in mires in northern Sweden (ROSSWALL et al., 1975b) and parasitic Nematoda are common in the invertebrates.

Of the Araneae species found in alpine southern Norway, 38% can be classified as arctic-alpine, 11% as subalpine and as much as 51% as lowland species, invading the alpine area from surrounding lowland districts (KAURI, pers. comm.). Input from surrounding districts may also be considerable for many winged insects in Fennoscandian tundra.

**Decomposers.** Most of the soil and litter invertebrate biomass in Fennoscandia belongs to species which feed mainly on organic matter and microorganisms. Often more than 90% of the soil invertebrate biomass consists of decomposer organisms (ØSTBYE et al., 1975), about half of which are saprovores and fungivores. The invertebrate activitiy is highest in the litter layer and the upper few cm in the soil. Nematodes are often the most abundant soil faunal group, in northwestern Sweden e.g. 4 mill. ind. $m^{-2}$ (LAGERLØF and MAGNUSSON, 1975) and an estimated biomass of 0.2–0.3 g dw $m^{-2}$ in mires. Rotifers are also found in high numbers in mires (about 900000 ind. $m^{-2}$ or somewhat below 0.1 g dw $m^{-2}$ in the Swedish mire). These groups are not studied in southern Norway. However, high densities are there often found for Enchytraeidae, Acari and Collembola in the litter and soil layers and more locally also some Diptera larvae. Collembola show highest densities in mesic areas (dry meadows), Acari often high values also in drier (lichen heath) areas (FJELLBERG, 1975; SOLHØY, 1975b), but the diversity of Collembola is highest in mires.

The density of Acari may exceed 200000 ind. $m^{-2}$ in the mesic areas at Hardangervidda, which is comparable to the density on grassland at Spitsbergen (BENGTSON et al., 1974) and in mires in northwestern Sweden (LOHM et al., 1973). The average biomass is found to be about 0.5 g dw $m^{-2}$ (max 2 g $m^{-2}$) in mesic eutrophic alpine meadows (SOLHØY et al., 1975), while 0.5 g dw $m^{-2}$ is the maximum biomass found in dry, oligotrophic heaths. There may also be a tendency to somewhat smaller species and/or individuals at these oligotrophic heaths than in wetter and richer habitats. Of the total acarids, about 80% may be classified as feeding on bacteria and fungi.

Although the density of Collembola is lowest in wet areas of the southern Norwegian tundra (about 60000 ind. $m^{-2}$), the biomass is equal or even higher than in drier communities (about 0.3 g dw $m^{-2}$) due to a greater percentage of larger species (FJELLBERG, 1975). The density of Collembola in wet meadows in alpine southern Norway is of the same order as found in tundra mires in Sweden (LOHM et al., 1973), while considerably higher densities (about 250000 ind. $m^{-2}$) are observed both in mires and grassland communities at Spitsbergen (BENGTSON et al., 1974).

The Enchytraeidae in the wettest areas in southern Norway reach a density and biomass averaging 38000 ind. $m^{-2}$ and 0.9 g dw $m^{-2}$ (SOLHØY, 1975a), while very few are found in dry lichen heaths. In northern Sweden, however, the population numbers seem to bee low for Enchytraeidae even in mires (cf. LUNDKVIST, 1974). One reason may be that postglacial colonization is still going on within the region (JONSSON, 1975).

Estimated annual production of soil invertebrates in Fennoscandian alpine tundra areas, about 0.3 g dw $m^{-2}$ in lichen heaths to about 0.6 g dw in eutrophic meadows (SOLHØY, unpubl.), or averagely 0.5 g dw $m^{-2}$ including Nematoda, is of the order 0.10–0.15% of the primary production and about 0.2% of the microbial production. Therefore, soil invertebrates are likely to play only a minor direct role in the decomposition processes in these ecosystems compared to microorganisms.

Both on biomass and production basis, however, invertebrates make up a larger part of the Fennoscandian tundra ecosystems than the vertebrates. As an average for Hardangervidda, we may use the biomass sum 850 mg dw $m^{-2}$ for the soil litter and vegetation invertebrate fauna sampled (KAURI, pers. comm.), and about 2 g dw $m^{-2}$ when Nematoda and Lumbricidae are included as well. For the whole mountain plateau, this makes about 20000 t dry weight or about 60000 t fresh weight which means about 10 times the biomass of small rodents in a peak year and 100 times the reindeer biomass. If the soil invertebrate biomass in birch forest tundra [above 1 g dw $m^{-2}$ may be estimated from data given by KALLIO (1975) from northern Finland] is added to the invertebrate biomass in the vegetation layer, even higher values may be found from the birch forest than estimated from the Hardangervidda alpine plateau.

## Microorganisms

Calculations of bacterial and fungal biomass and production are very uncertain and are based on several assumptions (CLARHOLM et al., 1975; HANSSEN and

Table 3. Decomposer production in $g\,m^{-2}\,yr^{-1}$ in Fennoscandian tundra areas. Based upon CLARHOLM et al. (1975); HANSSEN and GOKSØYR (1975); GOKSØYR (1975); ROSSWALL et al. (1975b)

| | Lichen heath | Dry meadow | Wet meadow | Oligotrophic mire |
|---|---|---|---|---|
| Fungi | 70 | 190 | 280 | 80 |
| Bacteria | 50 | 130 | 190 | 140 |
| Total | 120 | 320 | 465 | 220 |
| Prim. prod. | 270 | 530 | 830 | 160 |
| Producer/ Microorg. | 2.3 | 1.7 | 1.8 | 0.7 |

GOKSØYR, 1975). By the methods used, higher bacterial biomass was generally found (direct count) for the active soil layers than fungal biomass for different tundra communities in southern Norway, while the fungal biomass was highest in oligotrophic mires in northwestern Sweden. The annual production estimates of fungi and bacteria are given for some communities in Table 3. At the dry lichen heath on Hardangervidda, the total primary production is more than twice as high as the microbial production, while for the Swedish oligotrophic mire the situation is reversed. It seems as if the ratio between primary production and microbial production decreases as the wetness of a site increases. Moisture is, however, not the only factor influencing production rates, the production of both plants and microorganisms increases from oligotrophic to eutrophic sites. The main factor limiting bacterial production seems to be soil moisture, while nutrient conditions may be even more important to fungal production.

**Decomposition.** Most of the plant material produced goes directly through the decomposer chain (see Section on energy-flow of this paper); on the whole, considerably more than 90% (cf. GOKSØYR, 1975), except in years with small rodent or defoliating insect maxima. The decay rates vary strongly with plant species, plant compartment and environmental factors. There is an approximately 40% first-year decomposition as weight decrease of above-ground herbaceous nutrient-rich material, but only about 10% decomposition of lichens, and most bryophytes. The rate of decomposition decreases in the order forbs and graminoids > shrubs > bryophytes and lichens. For any one of these groups, moisture seems to be the most important environmental factor limiting decomposition, and low moisture often restricts the decay rate, especially of surface litter. Wind increases evaporation and causes rapid drying of litter. When litter is incorporated into cryptogam layers or peat in the mires (ROSSWALL et al., 1975b), mainly as a result of bryophyte growth, but also because of water movement in the mires, the moisture conditions become optimal in the upper peat layers. Below-ground material of mires is found to have a slower decay rate than above-ground parts, due to too high moisture and subsequent anaerobic conditions (ROSSWALL et al., 1975a). This contributes to the accumulation of organic matter found in both ombrotrophic and minerotrophic tundra mires in Fennoscandia. In southern Norway, a peat accumulation of 0.1 mm per year was estimated from radiocarbon dating in a wet, relatively eutrophic meadow. The slow decomposition also resulted in a high dead below-ground standing crop, more than 1300 g $m^{-2}$, which was separated from a similarly high living below-ground biomass by $^{14}C$ incorporation. Annual increments ten times higher were observed in an oligotrophic mire at Abisko. In addition to slow decomposition of below-ground material, this was also caused by only 5–10% first-year weight loss of dead bryophytes.

The rapid first-year weight loss often observed in vascular plant leaf litter is only partly caused directly by rapid microbial breakdown of easily metabolizable substances like amino acids and soluble carbohydrates. Leaching probably accounts for 40–50% of the weight loss in the first year after death (ROSSWALL et al., 1975a). Precipitation influences the leaching rates, and the high precipitation in fall may therefore be one reason for the particularly rapid weight losses often found in September in the Fennoscandian tundra systems. The higher soil-moisture content in fall, however, will also directly affect the microbial activity. In the summer, sufficient moisture for decomposition in drier communities occurs probably only in connection with rain, fog and dew formation. It is possible that the latter is the most important moisture factor for decomposition of surface litter during summer in many Fennoscandian tundra ecosystems (ROSSWALL et al., 1975b).

Temperatures are probably never over-optimal for decomposition in tundra areas if moisture conditions are favorable. Soil respiration shows, however, that microbial activity also continues at low temperatures, at least down to $-5°$ C (ROSSWALL et al., 1975a; SVENSSON et al., 1975). In soil with permafrost, the low temperatures down the profile even in summer will greatly decrease the decay rate compared to the rates at higher temperatures just below surface (ROSSWALL et al., 1975b). Particularly low soil respiration and microbial activity was observed at an ombrotrophic mire with permafrost in northwestern Sweden, even lower than at a poor, dry alpine lichen heath in southern Norway (SVENSSON et al., 1975). In addition to the temperature effect, this is probably also caused by anaerobic conditions in the peat, shown e.g. by considerable methane production which could account for up to 45% of the carbon lost to the atmosphere.

The influence of temperature is also shown by differences in decay rates of the same parent material in similar plant communities in alpine and subalpine regions. Decomposition of cellulose strips was found to be several times stronger in oligotrophic birch forests than in poor lichen heaths with about 3° C lower annual mean temperature (extremes 85% and 1% annual decomposition) (B. BERG et al., 1975). Both communities are dry, but the birch forest slightly moister. The decay is, however, also considerably higher in the subalpine birch forest than in a moister oligotrophic alpine snowbed for the same types of material (ROSSWALL et al., 1975a).

The effect of soil nutrients on decay rate is indicated by more rapid cellulose and barley straw decomposition in eutrophic than in oligotrophic communities. This is shown, for example, by about twice as high annual barley straw decomposition in an alpine dry meadow as in the oligotrophic snowbed (ROSSWALL et al., 1975a). The same indication is given by higher soil respiration rates at eutrophic than at oligotrophic tundra communities (SVENSSON et al., 1975).

## Conclusions

**Energy Flow.** Primary production in Fennoscandian tundra areas utilizes only about 0.7% of the global radiation within the wavelengths 400–700 nm (45% of the total global radiation) in lichen heaths, but about 2.4% in willow thickets. This results annually in from about $1000 \cdot 10^6$ to $4000 \cdot 10^6$ kcal in dry plant material produced per $km^2$ (WIELGOLASKI and KJELVIK, 1975). Values considerably lower than 1000 kcal $km^{-2}$ are, however, produced in snowbeds and under other extreme condition. An average of about $1400 \cdot 10^6$ kcal $km^{-2}$ is fixed annually in plants from different communities at Hardangervidda in southern Norway. Of this only an average of about $3 \cdot 10^6$ kcal $km^{-2}$ is consumed by reindeer on the Hardangervidda gross area basis (or about $10 \cdot 10^6$ kcal $km^{-2}$ on the reindeer home range basis). This intake is used for maintenance of about $0.075 \cdot 10^6$ kcal $km^{-2}$ assimilated in the reindeer population (including $0.025 \cdot 10^6$ kcal $km^{-2}$ new production) (GAARE and SKOGLAND, 1975), but more than twice that much plant material is spoiled by reindeer trampling and digging. On the other hand, 5–10% of annual primary production of plants available as winter forage is consumed or destroyed by reindeer. The production of these species is thus generally limiting the reindeer population.

In a maximum small rodent year, these animals (excluding lemmings) at Hardangervidda may consume in the order of $35 \cdot 10^6$ kcal $km^{-2}$ plant material per year (cf. HAGEN et al., 1975), for a small rodent biomass corresponding to about $0.75 \cdot 10^6$ kcal $km^{-2}$ (ØSTBYE, pers. comm.), i.e. 2–3% of the plant material produced, but locally the intake may be more than 10% of the plant material. In a minimum year, the intake by these animals is nearly negligible, and so is the herbivory by other vertebrates as well. The carnivory is also very limited in vertebrates in Fennoscandian tundra areas. As a rough estimate, probably less than 1% of the herbivores are consumed by predators.

Estimates of consumption by invertebrates in Fennoscandian tundra ecosystems also have to be very rough. Average total intake in an alpine region for all groups of invertebrates may also be about $35 \cdot 10^6$ kcal $km^{-2}$ $yr^{-1}$, Lumbricidae included, approximately twice that much in birch forest tundra. Most of this is consumption of detritus, although a relatively high consumption of live plant material is observed in some cases. In north Fennoscandian birch forest, for instance $10–35 \cdot 10^6$ kcal $km^{-2}yr^{-1}$ is consumed from the birch leaf biomass (HAUKIOJA and KOPONEN, 1975). At alpine Hardangervidda, only an average of about $0.3 \cdot 10^6$ kcal $km^{-2}$ live plant material seems to be consumed per year by an invertebrate biomass corresponding to approximately $0.15 \cdot 10^6$ kcal $km^{-2}$, when based on rough estimates from a few communities (SOLHØY, unpubl.). Mainly carnivorous invertebrates may consume somewhat more, i.e. about $0.5 \cdot 10^6$ kcal $km^{-2}yr^{-1}$ by a biomass corresponding to the same order of energy. Both consumption and biomass seem to be highest for mainly fungivorous organisms in the alpine region (consumption roughly of the order $20 \cdot 10^6$ kcal $km^{-2}yr^{-1}$ and energy in the biomass ten times lower, when based on unpublished data by SOLHØY for a few communities at Hardangervidda). The consumption by mainly saprovorous invertebrates is probably lower, although it is very difficult to classify Nematoda, Oribatei, Collembola, Enchytraeidae, Lumbricidae, Diptera larvae etc. either as fungivores or saprovores. The consumption of mainly saprovores may be approximately $3 \cdot 10^6$ kcal $km^{-2}yr^{-1}$ (biomass corresponding to $10^6$ kcal $km^{-2}$) in the alpine region of Hardangervidda when Lumbricidae is excluded, and sometimes probably more than five times higher than that when the latter group is included.

The energy in the biomass of the total invertebrate population may thus be of the order of 8 to $12 \cdot 10^6$ kcal $km^{-2}$ on an average in an alpine region (or about $4 \cdot 10^6$ kcal $km^{-2}$ excluding Lumbricidae and Nematoda), and may even be somewhat higher in Fennoscandian birch forest tundra. The annual production is generally somewhat lower than the biomass, and may be estimated at about $3.5 \cdot 10^6$ kcal $km^{-2}$ on an average for alpine Hardangervidda, excluding production of Lumbricidae (KAURI, pers. comm.).

Most of the energy fixed in plants and animals is used by microbes, normally more than 90%, and the decomposer chain is therefore more important than the grazing chain in the energy context. It is estimated that fungi often "consume" more than three times as much dead plants as the bacteria do in Fennoscandian tundra areas. Fungi also receive considerable amounts of energy from amorphic humus (organic matter), and at Hardangervidda total annual "consumption" by fungi is estimated at nearly $1500 \cdot 10^6$ kcal $km^{-2}$ of which nearly 1000 kcal $km^{-2}yr^{-1}$ is from dead plants. Energy input to bacteria is highest from humus and fungi, about two to three times higher than the input from dead plant material, resulting in a total energy input of the same order to bacteria as to fungi (GOKSØYR, 1975). Small amounts of energy are also supplied to the microbes from dead animals.

The biomass of microbes is difficult to estimate. By using the calculations in HANSSEN and GOKSØYR (1975) and in CLARHOLM et al. (1975), the average fungal biomass corresponds to about $500 \cdot 10^6$ kcal $km^{-2}$ for alpine southern Norway, but only half this value in some tundra regions of northern Sweden. Both fungal and bacterial biomass vary strongly with soil nutrients and soil moisture, and decrease rapidly with soil depth. In some cases higher biomass is estimated for bacteria (from direct counts which probably cause overestimates) than for fungi, while in other cases it is lower than for fungi, e.g. in some ombrotrophic Fennoscandian tundra mires. Rough estimates of microbial production show that the annual fungal production to biomass ratio may be somewhat above unity (HANSSEN and GOKSØYR, 1975), while it was more than 6 for bacteria in a mire in northern Sweden (ROSSWALL, 1975). The average production of fungi and bacteria together may be estimated to be of the order $1000–1500 \cdot 10^6$ kcal $km^{-2}$ $yr^{-1}$, cf. Table 3.

**Nutrients.** Both the energy content of plants (WIELGOLASKI and KJELVIK, 1975) and the content of some nutrients (N, P, K) (DOWDING et al., 1976) seem to increase with latitude, probably both because of the longer days and high incoming radiation and higher net assimilation at lower temperatures. Annual uptake of minerals compared to accumulation in plants is higher in Fennoscandian tundra than in forest areas of the world (WIELGOLASKI et al., 1975). Even on an area basis, the uptake may be somewhat higher (particularly N) than in forests and also than in grassland. Accumulation of minerals is usually relatively low in above-ground plant parts in the Fennoscandian tundra (except in birch forest areas), but the below-ground accumulation may be considerable due to the high biomass and slow decomposition rate. High amounts of N, P and Ca are found in green material particularly in spring. The N-content of birch leaf buds may in spring often be as high as in N-fixing legumes. Similarly very high N-content and other minerals are found in mushrooms (WIELGOLASKI et al., 1975). This is very important to the reindeer grazing. The consumption by reindeer as well as by other animals is selective. Thus, the functioning of the tundra ecosystems is not only dependent on the energy content, as expressed in calories, but also depends on the qualitative differences in composition of plants (KALLIO, 1975; WIELGOLASKI, 1975b).

Output of minerals in aqueous solution may be considerable for many of the Fennoscandian terrestrial tundra areas, as shown by chemical analyses of waters (VEUM, 1973). Particularly high amounts are leached out from eutrophic areas, and the nutrients are often deposited again along rivers and in flooded, lower elevated areas. This is particularly important if these are oligotrophic. High production may then occur because of availability of both water and nutrients. The output of nutrients is smaller from areas with permafrost due to its limitation of leaching.

Organic material from terrestrial organisms is also brought into the aquatic systems of tundra areas, and both nutrients and particularly the carbon in these organisms are of great importance to the aquatic productivity (LARSSON and TANGEN, 1975). This strongly reflects the interaction between terrestrial and aquatic systems, and indicates that nutrients may leave a system also in other ways than by leaching (see also ØSTBYE et al., 1975).

Compared to temperate ecosystems, however, the Fennoscandian tundra systems are usually conservative and losses of nutrients smaller, e.g. through leaching and, for nitrogen, through denitrification which is generally negligible. Many of the systems are accumulative and need an annual input of nutrients in order to sustain primary production.

Input of minerals to a terrestrial system takes place both through release from rocks, from decomposition of organic material and by precipitation. In addition, nutrients are supplied by animals e.g. by deposition of feces from mammals and birds, showing the great importance of all trophic levels in nutrient cycling, although a relatively small percentage of the energy goes through the grazing chain. For nitrogen, input by fixation in lichens and by microorganisms may also be significant. In ombrotrophic mires and at oligotrophic ridges, nutrient input through precipitation (KALLIO and VEUM, 1975) and migratory animals is particularly important because of low input from other sources, in mires probably also input through insects hatched in the wettest parts. Nutrient additions in the form of wet or dry deposition are important particularly in tundra areas with slow mineralization of organic matter. However, nutrients in eutrophic, mesic areas may be relatively rapidly released from organic material even in tundra (although slower than at higher temperatures), particularly the first year after death (DOWDING, 1974; DOWDING et al., 1976), and the recycling is also often rapid. Probably the release from newly dead organic material is the main source of nutrients for new growth in Fennoscandian tundra areas. In the case of nitrogen, it is also probable that at least bryophytes take up ammonia directly from the atmosphere.

Nitrogen is accumulated to a higher degree than other minerals in perennial tundra live plant parts (RODIN and BAZILEVICH, 1967); up to 20 t $km^{-2}$ may be estimated accumulated in live vegetation at Fennoscandian eutrophic wet areas (due to the high live root biomass) and in birch forest tundra (WIELGOLASKI et al., 1975). In mires even more nitrogen is accumulated in dead material than in live material, due to peat accumulation. The annual nitrogen uptake in plants is of the same order as the amounts accumulated in live and also in dead plants in dry areas, while the uptake is lower than the accumulation in wet areas.

The input of nitrogen through biological fixation and through precipitation seems generally to be of the same order in most communities in Fennoscandian tundra areas, both mostly varying from 0.10 to 0.25 t $km^{-2}yr^{-1}$. Fixation of N by some lichen species is most important in some dry areas. The lichen species *Stereocaulon paschale* in the driest areas and *Nephroma arcticum* and *Peltigera aphthosa* under somewhat moister conditions may fix 3–4 t N $km^{-2}yr^{-1}$ at 100% cover of the area, but in the field N-fixation by lichens is found to be lower, up to about 0.4 t N $km^{-2}yr^{-1}$ (KALLIO and KALLIO, 1975). In some of the mesic Fennoscandian tundra communities, bacterial N-fixation is relatively high (up to 0.1–0.2 t N $km^{-2}yr^{-1}$), mostly caused by free-living organisms, particularly aerobic bacteria, and is seldom dominated by symbiotic fixation in legumes (GRANHALL and LID-TORSVIK, 1975). Frequently, however, all types of nitrofying bacteria are absent in tundra soils (DUNICAN and ROSSWALL, 1974). Free-living blue-green algae, particularly in association with mosses, are the most important N-fixers in wet Fennoscandian tundra areas; in ombrotrophic mires this fixation is of the order 0.2 t N $km^{-2}yr^{-1}$, but it may be 50 times higher in minerotrophic mires (BASILIER, 1974; GRANHALL and SELANDER, 1973; KALLIO and KALLIO, 1975).

A nitrogen budget for an ombrotrophic subarctic mire in Sweden shows that while ca 1 g N $m^{-2}yr^{-1}$ was made available to the plants through mineralization of organic matter and litter, 0.2 g was added through biological nitrogen fixation, 0.1 g in precipitation and 0.7 g taken up by the bryophytes from the atmosphere and added through dry deposition (ROSSWALL et al., 1975b). Thus, as much nitrogen is added to the system annually as is mineralized. Ammonia concentrations in soil of tundra areas are low, and nitrate is rarely present, as most of what is made available is rapidly immobilized by plant roots and microorganisms.

Primary production, particularly of above-ground parts, increased strongly with mixed fertilization in relatively eutrophic dry meadow communities in alpine southern Norway (WIELGOLASKI, 1975a), but smaller influence was observed for supply by only nitrogen or phosphorus. Increased primary production by higher nutrient level of the soil is of course also of great importance to the other trophic levels. The input and cycling of nutrients is thus very important to the production of Fennoscandian tundra ecosystems together with soil moisture and snow cover, although the over-all limiting factor to production in these areas is temperature.

*Acknowledgements.* The author wishes to express his sincere thanks to the other members of the editorial board, Professors P. KALLIO and H. KAURI and Drs. T. ROSSWALL and E. ØSTBYE for their careful reading of the manuscript and for corrections proposed. Dr. T. SOLHØY is also thanked for valuable criticism of the paper.

## References


BASILIER, K.: Investigations on nitrogen fixation in moss communities. Progress report 1973. Swed. IBP Tundra Biome Project Tech. Rep. **16**, 83–97 (1974).

BENGTSON, S.-A., FJELLBERG, A., SOLHØY, T.: Abundance of tundra arthropods in Spitsbergen. Ent. Scand. **5**, 137–142 (1974).

BERG, A., KJELVIK, S., WIELGOLASKI, F. E.: Distribution of $^{14}C$ photosynthates in Norwegian alpine plants. In: WIELGOLASKI, F. E. (Ed.): Fennoscandian tundra ecosystems. Part 1: Plants and microorganisms, pp. 208–215. Berlin-Heidelberg-New York: Springer 1975.

BERG, B., KÄRENLAMPI, L., VEUM, A. K.: Comparisons of decomposition rates measured by means of cellulose. In: WIELGOLASKI, F. E. (Ed.): Fennoscandian tundra ecosystems. Part 1: Plants and microorganisms, pp. 261–267. Berlin-Heidelberg-New York: Springer 1975.

BLISS, L. C.: Arctic and alpine life cycles. Ann. Rev. Ecol. Systemat. **2**, 405–438 (1971).

CLARHOLM, M., LID-TORSVIK, V., BAKER, J. H.: Bacterial populations of some Fennoscandian tundra soils. In: WIELGOLASKI, F. E. (Ed.): Fennoscandian tundra ecosystems. Part 1: Plants and microorganisms, pp. 251–260. Berlin-Heidelberg-New York: Springer 1975.

DAHL, E.: Flora and plant sociology in Fennoscandian tundra areas. In: WIELGOLASKI, F. E. (Ed.): Fennoscandian tundra ecosystems. Part 1: Plants and microorganisms, pp. 62–67. Berlin-Heidelberg-New York: Springer 1975.

DOWDING, P.: Nutrient losses from litter on IBP tundra sites. In: HOLDING, A. J., HEAL, O. W., MACLEAN, S. F. JR., FLANAGAN, P. W. (Eds.): Soil organisms and decomposition in tundra, pp. 363–373. Stockholm: IBP Tundra Biome Steering Committee 1974.

DOWDING, P., CHAPIN, F. S. III, KILFEATHER, P., WIELGOLASKI, F. E.: Nutrients in tundra ecosystems. In: MOORE, J. J. (Ed.): Tundra: Comparative analysis of ecosystems. Cambridge: University Press 1976 (in press).

DUNICAN, L. K., ROSSWALL, T.: Taxonomy and physiology of tundra bacteria in relation to site characteristics. In: HOLDING, A. J., HEAL, O. W., MACLEAN, S. F. JR., FLANAGAN, P. W. (Eds.): Soil organisms and decomposition in tundra, pp. 79–92. Stockholm: IBP Tundra Biome Steering Committee 1974.

FJELLBERG, A.: Organization and dynamics of Collembola populations on Hardangervidda. In: WIELGOLASKI, F. E. (Ed.): Fennoscandian tundra ecosystems. Part 2: Animals and systems analysis, pp. 73–79. Berlin-Heidelberg-New York: Springer 1975.

GAARE, E., SKOGLAND, T.: Wild reindeer food habits and range use at Hardangervidda. In: WIELGOLASKI, F. E. (Ed.): Fennoscandian tundra ecosystems. Part 2: Animals and systems analysis, pp. 195—205. Berlin-Heidelberg-New York: Springer 1975.

GOKSØYR, J.: Decomposition, microbiology and ecosystem analysis. In: WIELGOLASKI, F. E. (Ed.): Fennoscandian tundra ecosystems. Part 1: Plants and microorganisms, pp. 230–238. Berlin-Heidelberg-New York: Springer 1975.

GRANHALL, U., LID-TORSVIK, V.: Nitrogen fixation by bacteria and free-living blue-green algae in tundra areas. In: WIELGOLASKI, F. E. (Ed.): Fennoscandian tundra ecosystems. Part 1: Plants and microorganisms, pp. 305–315. Berlin-Heidelberg-New York: Springer 1975.

GRANHALL, U., SELANDER, H.: Nitrogen fixation in a subarctic mire. Oikos **24**, 8–15 (1973).

HAFTORN, S.: Norges fugler, Oslo-Bergen-Tromsø: Universitetsforlaget 1971.

HAGEN, A.: Energibudsjett for en fjellrottepopulasjon [*Microtus oeconomus* (Pall.)] i den lav/mellomalpine sone på Finse, Hardangervidda. Univ. of Oslo: Unpubl. thesis 1973.

HAGEN, A., ØSTBYE, E., SKAR, H.-J.: Energy budget of a population of the root vole [*Microtus oeconomus* (Pall.)] in a high mountain habitat, Hardangervidda. In: WIELGOLASKI, F. E. (Ed.): Fennoscandian tundra ecosystems. Part 2: Animals and systems analysis, pp. 170–173. Berlin-Heidelberg-New York. Springer 1975.

HAGEN, Y.: The irruption of hawk-owls [*Surnia ulula* (L.)] in Fennoscandia 1950–51. Sterna, Opusc. Ser. Zool. **24**, 1–22 (1956).

HÅGVAR, S.: Energy budget and growth during the development of *Melasoma collaris* L. (Coleoptera). Oikos **26**, 140–146 (1975).

HÅGVAR, S., ØSTBYE, E.: Occurence and role of different invertebrate groups in alpine *Salix herbacea* snowbeds at Hardangervidda. In: WIELGOLASKI, F. E. (Ed.): Fennoscandian tundra ecosystems. Part 2: Animals and systems analysis, pp. 88–93. Berlin-Heidelberg-New York: Springer 1975.

HÄMET-AHTI, L.: Zonation of the mountain birch forests in northern-most Fennoscandia. Ann. Bot. Soc. "Vanamo" **34:4**, 1–127 (1963).

HANSSEN, J. F., GOKSØYR, J.: Biomass and production of soil and litter fungi at Scandinavian tundra sites. In: WIELGOLASKI, F. E. (Ed.): Fennoscandian tundra ecosystems. Part 1: Plants and microorganisms, pp. 239–243. Berlin-Heidelberg-New York: Springer 1975.

HAUKIOJA, E.: Weight development, consumption and egestion of *Dineura viridorsata* (Hym., Tenthredinidae) larvae. Rep. Kevo Subarctic Res. Stat. **10**, 9–13 (1973).

HAUKIOJA, E., HEINO, J.: Birch consumption by reindeer *(Rangifer tarandus)* in Finnish Lapland. Rep. Kevo Subarctic Res. Stat. **11**, 22–25 (1974).

HAUKIOJA, E., KOPONEN, S.: Birch herbivores and herbivory at Kevo. In: WIELGOLASKI, F. E. (Ed.): Fennoscandian tundra ecosystems. Part 2: Animals and systems analysis, pp. 181–188. Berlin-Heidelberg-New York: Springer 1975.

HELLAND, A.: Trægrændser og sommervarmen. Tidsskr. Skogbr. **1912**, 1–32 (1912).

HESJEDAL, O.: Vegetation mapping at Hardangervidda. In: WIELGOLASKI, F. E. (Ed.): Fennoscandian tundra ecosystems. Part 1: Plants and microorganisms, pp. 74–81. Berlin-Heidelberg-New York: Springer 1975.

HOFSVANG, T.: Life cycle and energy budget of *Tipula excisa* Schum. (Diptera, Tipulidae). In: WIELGOLASKI, F. E. (Ed.): Fennoscandian tundra ecosystems. Part 2: Animals and systems analysis, pp. 145–150. Berlin-Heidelberg-New York: Springer 1975.

HOGSTAD, O.: Structure of small passerine communities in subalpine birch forests in Fennoscandia. In: WIELGOLASKI, F. E. (Ed.): Fennoscandian tundra ecosystems. Part 2: Animals and systems analysis, pp. 94–103. Berlin-Heidelberg-New York: Springer 1975.

ISO-IIVARI, L.: Vertebrates of Inari Lapland. Kevo Notes **1**, 1–19 (1975).

JOHANSSON, L.-G.: The distribution and fate of $^{14}C$, photoassimilated by plants on a subarctic mire at Stordalen. Progress Report 1973. Swed. IBP Tundra Biome Project Tech. Rep. **16**, 165–172 (1974).

JOHANSSON, L.-G., LINDER, S.: The seasonal pattern of photosynthesis of some vascular plants on a subarctic mire. In: WIELGOLASKI, F. E. (Ed.): Fennoscandian tundra ecosystems. Part 1: Plants and microorganisms, pp. 194–200. Berlin-Heidelberg-New York: Springer 1975.

JONSSON, S.: Vertebrate fauna of the Torneträsk-area. Swed. IBP Tundra Biome Project Tech. Rep. **5**, 1–20 (1975).

KALELA, O.: On the fluctuations in the numbers of arctic and boreal small rodents as a problem of production biology. Ann. Acad. Sci. Fenn. Ser. A. IV Biol. **66**, 1–38 (1962).

KALLIO, P.: Kevo. In: ROSSWALL, T., HEAL, O. W. (Eds.): Structure and function of tundra ecosystems, pp. 193–223. Ecol. Bull. **20**. Stockholm: Swed. Nat. Sci. Res. Council 1975.

KALLIO, P., HEINONEN, S.: $CO_2$ exchange and growth of *Rhacomitrium lanuginosum* and *Dicranum elongatum.* In: WIELGOLASKI, F. E. (Ed.): Fennoscandian tundra ecosystems. Part 1: Plants and microorganisms, pp. 138–148. Berlin-Heidelberg-New York: Springer 1975.

KALLIO, P., KALLIO, S.: Nitrogen fixation by free-living microorganisms in the Kevo district Finnish Lapland. Rep. Kevo Subarctic Res. Stat. **12**, 28–33 (1975).

KALLIO, P., KÄRENLAMPI, L.: Photosynthesis in mosses and lichens. In: COOPER, J. P. (Ed.): Photosynthesis and productivity in different environments, pp. 393–423. Cambridge: University Press 1975.

KALLIO, P., LEHTONEN, J.: On the ecocatastrophe of birch forests caused by *Oporinia autumnata* (Bkh.) and the problem of reforestation. In: WIELGOLASKI, F. E. (Ed.): Fennoscandian tundra ecosystems. Part 2: Animals and systems analysis, pp. 174–180. Berlin-Heidelberg-New York: Springer 1975.

KALLIO, P., VALANNE, N.: On the effect of continuous light on photosynthesis in mosses. In: WIELGOLASKI, F. E. (Ed.): Fennoscandian tundra ecosystems. Part 1: Plants and microorganisms, pp. 149–162. Berlin-Heidelberg-New York: Springer 1975.

KALLIO, P., VEUM, A. K.: Analysis of precipitation at Fennoscandian tundra sites. In: WIELGOLASKI, F. E. (Ed.): Fennoscandian tundra ecosystems. Part 1: Plants and microorganisms, pp. 333–338. Berlin-Heidelberg-New York: Springer 1975.

KALLIO, S., KALLIO, P.: Nitrogen fixation in lichens at Kevo, North Finland. In: WIELGOLASKI, F. E. (Ed.): Fennoscandian tundra ecosystems. Part 1: Plants and microorganisms, pp. 292–304. Berlin-Heidelberg-New York: Springer 1975.

KÄRENLAMPI, L.: Comparisons between the microclimates of the Kevo ecosystem study sites and the Kevo meteorological station. Rep. Kevo Subarctic Res. Stat. **9**, 50–65 (1972).

KÄRENLAMPI, L., KAUHANEN, H.: A direct gradient analysis of the vegetation of the surroundings of the Kevo Subarctic Station. Rep. Kevo Subarctic Res. Stat. **9**, 82–98 (1972).

KJELLSEN, E. K.: Dynamics of Thysanoptera populations on Hardangervidda. In: WIELGOLASKI, F. E. (Ed.): Fennoscandian tundra ecosystems. Part 2: Animals and systems analysis, pp. 80–83. Berlin-Heidelberg-New York: Springer 1975.

KJELVIK, S., KÄRENLAMPI, L.: Plant biomass and primary production of Fennoscandian subarctic and subalpine forests and of alpine willow and heath ecosystems. In: WIELGOLASKI, F. E. (Ed.): Fennoscandian tundra ecosystems. Part 1: Plants and microorganisms, pp. 111–120. Berlin-Heidelberg-New York: Springer 1975.

KOPONEN, S.: On the mining insects of the mountain birch in northernmost Fennoscandia. Rep. Kevo Subarctic Res. Stat. **10**, 14–19 (1973a).

KOPONEN, S.: Herbivorous invertebrates of the mountain birch at Kevo, Finnish Lapland. Rep. Kevo Subarctic Res. Stat. **10**, 20–28 (1973b).

KOPONEN, S., OJALA, H.: On the mesofauna of the field layer of three subarctic habitats. Rep. Kevo Subarctic Res. Stat. **11**, 65–71 (1974).

LAGERLÖF, J., MAGNUSSON, C.: Investigation of nematodes at the Stordalen site. Swed. IBP Tundra Biome Project Tech. Rep. **18**, 1–21 (1975).

LARSSON, P., TANGEN, K.: The input and significance of particulate terrestrial organic carbon in a subalpine freshwater ecosystem. In: WIELGOLASKI, F. E. (Ed.): Fennoscandian tundra ecosystems. Part 1: Plants and microorganisms, pp. 351–359. Berlin-Heidelberg-New York: Springer 1975.

LIEN, L., ØSTBYE, E., HAGEN, A., SKAR, H.-J., SVALASTOG, D.: Density variations of bird populations in high mountain habitats, Hardangervidda. In: WIELGOLASKI, F. E. (Ed.): Fennoscandian tundra ecosystems. Part 2: Animals and systems analysis, pp. 105–110. Berlin-Heidelberg-New York: Springer 1975.

LOHM, U., LUNDKVIST, H., PERSSON, T.: Abundance and biomass of some soil animals at the Stordalen mire. Progress Report 1972. Swed. IBP Tundra Biome Project Tech. Rep. **14**, 90–92 (1973).

LUNDKVIST, H.: Occurrence of Enchytraeids at Stordalen. Progress Report 1973. Swed. IBP Tundra Biome Project Tech. Rep. **16**, 29–34 (1974).

ØSTBYE, E. (Ed.), BERG, A., BLEHR, O., ESPELAND, M., GAARE, E., HAGEN, A., HESJEDAL, O., HÅGVAR, S., KJELVIK, S., LIEN, L., MYSTERUD, I., SANDHAUG, A., SKAR, H.-J., SKARTVEIT, A., SKRE, O., SKOGLAND, T., SOLHØY, T., STENSETH, N. C., WIELGOLASKI, F. E.: Hardangervidda. In: ROSSWALL, T., HEAL, O. W. (Eds.): Structure and function of tundra ecosystems, pp. 225–264. Ecol. Bull. **20**. Stockholm: Swed. Nat. Sci. Res. Council 1975.

RESVOLL, T. R.: Om planter som passer til kort og kold sommer. Oslo. Arch. Mat. og Nat. vid. **35**, No. 6 (1917).

RODIN, L. E., BAZILEVICH, N. J.: Production and mineral cycling in terrestrial vegetation. Edinburgh-London: Oliver and Boyd 1967.

ROSSWALL, T.: Introduction, decomposer section. In: WIELGOLASKI, F. E. (Ed.): Fennoscandian tundra ecosystems. Part 1: Plants and microorganisms, pp. 227–229. Berlin-Heidelberg-New York. Springer 1975.

ROSSWALL, T., FLOWER-ELLIS, J. G. K., JOHANSSON, L.-G., JONSSON, S., RYDÉN, B. E., SONESSON, M.: Stordalen (Abisko). In: ROSSWALL, T., HEAL, O. W. (Eds.): Structure and function of tundra ecosystems, pp. 265–294. Ecol. Bull. **20**. Stockholm: Swed. Nat. Sci. Res. Council 1975b.

ROSSWALL, T., VEUM, A. K., KÄRENLAMPI, L.: Plant litter decomposition at Fennoscandian tundra sites. In: WIELGOLASKI, F. E. (Ed.): Fennoscandian tundra ecosystems. Part 1: Plants and microorganisms, pp. 268–278. Berlin-Heidelberg-New York: Springer 1975a.

SANDBERG, G.: Iagtagelser over arktiske sommerfugles metamorphoser. Ent. Tidsskr. **4**, 9–28 (1883).

SKARTVEIT, A., RYDÉN, B. E., KÄRENLAMPI, L.: Climate and hydrology of some Fennoscandian tundra ecosystems. In: WIELGOLASKI, F. E. (Ed.): Fennoscandian tundra ecosystems. Part 1: Plants and microorganisms, pp. 41–53. Berlin-Heidelberg-New York: Springer 1975.

SKRE, O.: $CO_2$ exchange in Norwegian tundra plants studied by infrared gas analyzer technique. In: WIELGOLASKI, F. E. (Ed.): Fennoscandian tundra ecosystems. Part 1. Plants and microorganisms, pp. 168–183. Berlin-Heidelberg-New York: Springer 1975.

SKRE, O., KJELVIK, S., BERG, A., WIELGOLASKI, F. E.: Nitrogen content in Norwegian alpine plants. In: VIK, R. (Ed.): IBP in Norway. Methods and results. Section PT-UM Grazing project, Hardangervidda. Botanical investigations, pp. 159–194. Oslo: Norwegian National IBP Committee 1975.

SKRE, O., WIELGOLASKI, F. E.: $CO_2$ exchange studies on single species of Norwegian alpine plants during the 1974 season. In: VIK, R. (Ed.): IBP in Norway. Methods and results. Section PT-UM Grazing project, Hardangervidda. Botanical investigations, pp. 297–321. Oslo: Norwegian National IBP Committee 1975.

SOLHØY, T.: Dynamics of Enchytraeidae populations on Hardangervidda. In: WIELGOLASKI, F. E. (Ed.): Fennoscandian tundra ecosystems. Part 2: Animals and systems analysis, pp. 55–59. Berlin-Heidelberg-New York: Springer 1975a.

SOLHØY, T.: Dynamics of Oribatei populations on Hardangervidda. In: WIELGOLASKI, F. E. (Ed.): Fennoscandian tundra ecosystems. Part 2: Animals and systems analysis, pp. 60–65. Berlin-Heidelberg-New York: Springer 1975b.

SOLHØY, T., ØSTBYE, E., KAURI, H., HAGEN, A., LIEN, L., SKAR, H.-J.: Faunal structure of Hardangervidda, Norway. In: WIELGOLASKI, F. E. (Ed.): Fennoscandian tundra ecosystems. Part 2: Animals and systems analysis, pp. 29–45. Berlin-Heidelberg-New York: Springer 1975.

SONESSON, M., WIELGOLASKI, F. E., KALLIO, P.: Description of Fennoscandian tundra ecosystems. In: WIELGOLASKI, F. E. (Ed.): Fennoscandian tundra ecosystems. Part 1: Plants and microorganisms, pp. 3–28. Berlin-Heidelberg-New York: Springer 1975.

STEIGEN, A. L.: Energetics in a population of *Pardosa palustris* (L.) (Araneae, Lycosidae) on Hardangervidda. In: WIELGOLASKI, F. E. (Ed.): Fennoscandian tundra ecosystems. Part 2: Animals and systems analysis, pp. 129–144. Berlin-Heidelberg-New York: Springer 1975.

SVENSSON, M., VEUM, A. K., KJELVIK, S.: Carbon losses from tundra soils. In: WIELGOLASKI, F. E. (Ed.): Fennoscandian tundra ecosystems. Part 1: Plants and microorganisms, pp. 279–286. Berlin-Heidelberg-New York: Springer 1975.

TAMBS-LYCHE, H.: Dynamics of Aphididae populations on Hardangervidda. In: WIELGOLASKI, F. E. (Ed.): Fennoscandian tundra ecosystems. Part 2: Animals and systems analysis, pp. 84–87. Berlin-Heidelberg-New York: Springer 1975.

TENOW, O.: The outbreaks of *Oporinia autumnata* Bkh. and *Operophthera* spp. (Lep., Geometridae) in the Scandinavian mountain chain and northern Finland 1862–1968. Zool. Bidr., Uppsala Suppl. **2**, 1–107 (1972).

VEUM, A. K.: Jordundersøkelser og nedbrytning av organisk materiale (Soil investigations and decomposition of organic matter). In: IBP i Norge Årsrapport 1972 (IBP in Norway Annual Report 1972), pp. 29–74. Oslo: Norwegian National IBP Committee 1973.

WIELGOLASKI, F. E.: Vegetation types and plant biomass in tundra. Arctic and Alpine Res. **4**, 291–305 (1972).

WIELGOLASKI, F. E.: Primary productivity of alpine meadow communities. In: WIELGOLASKI, F. E. (Ed.): Fennoscandian tundra ecosystems. Part 1: Plants and microorganisms, pp. 121–128. Berlin-Heidelberg-New York: Springer 1975a.

WIELGOLASKI, F. E.: Grazing by sheep. In: WIELGOLASKI, F. E. (Ed.): Fennoscandian tundra ecosystems. Part 2: Animals and systems analysis, pp. 216–228. Berlin-Heidelberg-New York: Springer 1975b.

WIELGOLASKI, F. E., KJELVIK, S.: Energy content and use of solar radiation of Fennoscandian tundra plants. In: WIELGOLASKI, F. E. (Ed.): Fennoscandian tundra ecosystems. Part 1: Plants and microorganisms, pp. 201–207. Berlin-Heidelberg-New York: Springer 1975.

WIELGOLASKI, F. E., KJELVIK, S., KALLIO, P.: Mineral content of tundra and forest tundra plants in Fennoscandia. In: WIELGOLASKI, F. E. (Ed.): Fennoscandian tundra ecosystems. Part 1: Plants and microorganisms, pp. 316–332. Berlin-Heidelberg-New York: Springer 1975.

# Subject Index


Abisco, location 4
Abundance, invertebrates 22–28, 37–43, 56, 57, 58, 60–64, 67–86, 90, 91, 133–135, 147, 154, 155, 313
—, small passerines 97–101
*Acanthis* spp. 15, 16
— *flammea* 95, 96, 98, 100–102
Acari 4, 22, 23, 33–39, 90–92, 152–156, 267, 269, 270, 312, 314, 315
Acclimatization 118, 120, 122, 127, 135, 307
Accumulation, minerals 319 – 321
—, organic matter 316, 317, 320
Acidification 234, 235
*Aclastus minutus* 39
*Aconitum* 233
*Acrocephalus schoenobaenus* 15, 16
Activity area, small mammals 276
—, Collembola 117–121
—, reindeer 206–215
Actual evapotranspiration 307
*Acyrthosiphon malvae* 84, 86
Adaptation 4, 12, 15, 16, 49, 121, 127, 156, 231
Aerial transport, invertebrates 4
Aggregation index 268
Air photo 237–242
— pollution 234–236
*Alces alces* 49, 311
*Alchemilla* 84
*Alopecosa aculeata* 25, 67, 69, 71, 72
*Alopex lagopus* 21, 52, 312
*Agyneta conigera* 69
*Amara* spp. 37
*Amauronematus* spp. 183
Amphibia 5, 13
*Amphorophora rubi* 85
*Anachrostichus lucidus* 51
*Anaerobic* 316, 317
*Anapothrips secticornis* 80–83, 124, 126, 127
*Anarta* spp. 50
*Ancylis myrtillana* 28
Animal behavior 271
— husbandry 233, 234
*Anthobium lapponicum* 50
*Anthoxanthum* 219, 220, 224, 225, 306
Anthus pratensis 22, 52, 92, 97, 106–108, 145, 160–169
— *trivialis* 95
Ants 4, 46
*Apatania stigmatella* 46
Aphididae 33, 36–38, 84–87, 183–186, 312
Apparent photosynthesis 252–257, 259–264, 303, 304, 306
*Aptinothrips stylifer* 80–82
Aquatic productivity 320
*Aquila chrysaëtos* 52, 213
Arachnoidea 38
Araneae 4, 19, 23–25, 27, 33, 35, 37–40, 51, 66–72, 89–91, 122–127, 129–144, 152–155, 161, 162, 269, 270, 314
*Araneus proximus* 27
*Arctella lapponica* 67
Arctic fox 4, 21, 30–32, 52, 312
*Arctorthecia cataphracta* 43
*Arctostaphylos* 303
*Argynnis* spp. 50
*Argyroploce hyperboreana* 28
Ash content, meadow pipit 160, 165, 167
— —, plants 219, 221, 222
*Asio flammeus* 51
— spp. 32
Assimilation, invertebrates 89, 137–143, 147–149, 189–191
—, root vole 172
*Astragalus* 219, 220, 222–224
Atlantic 4, 33
Atmospheric pollution 234–236
Auchenorrhyncha 36, 38
"Average rodent" 275–281
Aves, species number 5

Bacteria 315, 316, 319
Bacterial n-fixation 321
Bank vole 275
Bear 312
Beetle 50, 313, 314
Behavior, Vertebrates 210, 212, 271, 280
*Betula* 46, 66, 94, 174, 181, 190, 191, 197, 199, 204, 307, 311
Bibionidae 39, 42, 161, 162, 166
Bioelement-pools 235
Biological fixation, nitrogen 321
Biomass, bacteria 315, 316, 319
—, fungi 315, 316, 319

Biomass, invertebrates 23, 33, 34, 38, 39, 43, 60, 61, 74–76, 81, 82, 154, 155, 185, 186, 270, 314, 315, 318, 319
—, meadow pipit 161, 163, 164
—, plants 258, 260–263, 302, 304, 306–308
—, reindeer 196, 310, 318
—, small rodents 172, 309, 311, 318
Biotope value 241
Binomial distribution 268
Birch 5, 27, 181–188, 231–234, 307, 313, 318
— consumption, reindeer 310
—, damage 174–180
— defoliators 189–193
— forest 5, 19–27, 29, 32, 37, 46, 66–72, 94–103, 174–180, 181, 232, 234, 235, 300, 301, 302, 307, 308, 311–315, 317, 318, 320, 321
Bird density 22, 52, 312
— migrators 32, 49
— populations 105–110
Birth rate, small rodents 279, 280
Blackbacked gull 48, 50
Blackflies 212
Blood-suckers 51
Blue-green algae 307, 321
Body composition, meadow pipit 160–169
— —, small rodents 287, 288
Bog 307
*Bolyphantes index* 27, 37
Bombycimorpha 50
Borboridae 39, 42
*Boreaphilus henningianus* 50
*Boreus* sp. 162
Box and arrow diagram 246, 247, 253, 275
Brachychthoniidae 60
Brambling 21, 52
Breeding, small rodents 276, 279
Brown bear 312
Browsing 311
*Bryodrilus ehlersi* 55
Bryophytes 252, 253, 255, 256, 260–264, 303, 304, 307, 308, 316, 317, 320, 321
*Bubo bubo* 52
*Bufo* spp. 12, 13
Burrowing activity, small rodents 311
*Buteo lagopus* 21, 52
Butterflies 46, 50
Buzzard 52, 312
Byrrhidae, species 36
*Byrrhus* sp. 162

$^{14}$C 317
*Calcarius lapponicus* 15, 22, 52, 97, 106–108
Caledonian folding zone 301
*Calidris temminckii* 107
*Callophrys rubi* 50
Caloric content, invertebrates 89, 123, 126, 131, 132, 137–142, 147, 148, 151, 191
— —, meadow pipit 161, 162, 165, 167, 168
— —, plants 219, 221, 222
— —, small rodents 172, 287, 288
*Calyptozetes sarekensis* 61, 62
Cambro-Silurian 301
*Canis lupus* 52, 312
Carabidae 36, 37, 41, 90, 91, 152, 267, 270, 314
*Carabodes labyrinthicus* 61–63, 112–116
Carapace 71, 131, 132
Carbohydrates, plants 219, 221, 223–225
Carbon loss 317
Carcass, meadow pipit 160
*Carduelis flammea* 21, 49
*Carex* 50, 197, 198, 219–221, 223, 224, 304, 306
Carnivores, invertebrates 314, 318
—, vertebrates 312, 318
Carrying capacity 157, 233
Caterpillar 186, 232
Cation pool, mobile 235
Cattle 233
Cecidomyiidae 27, 39, 42, 183–187, 313
Cellulose strips, decomposition 317
Centipedes 4
Ceraphronidae 38, 39, 42
*Ceratinella oculatissima* 67
*Cernosviloviella* sp. 55
*Cetraria* 197, 204, 303, 304, 311
Chamaephytes 301
Chemical content, meadow pipit 160–168
— —, plants 219, 221–225
— —, reindeer diet 199, 200
Chilopoda 19, 24, 33, 50, 312
Chionophilous 305
Chironomidae 39, 42, 166, 314
*Chloroperla borealis* 28
Chloropidae 39, 42
Chrysomelidae 37, 89, 183, 185, 313
*Chrysops nigripes* 51
Cicadina 19, 24–26, 162, 183–185
*Cicadula intermedia* 50
*Cidaria caesiata* 25, 28
Cimbicidae 185
*Circus cyaneus* 48
*Cladonia* 197, 204, 232, 303, 304, 310, 311
Classification, ecological 241
*Clethrionomys* spp. 21, 49, 275, 311
Cloudberry 50
C/N ratio 301
Coal tit 52
Coccina 36, 162, 183
*Coeliodes* spp. 27, 183, 184
*Coenonympha tullia* 50

*Cognettia lapponica* 51
— spp. 55
Coleoptera 19, 23–25, 27, 28, 33, 37–39, 50, 51, 89, 91, 152–155, 161, 162, 167, 183–185, 187, 312, 314
*Colias palaeno* 50
*Colladonus torneellus* 25, 183
Collembola 4, 22, 23, 33–39, 51, 73–79, 90–92, 117–121, 123, 140, 152–156, 267, 269, 270, 312, 314, 315, 318
Colonization 10
Combustion 234
Compartment model 246, 251–266
Competition 11, 12
*Conigerella borealis* 37
Conservation, remote sensing 237–242
"Consumption", fungi 319
—, herbivores 186–188
— hypothesis, small rodent cyclisity 278, 279
—, invertebrates 137–143, 189–193, 257, 260, 313, 318
—, reindeer 201–204, 310, 318, 320
—, sheep 216–225
—, small rodents 172, 173, 290–292, 295, 311, 318
Continental 4, 33, 50, 195, 233, 304
*Cornus* 177
*Corvus corax* 49, 213
— *corone cornix* 53
*Corymbites costalis* 25, 183
Craneflies 145–150, 314
Crude protein 219, 221, 223, 225
Crustacea 43
Cryptogams, production 302, 303, 308
Cryptostigmata 22
Culicidae 51, 212
*Curculionidae* 27, 36, 89, 152, 162, 183–185
"Curve fitting" 277, 281
Cycles, birds 105–110
—, invertebrates 313
—, *Oporinia* 178, 183
—, predators 277, 278
—, small rodents 30, 107, 108, 183, 232, 273–281, 283–286
—, vertebrates 30, 51, 309
*Cyphon variabilis* 50

Day length 301
Death rate, invertebrates 257
— —, plants 256, 257, 259, 263, 264
— —, small rodents 277, 279
Decay rate 316, 317
Decomposers, invertebrates 149, 314, 315, 318
Decomposition 257, 259, 316, 317, 319, 320
*Deductive* model 286
Defoliation of birch 174–188, 190–193
*Delphacodes discolor* 50
*Dendrobaena octaedra* 24, 51
*Dendrobaena* spp. 42
Denitrification 320
Densitometry 241
Density, birds 22, 52, 312
—, invertebrates 24, 90, 314, 315
— reindeer 202, 208, 310, 318
—, small passerines 94, 98, 100—102
—, small rodents 30, 107, 108, 276, 281, 311
Depressions, photosynthesis 303
*Deschampsia* 177, 197, 219, 220, 223, 224, 233, 306
Description, model 252, 273–277, 286
Deterministic model 246, 251–266, 271, 286
Diapause 120
Diaspores 160–162, 167
Diet, reindeer 195–202
—, sheep 216–225
Difference equation model 246, 270, 277
Differential equation model 246, 251–266
Differentiation, flower 302
Digestibility 203, 207, 222, 223, 225, 311
Digital computer simulation 273
*Dineura virididorsata* 27, 183–187, 190, 191, 313
*Diplocolenus limbatellus* 25
Diplopoda 24, 33, 50, 312
*Diprion* spp. 27
Diptera 4, 23–25, 27, 33, 39, 42, 51, 140, 145, 153–155, 161, 162, 183, 187, 212, 269, 270, 312–314, 318
Dispersal capacity 10
Distribution, fauna 4, 8, 12, 32, 46, 99
Diversity 4, 9, 10, 14–16, 19, 33, 37, 43, 67, 68, 97, 99–102, 309, 314
Dorylaimoid nematodes 51
Dotterel 22
Driving variables 252, 254
Ducks 49
Dwarf shrubs 303, 305
Dynamic model 246
*Dyschirius norvegicus* 50

Eagle 52
*Eana osseana* 28
Earthworms 4
Ecocatastrophe 174–180, 240
Ecological efficiency 137–143
— models 246–249, 251–266, 267–272, 283–296
— plasticity 16
— survey 240
Ecosimulation 264
Ecosystem models 246–249, 251–266

*Edwardzetes edwardsi* 61, 62
*Eiseniella tetraedra* 51
*Elaphrus lapponicus* 34
Elateridae 183
Electromagnetic spectrum 237
*Emberiza schoeniclus* 15, 16, 95, 96, 100, 101
Emberizidae 8–16
*Empetrum* 177, 204, 303, 307, 311
Empididae 33
Enchytraeidae 4, 35–39, 51, 55–58, 155, 267, 269, 270, 312–315, 318
Energy budget 136–142, 145–150, 190, 191, 213
— circuit language 267
— flow 122, 189–191, 267, 270, 271, 318, 319
— metabolism 123, 125–127, 130, 135, 137, 139, 158, 167
— range, invertebrates 156
—, small rodents 290, 291, 295, 296
Entropy 271
Ephemeroptera 46
*Equisetum* 311
*Erebia disa* 50
*Eremophila alpestris* 106–108
*Erigone* spp. 37
Erigoninae 66
*Eriocrania* spp. 27, 183–185, 193, 313
*Eriogaster* spp. 50
*Eriophorum* 197, 198, 307
Eriphyidae 27
Erosion 231, 232, 240
Esker ridges 303
Esophageal fistula 196, 197, 199, 200, 203
*Euceraphis punctipennis* 183
*Euconulus fulcus* 25, 42
*Eudromias morinellus* 22
Eulophidea 39, 42
*Eupelops plicatus* 61, 62
*Euphrasia* 84, 162
*Eustigmaeus* sp. 43
Eutrophic 94, 252, 302, 305–307, 315–317, 320, 321
Evaluation 285
Evapotranspiration 306, 307
Everlasting days 304

*Falco* spp. 51
Falcon 51, 52, 312
False color 237–240
Fat content, small rodents 287, 288
Faunistic similarity 73
Feedback loop 260
Feeding habit 5
Fen 305
Fertilization 306, 321
*Festuca* 220–222, 224, 225, 233, 303
Fetuses, small rodents 287, 288, 290, 291, 296
*Ficedula hypoleuca* 21, 52
Field control, remote sensing 238–241
— vole 275
Fieldfare 52
Film cameras 237
Fire 232
Flies 212
Flow diagram 247, 253, 267, 275
Fluctuation, birds 94, 98, 100–102, 105–110
—, invertebrates 56–58, 60–64, 68–79, 81, 85, 178, 183, 269, 313
—, predators 277, 278
—, small rodents 107, 108, 273–281, 283, 284–286
Foliage feeders 181–188
*Folsomia* spp. 73–77
Food composition, meadow pipit 160–169
— habits, reindeer 195–205
— —, reindeer 213, 214, 310
— —, sheep 216–225
— —, small rodents 277, 291
— limitations, small rodent cyclisity 278, 279
— resources 16, 102
— selection, reindeer 311, 320
— —, small rodents 296
Forage quality 217, 219, 221–225
Forbs 197–199, 201, 220, 222, 252, 254, 255, 259–264, 305, 306, 311, 316
Forest tundra 309, 312–315, 318, 321
*Formica gagatoides* 50
Formicidae 33, 312
Fox 21, 30, 312
*Fredericia bulboides* 55
Frequency, Collembola 78
—, Thysanoptera 83
*Fringilla montifringilla* 15, 16, 21, 52, 95, 96, 98–102
Fringillidae 8–16, 102
Frog 22
*Fumea casta* 50
Fungi 315, 316, 319
Fungivores, invertebrates 314, 318

Gamasid mites 51
Gamasina 152
*Garrulus glandarius* 48
Gastropoda 19, 33, 35, 38, 42, 312
Geometridae 174, 183, 232, 312
Gestation period, small rodents 290
Global radiation 318
*Gnaphosa* spp. 67, 69
Gnaphosidae 66, 67, 69
Gneiss 301
Goats 233

Golden eagle 213
*Gonatium rubellum* 27
— *rubens* 67
Graminoids 197–199, 201, 311, 316
Granite 301
Grasses 220, 223, 252, 254, 304–307, 311
Grazing, invertebrates 174–180, 189–191, 313
—, birch 5, 174–194
—, reindeer 196–205, 206–214, 304, 310, 318, 320
—, sheep 216–225, 306
—, small rodents 311
Grazing-land model 247
Grey owl 48
Greysided vole 4, 275
Growing season 301, 303
Growth, plants 255, 258–264
—, sheep 218
—, small rodents 277, 285, 287, 288, 292, 293, 295, 296
Gull 32, 213
*Gulo gulo* 21, 52, 312
*Gymnusa brevicollis* 50
Gyr falcon 51, 52

Habitat diversity 9, 15
— resources 99, 102
*Haematopus ostralegus* 48
*Hahnia ononidum* 25, 67, 69
Hahniidae 67, 69
Half-saturation constant 292, 293, 295
*Haliaëtus albicilla* 52
*Haplodrassus soerenseni* 67, 69
Hardangervidda, location 3
Hare 21, 49, 311
Harpacticidae 43
Hawk owl 21
Heath 19–26, 27, 60–63, 73–77, 80, 81, 88, 94–102, 111, 154, 155, 195, 200, 201, 203, 204, 232–235, 248, 301–305, 311, 314, 315, 316–318
Heavy metal 235
Hedge sparrow 52, 53
*Helophorus glacialis* 161, 162
Hemerobidae 27
Hemiptera 23–25, 27, 33, 37–39, 43, 153, 154, 155, 162, 185, 267, 269, 270, 312, 314
Hen harrier 48
*Henlea nasuta* 55
Herbage intake 195–205, 216–225
Herbivory, invertebrates 174–180, 189–193
Herd, reindeer 202, 203, 206–215, 310
Hesperoidea 50
Heteroptera 36, 38
Hibernation 132
*Hilaira frigida* 37
*Hirundo rustica* 48
Home range, reindeer 200–202, 310
Homoptera 25–28, 38, 50, 91, 122, 124, 152, 183–185, 187, 314
Hooded crow 53
Horseflies 212
Horses 233
House sparrow 32
Huddling 289
Humidity factor 254, 255
Hunting, reindeer 196, 213, 214
Hydrophilidae 36, 152, 162
*Hydrophorus alpinus* 51
Hymenoptera 24–27, 50, 89, 153–155, 161, 162, 182, 185, 312, 314
— parasitica 27, 38, 39, 42, 43, 314
*Hyperomyzus* sp. 84
*Hyphoraia lapponica* 50
*Hypnoides rivularius* 25

IBP tundra research areas 3, 4,
Ichneumonidae 4, 33, 38, 39, 43
Immision damage 240
*Inductive* model 286
Infrared photography 237–240
Ingestion, small rodents 292
— efficiency, *Pardosa palustris* 137–140
Insect harassment, reindeer 212, 213
— infestation 240
Intake, invertebrates 313, 318
—, reindeer 202, 203, 213, 214, 310, 318
—, sheep 216–225
Interaction, rodent-vegetation 278, 279
Intraspecific competition 12
Inventory, vegetation 237–241
Invertebrates, abundance, IBP tundra research areas 22–28, 37–43
—, aerial transport 4
—, biomass 23, 33, 34, 38, 39, 43, 154, 155, 185, 186, 315
—, caloric content 151–159
—, carrying capacity 157
—, consumption 182–188, 190–193, 257, 260
—, ecological efficiency 4
—, life cycle 4, 33
—, production 186, 314, 315, 319
—, respiration 122–127
—, species IBP tundra research areas 19, 20–28, 33–43, 50–51
—, systems analysis 267–272
Isophlet diagram, temperatures 117
Isopoda 4, 24, 33, 312
*Isotoma* spp. 74–78
— *viridis* 117–120, 124–127

Jaeger 312
Jay 48, 52
Juniper 233, 234

Kevo, location 3

Labiata 38
*Lacerta vivipara* 22
Lactation, small rodents 289–291, 296
*Lagopus lagopus* 49, 51, 311
– *mutus* 22, 311
Lapland bunting 15, 16, 22, 52, 106–108
Lapwing 50
*Larus canus* 213
— *fuscus* 50
— *marinus* 48
*Lathrobium punctatum* 50
Leaching 232, 234, 317, 320
Leaf grazers 5, 182–184, 313
Leafhopper 43, 184
Legumes, nitrogen fixation 321
Lemming 21, 49, 51, 173, 232, 275, 283, 290, 313
*Lemmus lemmus* 21, 49, 173, 275, 290, 311
Lepidoptera 19, 26–28, 37–39, 43, 46, 89, 153–155, 162, 182–185, 187, 267, 269, 270, 312–314
*Leptorhoptrum robustum* 37
*Lepus timidus* 21, 49, 311
Lichen heath 34–43, 60–63, 73–77, 80, 81, 111, 154, 155, 196, 200, 201, 203, 204, 234, 235, 248, 302–304, 314–318
Lichens 177, 197–204, 218, 232–234, 302, 303–305, 307, 308, 310–312, 316, 321
Life cycle, invertebrates 4, 33, 62–64, 77, 78, 79, 82, 87, 132, 133, 136, 268–270, 312, 313
— form 301
Light compensation point 254
— saturation 303
Lignin 219, 221, 223, 225
Likelihood fitting 268
*Limotettix transversus* 25
Linyphiidae 24, 25, 66, 67, 69
*Lispocephala erythrocera* 51
*Lithobius curtipes* 24
*Lithocolletis* 183
Lizard 22
Location, IBP tundra research areas 3
Long-eared owl 32
*Loxia pytyopsittacus* 52
Lumbricidae 24, 33, 35, 38, 42, 51, 312, 315, 318, 319
*Lumbricus rubellus* 42
*Luscinia svecica* 95, 96, 100–102
Lycosidae 66–72, 129–144
*Lynx lynx* 32, 312

*Macrargus multesimus* 67, 69
*Macrosiphum avenae* 84
Magpie 32, 49
Maintenance energy, reindeer 213, 214, 318
— metabolism, small mammals 292
— respiration, plants 255
*Malthodes* spp. 27, 28
Mammalian predators 309, 311
*Mannerheimia arctica* 25
Mapping, vegetation 237–241
Maritime 4, 50
Markov property 287
*Massalongia rubra* 27, 183
Mathematical models 246–249, 251–266, 273–281, 283–296
Meadow 34–48, 55–58, 60–63, 73–77, 80, 81, 84, 86, 94–102, 117–119, 129, 153–158, 201, 216–225, 233, 235, 248, 251, 253, 258, 260–264, 269, 270, 293, 305–307, 311, 315, 317, 321
Mechanistic model 246, 248
*Mecomma ambulans* 25
Mecoptera 36, 38, 153, 154, 162
— pipit 22, 52, 92, 106–108, 145, 160–169
*Meioneta* spp. 67
*Melasoma collaris* 37, 88–92, 313
Mesofauna 23–28, 37–43, 66
Mesostigmata 43, 314
*Messa nana* 183
Metabolism, invertebrates 120, 121, 123, 125–127
—, meadow pipit 167
—, *Pardosa palustris* 129–144
—, small rodents 170–173, 277, 288, 290, 292, 294–296
—, *Tipula excisa* 145–150
Methane 317
*Micaria alpina* 67, 69
Microarthropods 4
Microbial activity 317
— breakdown 317
— production 316, 319
Microfauna, IBP tundra research areas 22, 23, 37–39, 51
Micropredators 314
*Microtus oeconomus* 21, 170–173
— spp. 275, 283, 290, 311
Midnight sun 301
Migration, birds 32, 49, 311
—, reindeer 195, 310
Millipedes 4
Mineral content, meadow pipit 160–169
— —, plants 221–225, 310, 311
Mineralization 320
Minerals, input and output 320, 321
Minerothropic 50, 305–307, 316, 321

Mining invertebrates 5, 182–184, 187, 193, 313
Mink 32
Mire 46, 49–51, 201, 220, 302, 303, 305–307, 311, 314–316, 319, 321
Mites 27, 43, 60–64, 111–116
*Mitopus morio* 24, 43, 91, 123–127, 161, 267, 270
Moisture 303, 304, 306, 308, 314–317, 319, 321
Mollusca 24, 38, 51
Monocormous 179
Monocotyledons 252, 255, 256, 258–264, 303, 304, 306
Monte Carlo technique 283
Moose 32, 49, 53, 311
Moraine 303
Mortality rate, small rodents 277, 279, 280
Mosses 197–199, 212, 235, 259, 304, 307, 311, 321
*Motacilla alba* 97
Moth 232
Multispectral recording 237–239
*Muscicapa striata* 53
Muscidae 267, 270
Mushroom 197, 198, 204, 311, 320
*Mustela* spp. 21, 312
Mustelid 30
*Mycetophorus brunneus* 25
*Mycetophorus nigrans* 37
*Myllaena gracilicornis* 50
Myrmaridae 39, 42

*Nardus* 233, 304, 306
Natality, *Pardosa palustris* 135
—, small rodents 279, 280
Nematocera 27, 46, 267, 270
Nematoda 22, 23, 43, 51, 314, 315, 318, 319
*Nephroma* 304, 321
*Nepticula* 183
Neuroptera 36, 38
Niche 8–16, 67, 68, 99
Nitrogen budget 321
— fixation 304, 307, 320, 321
Nostril flies 212
*Notiophilus aquaticus* 37
— *germinyi* 25
*Notodonta dromedarius* 183
*Nucifraga caryocatactes macrorhynchos* 52
Nutcracker 52
Nutrient recovery, small rodent cyclicity 278
— stress 263
Nutrients 303, 304, 306, 307, 311, 312, 316, 317, 319–321
*Nyctea scandiaca* 51

Oceanic 195, 306, 307
*Ocyusa nivicola* 37
*Oenanthe oenanthe* 22, 92, 97, 106–108
*Oeneis norna* 50
*Oestridae* 212
*Okeniella caudata* 51
Oligochaeta 38
Oligotrophic 94, 235, 302–305, 307, 308, 311, 315–317, 320
Ombrotrophic 50, 301, 303, 307, 316, 317, 319–321
Omnivore 309
*Oncopsis flavicollis* 183, 184
*Operophthera brumata* 51
Opiliones 36, 38, 43, 90, 91, 122–127, 152, 153–155, 161, 162
*Oporinia autumnata* 28, 46, 51, 174–180, 183, 190, 191, 232, 312, 313
*Oppia translamellata* 50
Optimization 246
Oribatei 35–39, 43, 60–64, 152, 318
Oribatidae 50, 111–116
*Oromurcia bicuspidata* 60–63
Osprey 52
Ostracoda 43
*Otiorrhynchus dubius* 89, 162
Outfarm 233
Output, model 259–261, 269, 270, 278–281
Overgrazing 233, 240
Owl 30, 51, 52, 312
Oxen 233
Oxygen consumption, invertebrates 111, 112–127, 130, 135, 136, 148
*Oxyptila rauda* 25, 67
Oystercatcher 48, 50

Palatability 225
Panchromatic photo 238–240
*Pandion haliaëtus* 52
Parameter values 275, 277–280, 292, 293
*Pararamphomyia obscura* 51
Parasitic nematodes 51
*Paraswammerdamia conspersella* 28
*Pardosa palustris* 4, 129–144
– spp. 25, 37, 50, 51, 67, 72, 123–127
*Parornix* 183
Parrot crossbill 52
*Parus ater* 52
— spp. 21, 49
Passerines 16, 30, 50–52, 94–103, 106, 168, 309, 311
Pasture plants 217, 220, 222
*Patrobus septentrionis* 50
— spp. 37
Pawing, reindeer 204
Peat 301, 306, 307, 316, 317, 321

*Pelecopsis mengei* 25
*Pelophila borealis* 120
*Peltigera* 304, 321
Peregrine 51, 52
Periodic length, small rodent cyclisity 280
Permafrost 301, 317, 320
pH 235, 301, 306
Phalangida 19, 24, 25
Phenology 69, 184
Phenotypic variability 15
*Philodromus cespitum* 27
*Phleum* 220, 221
*Phoenicurus phoenicurus* 52, 95, 96, 100–102
Photo interpretation, vegetation 239–241
Photoperiod, invertebrates 120, 123
Photosynthesis, maximum 254, 255
Photosynthetic efficiency 308
*Phthiracarus*, caloric content 151, 152
*Phygadeuon leviventris* 39
Phyllite 301
*Phyllodecta polaris* 183
*Phylloscopus* spp. 15, 16
– *trochilus* 21, 52, 95, 96, 98–102
Phytophageous insects 192
*Pica pica fennorum* 49
Pied flycatcher 21, 52
Pine forest 19–26, 72, 232–235
Pisces, species number 5
Plant communities 200, 201, 218, 220, 239, 302–308, 311–313, 315–318
— intake 310, 311, 318
— preferences, reindeer 195–204
— —, sheep 216–225
Plecoptera 46
*Plectrophenax nivalis* 15, 22, 92, 106
*Pleurozium* 232, 308
Plovers 312
*Poa* 219, 220, 306
*Podistra pilosa* 28, 50
Podzolization 301, 306
Poisson distribution 268
Polar days 301
Pollution 234–236, 306
Polycormous 179
*Polydrosus ruficornis* 25, 27, 183
Polymorphic behavior, small rodent cyclisity 279, 280
*Polyploca flavicornis* 183
Population, reindeer 183, 196, 202–204, 208, 213, 310, 318
— density, birds 22, 51, 105–110
— —, invertebrates 22–28, 38–43, 51, 56, 57, 58, 60–64, 67–86, 91, 133–135, 147, 174, 185, 314, 315
— —, small passerines 97–102
— —, small rodents 30, 172, 173, 183, 311
— — regulators, small rodents 276–281
Population dynamics, small rodents 273, 283–286, 295
*Potentilla* 84
Predation, small rodents cyclisity 277–280
Predatory birds 52, 309, 312
– invertebrates 66, 89–92, 129–144, 314
— mammals 21, 51, 232, 309, 312
Prediction, model 285
Preference of food, reindeer 311, 320
Primary production 258, 260–264, 301, 302–308, 316, 318, 320, 321
*Pristiophora mollis* 25
Probabilistic behavior, variables 246
Probability functions 267, 268
Production, bacteria 315, 316, 319
—, fungi 315, 316, 319
–, invertebrates 89, 137–142, 147–149, 186, 187, 189–193, 314, 315, 319
—, plants 258, 260–264, 301–308
—, reindeer 203
—, root vole 171, 172
Programming 252
Prostigmata 43, 314
Protozoa 43
*Prunella modularis* 52, 95
*Prunus* 85
*Psammotettix frigidus* 43
Pseudoscorpiones 24, 50
Psocoptera 24, 25, 27, 33, 36, 38, 153, 154
—, species 36
*Psylla betulae-nanae* 50
— *hartigi* 27, 183
Psyllina 26, 36, 38, 182–186
Ptarmigan 22, 30, 32, 309, 311
*Pterocomma* sp. 84
*Pteronidea* spp. 183
*Prerostichus diligens* 50
*Pyrgus centaureae* 50

$Q_{10}$, respiration invertebrates 123, 125–127
*Quedius boops* 37

Radiation 237
Radioautographs 304
Radiocarbon dating 317
*Rana arvalis* 12, 13
— *temporaria* 12, 13, 22
Random niche 67, 68
Range size 8–16
— use, reindeer 195–205, 207–209
*Rangifer tarandus* 21, 49
*Ranunculus* 219–221, 224, 225
Raptorial birds 309, 311
Raven 49, 213
Recreation 239, 240

Recycling, nutrients 320
Red fox 53, 213
Redpoll 21, 49, 52
Redstart 52
Redwing 21, 32, 52
Reforestation, *Oporinia* damage 174–180
Regression model 245
Regulating factors, small rodent cyclisity 276–281
Reindeer 5, 21, 49, 182, 183, 187, 195–205, 232, 233, 240, 309–312, 315, 318
— grazing 304
Reproduction, small rodents 276–279, 289, 290, 291, 293, 295, 296
Reptilia, species number 5
Resolution, model 247, 248
—, photo interpretation 238
Resource, habitat 99, 102
— diversity 14
— gradient 8
— spectrum 13, 16
Respiration, invertebrates 4, 111–116, 129, 130, 135–139, 148, 149, 189–193
—, plants 255, 259, 264
—, root vole 172
*Rhacomitrium* 303, 304
*Rhaebothorax sphagnicola* 50
Rhopalocera 50
*Rhopalosiphum padi* 85
Ring-ouzel 32
Rodent-vegetation interactions 277–280
Rodents 21, 49, 275–281
Roe deer 32
Root respiration 255, 259–264
— vole 170–173, 275, 283–296
Rotifera 43, 314
Rough-legged buzzard 21
*Rubus* 50, 85, 307
Rumen samples, reindeer 196–200, 203
Rumination, reindeer 207

Saccharides 219, 221, 223–225
*Salix* 84, 88–92, 197, 198, 219–222, 232, 301, 302, 305, 311, 313
*Saloca strandi* 67
Sap suckers 5, 182
Saprovores, invertebrates 314, 318, 319
Satellite recording 237, 238
Sawfly 183–185, 187, 313
Scale bugs 43
Scarabaeidae, species 36
Sciaridae 39, 42
*Scolytidae* 27
Secondary production, invertebrates 314, 315, 319
Sedge 252, 254, 307, 311
Selective food intake, reindeer 195–205, 211, 320
— — intake, sheep 216–225
— grazing, *Oporinia* 179
Senescence 256
Sensitivity, ecosystems 234–236
— analysis 260, 261, 280, 285
—, models 278–281
Sewage 239
Sex ratio, spiders 71
Shading factor 254, 255
Sheep 233
—, grazing 216–228, 306
—, growth 218
Shore lark 106–108
Short-eared owl 32, 51
Shrew 21, 30
Siberian tit 21
Similarity, faunistic 73
Simulation 246, 251–266, 269–271, 283–296
Simuliidae 212
Skylark 32
Small passerines 94–103
— rodents 21, 30, 49, 107, 108, 173, 183, 232, 273–282, 283–299, 309, 311, 315, 316, 318
Snail 4
Snow bunting 15, 16, 22, 92, 106–108
— cover 34, 58, 88, 106, 109, 146, 204, 212, 232, 240, 241, 288, 293, 294, 301–304, 306, 308, 321
Snowbed 88–92, 201, 204, 235, 302, 313, 317, 318
Snowy owl 4, 30–32, 51
Social behavior, reindeer 212
Soil fauna 314, 315
— moisture, vegetation structure 303, 304, 306, 308
— respiration 317
Solar radiation 252, 254, 255
*Solidago* 177
Solifluction 88, 232
*Sorex* spp. 21
Species number, invertebrates 19–28, 37, 38, 39, 46, 50, 51
— —, vertebrates 5, 19–28, 29–32, 47–53
Spectral band 237
*Sphagnum* 50, 52, 307
Sphinges 50
Spiders 4, 24, 25, 27, 33, 37, 46, 50, 90, 123–127, 129–144, 267, 268, 314
Spillage factor 203, 204, 310
*Spilogona* spp. 51
Spotted flycatcher 53
Springtails 4, 122–127, 314
Stability, models 278–281

Staphylinidae 36, 37, 41, 90, 91, 152, 162, 267, 270, 314
Startling 53
State variables 247, 252, 253, 258
State-oriented model 246, 286
Static model 245
Steady state 263
*Stenomacrus affinis* 39
*Stenus* spp. 50
*Stereocaulon* 304, 321
Stochastic model 246, 264, 267, 268, 270, 271
Stomach content, meadow pipit 160–164, 167
Stoneflies 4, 33
Stress, small rodent cyclisity 279
*Strix nebulosa* 48
Structure, model 284, 286
*Sturnus vulgaris* 53
Submodels 248, 285, 286
Subnivean space 275, 288, 295
Succession 232
Suckling, reindeer 210
—, small rodents 287, 290, 291
Sugar 219, 221, 223–225, 310
Sulphur 234
Sunshine hours 263
*Surnia ulula* 21
Survival, small rodents 278, 294
Swallow 48
Sylviidae 8–16
Symbiotic bacteria 306, 321
Symphyta 25, 89, 183–185, 187, 313, 314
Systems analysis 245, 251, 283

Tabanidae 212
*Tachinus elongatus* 25
Tachydrominae 33
Tardigrada 43
*Tectocepheus velatus* 60–64
Temminck's stint 107, 109
Temperature factor 254
—, isophlet diagram 117
—, respiration rates invertebrates 111–116, 122–127
—, tolerance 12, 13
—, vegetation structure 301, 308, 321
Territories, birds 105–110
*Tetracanthella britannica* 117–120
— spp. 74, 75, 77, 78
*Thamnotettix confinis* 25, 28
Theridiidae 24
Thermal inversion 178
Thermopause 120
*Theridion ohlerti* 27
— *varians* 27
Thomisidae 66, 67, 69
Thrips 80–83
Thysanoptera 33, 36–38, 80–83, 91, 122, 123–127, 152–154, 184, 185, 267, 269, 270, 314
*Tibioplus diversus* 67
*Tipula excisa* 39, 145–150
Tipulidae 33, 145, 161, 162, 166, 267, 269, 270
*Tiso aestivus* 37, 67
Top/root ratio 307
Tortricidae 89
Trampling 203, 231, 303, 304, 310, 318
Translocation 253, 255–257, 259–263
*Trichiura crataegi* 183
Trichoptera 28, 46
*Trientalis* 177
Trombididae 152
Trophic indices, *Parposa palustris* 140, 143
— level 309
— niche 13
Turdidae 102
*Turdus iliacus* 21, 95, 96, 100–102
— spp. 52
Turnover rate, root vole 171, 172
— time 263
Twite 15, 16

*Vacciniina optilete* 50
*Vaccinium* 46, 50, 301, 303, 304, 307, 313
Validation 280
*Vanellus vanellus* 50
Vascular plants, production 301–308
Vegetation inventory, remote sensing 237, 238–242
– pattern 231, 232
Vegetation-rodent interation 278, 279
Vertebrates, biomass 310, 318
—, caloric content 161, 163, 165, 168, 172, 287, 288, 318
—, consumption 173, 182, 187, 188, 201, 202–204, 290–292, 295, 310, 311, 318, 320
—, cyclicity 30, 309, 311, 312
—, distribution range 4, 32
—, production 172, 203, 218, 287, 294, 318
—, species 5, 19–28, 29–32, 47–53, 308–312, 318
*Vertigo arctica* 42
*Vespula norvegica* 51
Visible radiation 237
*Vitrina pellucida* 42
Vole 5, 182, 183, 187, 283
*Vulpes vulpes* 21, 53, 213, 312

Waders 30, 49, 312
Warble flies 212
Water content, meadow pipit 160, 164, 167
Waterpower regulations 240

Weaning, reindeer 211
—, small rodents 288, 289
Weasel 49, 312
Weight changes, small rodents 293
Wheatear 22, 92, 106–108
Wildlife 239, 241
Willow grouse 49, 51, 309, 311
— thicket 220, 318
— tit 21, 49
— warbler 21, 52
Willows 301, 306
Winter dormancy, Collembola 120
Wolf 52, 232, 312
Wolf spiders 67
Wolverine 21, 32, 52, 232, 312
Woody plants 252, 254, 255, 259–264

*Xenylla maritima* 74, 75, 77
*Xysticus obscurus* 67, 69

*Zora nemoralis palmgreni* 67
Zoridae 67, 69
*Zygaena exulans* 43, 50

**Ecological Studies**
**Analysis and Synthesis**
**Editors: W.D.Billings, F.Golley, O.L.Lange, J.S.Olson**
**Vol. 16**

# Fennoscandian Tundra Ecosystems

**(In 2 Parts)**

**Editor: F.E.Wielgolaski**
**Editorial Board: P.Kallio, H.Kauri, E.Østbye and T.Rosswall**

**Part 1: Plants and Microorganisms**
90 figures. 96 tables. XV, 366 pages. 1975.

Contents: Site Description. — Abiotic Variables. — Primary Producers: Plant Structure. Biomass and Production. Plant Processes. — Decomposers. — Nutrient Cycling.

The two volumes on the Fennoscandian tundra ecosystems, published as Ecological Studies, Vols. 16 and 17, present new findings on the productivity (carbon and energy flows and nutrient cycling) and synthesis of these ecosystems. These two volumes give the first relatively detailed analysis of the whole system. This has been possible thanks to teamwork between specialists in various fields working within the International Biological Programme (IBP) on Tundra Biome projects in the Fennoscandian countries. Part 1 opens with a functional description of abiotic factors at specific tundra areas in Norway, Sweden, and Finland: the IBP sites at Hardangervidda in southern Norway, Abisko in northern Sweden, and Kevo in northern Finland. Aspects dealt with are the structure of the Fennoscandian tundra plant communities, primary production, and plant physiology, also problems concerning microorganisms, decomposition, and nutrient cycling within these ecosystems.

## Ecological Studies
## Analysis and Synthesis

Editors: W.D.Billings, F.Golley, O.L.Lange, J.S.Olson

**Vol. 1: Analysis of Temperate Forest Ecosystems**
Editor: D.E.Reichle
1st corrected reprint
91 figures. XII, 304 pages. 1973

**Vol. 2: Integrated Experimental Ecology.**
Methods and Results of Ecosystem Research in the German Solling Project
Editor: H.Ellenberg
53 figures. XX, 214 pages. 1971

**Vol. 3: The Biology of the Indian Ocean.**
Editor: B.Zeitschel in cooperation with S.A.Gerlach
286 figures. XIII, 549 pages. 1973

**Vol. 4: Physical Aspects of Soil Water and Salts in Ecosystems**
Editors: A.Hadas, D.Swartzendruber, P.E.Rijtema, M.Fuchs, B.Yaron. 221 figures, 61 tables. XVI, 640 pages. 1973

**Vol. 5: Arid Zone Irrigation**
Editors: B.Yaron, E.Danfors, Y.Vaadia. 181 figures. X, 434 pages. 1973

**Vol. 6: K.Stern, L.Roche Genetics of Forest Ecosystems**
70 figures. X, 330 pages. 1974

**Vol. 7: Mediterranean Type Ecosystems.** Origin and Structure
Editors: F.di Castri, H.A.Mooney. 88 figures. XII, 405 pages. 1973

**Vol. 8: Phenology and Seasonality Modeling**
Editor: H.Lieth
120 figures. XV, 444 pages. 1974

**Vol. 9: B.Slavík Methods of Studying Plant Water Relations**
181 figures. XVIII, 449 pages. 1974

**Vol. 10: Coupling of Land and Water Systems**
Editor: A.D.Hasler
95 figures. XVII, 309 pages. 1975

**Vol. 11: Tropical Ecological Systems**
Trends in Terrestrial and Aquatic Research
Editors: F.B.Golley, E.Medina. 131 figures. XV, 398 pages. 1975

**Vol. 12: Perspectives of Biophysical Ecology**
Editors: D.M.Gates, R.B.Schmerl. 215 figures. XIII, 609 pages. 1975

**Vol. 13: Epidemics of Plant Diseases**
Mathematical Analysis and Modeling. Editor: J.Kranz
46 figures. X, 170 pages. 1974

**Vol. 14: Primary Productivity of the Biosphere**
Editors: H.Lieth, R.H.Whittaker. 67 figures. VIII, 339 pages. 1975

**Vol. 15: Plants in Saline Environments**
Editors: A.Poljakoff-Mayber, J.Gale. 54 figures. VII, 213 pages. 1975

**Vol. 16: Fennoscandian Tundra Ecosystems**
Part 1: Plants and Microorganisms.
Editor: F.E.Wiegolaski
Editorial Board: P.Kallio, T.Rosswall. 90 figures, 96 tables. XV, 366 pages. 1975

**Vol. 17: Fennoscandian Tundra Ecosystems**
Part 2: Animals and Systems Analysis
Editor: F.E.Wiegolaski
Editorial Board: P.Kallio, H. Kauri, E. Østbye. T.Rosswall
81 figures, 97 tables. Approx. 370 pages. 1975

**Vol. 18: Remote Sensing for Environmental Sciences**
Editor: E.Schanda.
178 figures. 31 tables. Approx. 370 pages. 1976

**Vol. 19: Water and Plant Live. Problems and Modern Approaches**
Editors: O.L.Lange, L.Kappen, E.D.Schulze
In Preparation

MIX
Papier aus verantwortungsvollen Quellen
Paper from responsible sources
FSC® C105338

If you have any concerns about our products,
you can contact us on
**ProductSafety@springernature.com**

In case Publisher is established outside the EU,
the EU authorized representative is:
**Springer Nature Customer Service Center GmbH**
**Europaplatz 3, 69115 Heidelberg, Germany**

Printed by Libri Plureos GmbH
in Hamburg, Germany